AF784866

PSYCHIATRY - THEORY, APPLICATIONS AND TREATMENTS

NEUROBIOLOGY OF MEN'S MENTAL HEALTH

PSYCHIATRY - THEORY, APPLICATIONS AND TREATMENTS

Additional books in this series can be found on Nova's website under the Series tab.

Additional e-books in this series can be found on Nova's website under the e-book tab.

PSYCHIATRY - THEORY, APPLICATIONS AND TREATMENTS

NEUROBIOLOGY OF MEN'S MENTAL HEALTH

LEO SHER, MD,

AND

TIMOTHY R. RICE, MD

EDITORS

NOVA BIOMEDICAL

New York

Copyright © 2015 by Nova Science Publishers, Inc.

All rights reserved. No part of this book may be reproduced, stored in a retrieval system or transmitted in any form or by any means: electronic, electrostatic, magnetic, tape, mechanical photocopying, recording or otherwise without the written permission of the Publisher.

We have partnered with Copyright Clearance Center to make it easy for you to obtain permissions to reuse content from this publication. Simply navigate to this publication's page on Nova's website and locate the "Get Permission" button below the title description. This button is linked directly to the title's permission page on copyright.com. Alternatively, you can visit copyright.com and search by title, ISBN, or ISSN.

For further questions about using the service on copyright.com, please contact:
Copyright Clearance Center
Phone: +1-(978) 750-8400 Fax: +1-(978) 750-4470 E-mail: info@copyright.com.

NOTICE TO THE READER

The Publisher has taken reasonable care in the preparation of this book, but makes no expressed or implied warranty of any kind and assumes no responsibility for any errors or omissions. No liability is assumed for incidental or consequential damages in connection with or arising out of information contained in this book. The Publisher shall not be liable for any special, consequential, or exemplary damages resulting, in whole or in part, from the readers' use of, or reliance upon, this material. Any parts of this book based on government reports are so indicated and copyright is claimed for those parts to the extent applicable to compilations of such works.

Independent verification should be sought for any data, advice or recommendations contained in this book. In addition, no responsibility is assumed by the publisher for any injury and/or damage to persons or property arising from any methods, products, instructions, ideas or otherwise contained in this publication.

This publication is designed to provide accurate and authoritative information with regard to the subject matter covered herein. It is sold with the clear understanding that the Publisher is not engaged in rendering legal or any other professional services. If legal or any other expert assistance is required, the services of a competent person should be sought. FROM A DECLARATION OF PARTICIPANTS JOINTLY ADOPTED BY A COMMITTEE OF THE AMERICAN BAR ASSOCIATION AND A COMMITTEE OF PUBLISHERS.

Additional color graphics may be available in the e-book version of this book.

Library of Congress Cataloging-in-Publication Data

Library of Congress Control Number: 2015942938
ISBN: 978-1-63483-191-8

Published by Nova Science Publishers, Inc. † New York

CONTENTS

PREFACE

Men frequently suffer from adversity and experience psychological problems irrespective of how physically strong or economically well off they are. Men suffer from physical, sexual and emotional abuse, and may be victims of domestic violence. A study in the U.S. showed that about 7% of men were physically assaulted by a wife or girlfriend. Millions of men suffer from depression. Many men try to deal with it on their own, but depression symptoms can make them chronically miserable. Depression puts men at risk for suicide. Over the past several decades, three to four times more men have taken their own lives than women, and at no point during this period has the rate of suicide in women been higher than that of men. There is much more data indicating that men's mental health needs more attention. This raises questions about why we are not thinking about the psychological needs of men and boys.

I was involved in the activities of the World Federation of Societies of Biological Psychiatry (WFSBP) since 2009. In Fall 2013, I proposed to create a new WFSBP Task Force. I suggested to establish the Task Force on Men's Mental Health. My proposal was supported by the WFSBP leadership. The new Task Force was created. Zoltan Rihmer, M.D., Ph.D. (Budapest, Hungary), an outstanding expert on mood disorders and suicide and one of the best-known psychiatric researchers in the world has accepted my invitation to be the Vice-Chair of the Task Force. Timothy R. Rice, M.D. (New York, USA), a young, energetic, enthusiastic and obviously gifted physician and scholar has accepted my invitation to be the Secretary of the newly created Task Force on Men's Mental Health. Outstanding, internationally known scholars have become members of the Task Force: Mikkel Arendt, Ph.D. (Aarhus, Denmark), M. Dolores Braquehais, M.D., Ph.D. (Barcelona, Spain), Javier Didia-Attas, M.D. (Buenos Aires, Argentina), Masahito Fushimi, M.D., Ph.D. (Akita, Japan), Julia Golier, M.D. (New York, USA), Jose de Leon, M.D. (Kentucky, USA), Shih-Ku Lin, M.D. (Taipei, Taiwan), J. John Mann, M.D. (New York, USA), Anne-Maria Möller-Leimkühler, Ph.D. (Munich, Germany), Alexander Neumeister, M.D. (New York, USA), Jorge Ospina-Duque, M.D. (Bogota, Colombia), Carlos Roncero, M.D. (Barcelona, Spain), Wolfgang Rutz, M.D., Ph.D. (Uppsala, Sweden), Robert G. Stern, M.D. (New Jersey, USA), and Nestor Szerman, M.D. (Madrid, Spain). The Task Force concentrates on issues which are important to men's health as well as for general public health and greater society.

This book is dedicated to research on men's mental health. Timothy R. Rice and I designed this book to integrate knowledge in the field of men's mental health and to stimulate research and clinical work in this field.

The contributors to this book are the leading international experts in psychiatry, medicine, and psychology. We would like to acknowledge and thank all the contributors. Our task as the editors was greatly facilitated by their swift and positive response to our initial inquiry, and thereafter by producing their manuscript diligently.

Leo Sher, MD
New York, February 2015

Dr. Timothy R. Rice, MD
Icahn School of Medicine at Mount Sinai
1240 Park Avenue, New York, NY 10029
timothy.rice@mssm.edu

Dr. Leo Sher, MD
Icahn School of Medicine at Mount Sinai
James J. Peters Veterans Administration Medical Center
drleosher@gmail.com

SECTION ONE: MEN'S MENTAL HEALTH IN CHILDHOOD, ADOLESCENCE, AND NEURODEVELOPMENT

In: Neurobiology of Men's Mental Health
Editors: Leo Sher and Timothy R. Rice
ISBN: 978-1-63483-191-8
© 2015 Nova Science Publishers, Inc.

Chapter 1

TESTOSTERONE AND MALE DEVELOPMENT

***Timothy R. Rice*, MD**
Department of Psychiatry, Icahn School of Medicine at Mount Sinai,
New York, NY, US

ABSTRACT

An understanding of the neurobiology of men's mental health begins with a solid understanding of normative development and physiology. This chapter introduces the perspective of this collection of works by focusing upon the current data concerning putative mechanisms of testosterone-mediated effects upon the developing central nervous system. Topics discuss include the activity of testosterone in the gestational and early neonatal periods, the concept of organizational and activational effects, the emergence of puberty, and the complications of declining testosterone levels in the elderly. Reference to recent data concerning a molecular mechanism of the role of testosterone in influencing male fear circuitry among several other mechanistic models are advanced. The conclusion narrows the focus to the emotion regulation system, to be developed in the following chapter and further in relation to therapeutics in chapter nine.

Keywords: testosterone, development, neuropsychiatric disorders of childhood, attention deficit hyperactivity disorder, autism spectrum disorder

INTRODUCTION

The influence of a biological agent – testosterone – on male development and behavior is the focus of this chapter. This interaction between biology and other determinants in the formation of the outcome product of male experience is a complex process. It is one for which the current scientific knowledge base is only just nascent. Yet, the further study of this

* Corresponding author: Timothy R. Rice, M.D. Icahn School of Medicine at Mount Sinai, Box 1230, One Gustave L. Levy Place, New York, NY 10029, USA, Email: timothy.rice@mssm.edu

process is highly promising for a vastly improved capability to target issues of health and sickness that are specific to men.

As the 21st century ushers in an anticipated greater transition to brain-based models of human behavior and health, most recently spearheaded by the principles of the Research Domain Criteria (RDoC) Project of the American National Institute of Mental Health (NIMH) [1, 2], the relevance and importance of such study will gain additional weight. At this point, it is timely to introduce the chapters that are to follow in this collection by a biological-developmental perspective which incorporates dimensional models of both childhood and adult psychopathology. In fact, in the pages which are to follow, the stress will be laid upon not only childhood and adult differentiation, but also the variations in brain and behavior which occur in infancy, toddlerhood, pre-school age, school age, pre-adolescent, adolescent, young adult, adult, and the elderly. A perspective which investigates the actions of testosterone in the context of each of these developmental ages establishes a functional model for better understanding the nuances of the male experience across the life span.

To be capable of taking such a perspective, the actions of testosterone will be simplified in order to permit a breadth which would otherwise be impossible. We must also accept the limitations of the current evidence base. There is value to such an approach, as without it there would be no platform from which to begin our dive into the study of this field. The significant disclaimer to such an approach is: do not lose sight of the very significant limitations of our knowledge concerning the role of testosterone in human development, as well as of the significant hypothetical nature of some of these discussions. Where possible, evidence from the scientific literature will be cited to support the model, but our data in this field remains young and underdeveloped.

Nonetheless, we must use this data to begin to piece together an increasingly advanced understanding of the neurobiology of men's mental health. In the chapter which will follow, this perspective will be drawn out further into a discussion of endophenotypes as a useful heurisitc in this study. The concept of emotion regulation will organize the material. It is the goal of this chapter to first lay out what we know of testosterone, its chronological presences and absences throughout the lifespan, and its functions on the human central nervous system (CNS).

Normative Physiology of Testosterone

Male development begins with the inheritance of the Y chromosome at fertilization. The sex-determining region Y (*SRY*) gene and its downstream products induce differentiation of pluripotent gonadal cells into fetal Leydig cells. Differentiation is effected as early as at seven weeks of gestation [3] and small amounts of testosterone are produced. Shortly thereafter the hypothalamic-pituitary-gonadal (HPG) axis matures sufficiently to stimulate the Leydig hormones through luteinizing hormone to produce testosterone.

Testosterone is one of a number of androgens, or C19 steroid hormones, that include testosterone, and rostenedione, a biochemically reduced form of testosterone 5α- and 5β-dihyrotestosterone (DHT), and dehydroepiandrosterone (DHEA). Its effects are mediated through multiple factors including its bioavailability and its metabolization. Testosterone is present in the blood as free (unbound) testosterone, as albumin-bound testosterone, and as sex hormone-binding globulin (SHGB)-bound testosterone, and in some tissues it is metabolized to a more active metabolite, 5α-dihydrotestosterone (DHT).

Contrary to the common belief that testosterone and sex hormones only begin to play a role in male development at puberty, by the start of the second trimester serum testosterone levels are approximately 2.5 times higher in males than in females [4]. This period forms the first critical period at which male CNS differentiation is believed to occur. At the end of the second trimester the fetal Leydig cells begin to degenerate [5], the HPG axis becomes suppressed through rising levels of placental hormones, and male fetal testosterone levels decline significantly and return to levels comparable to those female fetuses.

With birth and the release of the newborn HPG axis from placental suppression a second rise in testosterone occurs through HPG activation [6]. This second critical period begins at birth, peaks at approximately three months, and ends at approximately six months [7–12]. Putative mechanisms for this physiologic period include immaturity of hypothalamic feedback mechanisms as well as the activities of residual degenerating fetal Leydig cells. It ceases when the HPG axis feedback mechanisms mature and a new population of neonatal Leydig cells that are less responsive to LH replace the fetal Leydig cells [3]. Whatever the mechanism, during this time testosterone levels are effectively at pubertal concentrations, leading to both measurable CNS and extra-CNS effects.

There are some limitations to this model of two critical windows in utero and in early infancy. The testosterone produced during this interval may have low bioavailability, particularly so in the early infancy window owing to the rise in sex hormone binding globulin (SHBG) at this time [13–15]. However, cellular-level biological assays do reveal biologically-active androgens in male infants at 3 months of age [16]. More studies will be needed to clarify the impact of these considerations of bioavailability, as well as potential metabolic derangements in the gestational fetus and newborn infant which may deviate significantly from adult physiology.

Nevertheless, this data enables us to consider some interesting ideas concerning the biological basis of male experience across development. It is the case that testosterone and androgens are virtually absent from the late infant, toddler, pre-school aged, school aged, and pre-adolescent man. The imprint of the early testosterone exposure is instead understood to be the important factor in these ages, the effects of which we will develop below. These are termed organizational effects, in contrast to the direct activation effects which testosterone is believed to produce while actively present in the blood stream.

Probably more likely than not, activation effects are indeed present in the early infant, followed by their disappearance as the child enters the mid infancy years. This may have important ramifications to those who study and theorize on infant development. For example, from the historical psychoanalytic perspective, Melanie Klein hypothesized that the first four to six months of life come to a conclusion upon a significant shift in the child's way of understanding the world, from the to the paranoid schizoid to the depressive position. Knowing the effects that testosterone may have upon the emotion regulation system, as we will discuss in greater detail in chapters 2 and 9, may offer a unique correlate between biology and psychoanalytic theory in the development of male infant. Certainly, more recent and evidence based considerations of infant development could also be furthered by integrating this biological data.

The supposedly "quiescent" period of childhood in relation to activation effects comes to a significant and life-changing end upon the pre-adolescent's arrival to puberty. There is no single biological event that singles out the beginning or end to puberty, but rather it is the process of maturation set about by the rise in male androgens which have widespread biological, psychological, and social products. Puberty is such a significant event that adolescent specialists have found it of utility to separate the period into early, middle, and late adolescence, with the following stage of early, young, or emerging adulthood sharing many of the characteristics of the earlier phases of this critical developmental period [17–19]. Thereafter, normative serum testosterone levels will remain biologically active and fundamental in the male neurobiology throughout most of adult life followed by the gradual decline in these levels into old age.

Among American boys, puberty takes place on average at a slightly later date than among girls [20]. Boys on average generally begin puberty around age 12 and complete the traditional pubertal period around age 17. Many factors, such as nutritional status, advance or delay the onset of puberty. There is also great individual variation. As authors Varlinskaya, Vetter-O'Hagen, and Spear note [21], this process coincides with a great diversity of processes in the CNS including considerable remodeling in cortical and limbic brain areas, decreases in brain energy utilization, selective synaptic pruning, and myelination of axons with maturation of neuronal tracks and supraordinate systems [22–24]. These authors go on to succinctly review some of the interesting androgen-brain-behavior correlates that have long been noted in adolescence, including the effects of testosterone upon not only risk taking, pleasure seeking, and alcohol consumption, but also including detailed nuances of drinking behaviors, or the microstructure of social behavior [21].

Testosterone levels decline with male aging [25–27]. As reviewed by Sher [28], age-related plasma testosterone decrease is a result of different biological alterations including primary structural gonadal damage, age-related degenerative changes of the pituitary gland, insufficiencies of the neurohypothalamic system, and primary peripheral metabolic abnormalities, such as the age-associated increase in the concentration of serum SHBG and consequent decrease in free testosterone [26]. In the aging man, there is approximately a 1%–2% decrease of total testosterone levels per year. Because of this gradual decrease in testosterone levels, the androgen deficiency of the elderly man is defined as partial androgen deficiency of the aging male (PADAM) or late onset hypogonadism (LOH). As Sher reviews, the prevalence of testosterone deficiency may be as high as 30% in men aged 40–79 years [29, 30]. Of course, as he suggests, there may be significant clinical correlations to these data.

Preliminary Evidence Regarding the Role of Testosterone in Male CNS Development

Rather than serve as a review of the myriad androgen-brain-behavior correlation studies and evidence regarding androgen effects upon psychopathology, form the perspective of this chapter we will attempt to limit ourselves to a consideration of what we know of the neural structure and mechanisms underlying these behavioral and pathologic consequences. Molecular, cellular, and tissue- and pathway-specific differences have long been known to exist between men and women. For example, both dopaminergic [31, 32] and serotonergic [33] function varies significantly between males and females. We have long been only able to presumably infer that the origin of these differences exist, at least partially, as a function of sex hormones and other biological factors.

However, such inferrances are uncertain. Among the biological factors, differences in the genetic code between men and women could theoretically account for all of the biological determination of CNS determination, rather than any direct role from sex hormones. Additionally, it could be said that biological factors play no role in comparison to the effects of the differing psychosocial environment between men and women. For example, the gender differential in expectations for young children – such as the increased tolerance for physical play and aggression in boys over girls – may preferentially reinforce pathways that later determine future behavior. Yet, is has now been shown through multiple studies that "rough and tumble play" is directly correlated with increased prenatal testosterone exposure [26].

Determining the brain-based mechanisms for these demonstrated associations between sex hormone levels and behavior remains the challenge. Much as it may be interesting to some to see observational evidence that, for example, there is an association in male infants between Klein's Paranoid Schizoid phase and circulating testosterone levels, it would be of far greater value to have biological data demonstrate how testosterone effects the brain. This brain-based approach enables a jettison of a reliance upon psychological theories of the mind which do not always generate agreement among providers. The biological mechanistic data can then be used to explain various psychological theories. This bottom-up rather than a top-down approach is of key relevance to an approach such as that contained in this collection of beginning with a neurobiological foundation to men's mental health.

Fortunately for those of us with a trust and faith in this perspective, evidence appears to be accumulating regarding a mechanism for the hypothesis that sex hormones play a large role in CNS development that is directly relevant to male-female differences in psychiatric pathology. It is these mechanisms in which we are interested here, understanding that all human experience is a product of the brain structure. Though the majority of animal research has investigated sex hormone alterations in brain regions that are related to reproductive behaviors, studies are now investigating reasons that may play significant roles in a large array of neuropsychiatric conditions of childhood.

For example, G-protein-coupled receptor 30 (GPR30) modulates the ability of sex hormones to effect promote neuronal proliferation and differentiation activities. There is a sexually dimorphic distribution pattern of this receptor in select brain areas in animal models, including in the amygdala [35]. These receptors appear to produce different effects on anxiety in male versus female mice [36]. Thus, sex hormones indeed appear to be at the source of our ability to differentiate between male and female biologies of fear and its extinction [37]. These molecular differences may have a profound effect on synaptic plasticity and integration of prefrontal-limbic integrated networks that modulate fear pathways as well as the full experience of emotion and emotion regulation. As multiple disorders may be conceptualized as products of aberrant fear and arousal pathways, including anxiety disorders like post-traumatic stress disorder (PTSD), generalized anxiety disorder, and panic disorder where there is an approximate 2:1 female:male sex ratio [38, 39], as well as mood disorders and others.

We know more generally as well that fetal testosterone levels correlate negatively and positively with future prefrontal and amygdala volumes, respectively [40]. This suggests that there may be developmental functional correlates of fetal sex hormone levels and future prefrontal-mediated attenuation of limbic regions through the emotion regulation circuitry

[41]. This is certainly related to the above specific molecular findings underlying fear and arousal pathways.

Sexually dimorphic effects driven by the male androgen outside of the emotion regulation system have also been described. Impulsive aggression may also be a secondary product of slowed structural development wrought by organizational effects. Fetal testosterone slows normal structural development in the left hemisphere of the brain, modifying cerebral lateralization and making males vulnerable for a longer period of time to insult and to structural abnormalities in the left hemisphere. These insults may predispose to neuropsychiatric deficits and could account for the increased frequency of attention-deficit hyperactivity disorder and learning disorders among males [42]

As we as a field develop our understanding of the neurobiology of men's mental health, attention to these mechanistic understandings regarding the role of testosterone in central nervous system development will be crucial. A model which integrates both organizational and activational effects, one which by necessity must include a detailed integration of the gestational and early neonatal androgen activity critical windows, will be required to develop an advanced understanding of mental health issues as they pertain to men. The emerging evidence that fear circuitry and other top-down prefrontal-limbic inhibitory and excitatory networks are directly influenced by androgens may serve to be a finding of fundamental importance. The value of the concept of emotion regulation (ER) and the ER system, in particular as it relates to child and adolescent brain development where there is a deficit of prefrontal development relative to lower brain structure development [43], will be developed in relation to this finding in the following chapter.

REFERENCES

[1] Insel TR. The NIMH Research Domain Criteria (RDoC) Project: Precision Medicine for Psychiatry. *Am J Psychiatry* (2014) 171:395–397. doi:10.1176/appi.ajp. 2014. 14020138.

[2] Insel T, Cuthbert B, Garvey M, Heinssen R, Pine DS, Quinn K, Sanislow C, Wang P. Research domain criteria (RDoC): toward a new classification framework for research on mental disorders. *Am J Psychiatry* (2010) 167:748–751. doi:10.1176/appi.ajp.2010. 09091379.

[3] Svechnikov K, Landreh L, Weisser J, Izzo G, Colón E, Svechnikova I, Söder O. Origin, development and regulation of human Leydig cells. *Horm Res pædiatrics* (2010) 73:93–101. doi:10.1159/000277141.

[4] Beck-Peccoz P, Padmanabhan V, Baggiani AM, Cortelazzi D, Buscaglia M, Medri G, Marconi AM, Pardi G, Beitins IZ. Maturation of hypothalamic-pituitary-gonadal function in normal human fetuses: circulating levels of gonadotropins, their common alpha-subunit and free testosterone, and discrepancy between immunological and biological activities of circulating follicle-s. *J Clin Endocrinol Metab* (1991) 73:525–32. doi:10.1210/jcem-73-3-525.

[5] Codesal J, Regadera J, Nistal M, Regadera-Sejas J, Paniagua R. Involution of human fetal Leydig cells. An immunohistochemical, ultrastructural and quantitative study. *J Anat* (1990) 172:103–14.

[6] Mann DR, Fraser HM. The neonatal period: a critical interval in male primate development. *J Endocrinol* (1996) 149:191–7.

[7] Forest MG, Cathiard AM, Bertrand JA. Evidence of testicular activity in early infancy. *J Clin Endocrinol Metab* (1973) 37:148–51. doi:10.1210/jcem-37-1-148.

[8] Forest MG, de Peretti E, Bertrand J. Testicular and adrenal androgens and their binding to plasma proteins in the perinatal period: developmental patterns of plasma testosterone, 4-androstenedione, dehydroepiandrosterone and its sulfate in premature and small for date infants as compared wit. *J Steroid Biochem* (1980) 12:25–36.

[9] Winter JS, Faiman C, Hobson WC, Prasad A V, Reyes FI. Pituitary-gonadal relations in infancy. I. Patterns of serum gonadotropin concentrations from birth to four years of age in man and chimpanzee. *J Clin Endocrinol Metab* (1975) 40:545–51.

[10] Winter JS, Hughes IA, Reyes FI, Faiman C. Pituitary-gonadal relations in infancy: 2. Patterns of serum gonadal steroid concentrations in man from birth to two years of age. *J Clin Endocrinol Metab* (1976) 42:679–86.

[11] Schmidt H, Schwarz HP. Serum concentrations of LH and FSH in the healthy newborn. *Eur J Endocrinol* (2000) 143:213–215. doi:10.1530/eje.0.1430213.

[12] Kuiri-Hänninen T, Seuri R, Tyrväinen E, Turpeinen U, Hämäläinen E, Stenman U-H, Dunkel L, Sankilampi U. Increased activity of the hypothalamic-pituitary-testicular axis in infancy results in increased androgen action in premature boys. *J Clin Endocrinol Metab* (2011) 96:98–105. doi:10.1210/jc.2010-1359.

[13] Bolton NJ, Tapanainen J, Koivisto M, Vihko R. Circulating sex hormone-binding globulin and testosterone in newborns and infants. *Clin Endocrinol* (Oxf) (1989) 31:201–7.

[14] Huhtaniemi I, Dunkel L, Perheentupa J. Transient increase in postnatal testicular activity is not revealed by longitudinal measurements of salivary testosterone. *Pediatr Res* (1986) 20:1324–7. doi:10.1203 /00006450-198612000-00028.

[15] Knickmeyer RC, Woolson S, Hamer RM, Konneker T, Gilmore JH. 2D:4D ratios in the first 2years of life: Stability and relation to testosterone exposure and sensitivity. *Horm Behav* (2011) 60:256–63. doi:10.1016/j.yhbeh.2011.05.009.

[16] Main KM, Schmidt IM, Skakkebaek NE. A possible role for reproductive hormones in newborn boys: progressive hypogonadism without the postnatal testosterone peak. *J Clin Endocrinol Metab* (2000) 85:4905–7. doi:10.1210/jcem.85.12.7058.

[17] Arnett JJ. Emerging adulthood: The winding road from the late teens through the twenties. (2004). doi::10.1093/acprof:oso/9780195309379. 003.0001.

[18] Arnett JJ. Emerging Adulthood: What Is It , and What Is It Good For? *Child Dev Perspect* (2007) 1:68–73. doi:10.1111/j.1750-8606.2007. 00016.x.

[19] Arnett JJ. Emerging adulthood. A theory of development from the late teens through the twenties. *Am Psychol* (2000) 55:469–480. doi:10.1037/0003-066X.55.5.469.

[20] Sun SS, Schubert CM, Chumlea WC, Roche AF, Kulin HE, Lee PA, Himes JH, Ryan AS. National estimates of the timing of sexual maturation and racial differences among US children. *Pediatrics* (2002) 110:911–919. doi:10.1542/peds.110.5.911.

[21] Varlinskaya EI, Vetter-O'Hagen CS, Spear LP. Puberty and gonadal hormones: Role in adolescent-typical behavioral alterations. *Horm Behav* (2013) 64:343–349. doi:10.1016/j.yhbeh.2012.11.012.

[22] Bava S, Tapert SF. Adolescent brain development and the risk for alcohol and other drug problems. *Neuropsychol Rev* (2010) 20:398–413. doi:10.1007/s11065-010-9146-6.

[23] Brenhouse HC, Andersen SL. Developmental trajectories during adolescence in males and females: A cross-species understanding of underlying brain changes. *Neurosci Biobehav Rev* (2011) 35:1687–1703. doi:10.1016/j.neubiorev.2011.04.013.

[24] Sturman DA, Moghaddam B. The neurobiology of adolescence: Changes in brain architecture, functional dynamics, and behavioral tendencies. *Neurosci Biobehav Rev* (2011) 35:1704–1712. doi:10.1016/j.neubiorev.2011.04.003.

[25] Valenti G. The pathway of partial androgen deficiency of aging male. *J Endocrinol Investig* (2005) 28:28–33.

[26] NardozzArchimedes Nardozza, Szelbracikowski S dos S, Nardi AC, de Almeida JC. Age-related testosterone decline in a Brazilian cohort of healthy military men. *Int Braz J Urol* (2011) 37:591–597. doi:10.1590/S1677-55382011000500004.

[27] Araujo AB, O'Donnell AB, Brambilla DJ, Simpson WB, Longcope C, Matsumoto AM, McKinlay JB. Prevalence and incidence of androgen deficiency in middle-aged and older men: Estimates from the Massachusetts male aging study. *J Clin Endocrinol Metab* (2004) 89:5920–5926. doi:10.1210/jc.2003-031719.

[28] Sher L. Low testosterone levels may be associated with suicidal behavior in older men while high testosterone levels may be related to suicidal behavior in adolescents and young adults: a hypothesis. *Int J Adolesc Med Health* (2013) 25:263–8. doi:10.1515/ijamh-2013-0060.

[29] Allan CA, McLachlan RI. Age-related changes in testosterone and the role of replacement therapy in older men. *Clin Endocrinol* (Oxf) (2004) 60:653–70. doi:10.1111/j.1365-2265.2004.02002.x.

[30] Liu CC, Wu WJ, Lee YC, Wang CJ, Ke HL, Li WM, Hsiao HL, Yeh HC, Li CC, Chou YH, et al. The prevalence of and risk factors for androgen deficiency in aging Taiwanese men. *J Sex Med* (2009) 6:936–946. doi:10.1111/j.1743-6109.2008.01171.x.

[31] Xiao L, Becker JB. Quantitative microdialysis determination of extracellular striatal dopamine concentration in male and female rats: effects of estrous cycle and gonadectomy. *Neurosci Lett* (1994) 180:155–8.

[32] Andersen SL, Teicher MH. Sex differences in dopamine receptors and their relevance to ADHD. in *Neuroscience and biobehavioral reviews,* 137–41. doi:10.1016/S0149-7634(99)00044-5.

[33] Cosgrove KP, Mazure CM, Staley JK. Evolving knowledge of sex differences in brain structure, function, and chemistry. *Biol Psychiatry* (2007) 62:847–55. doi:10.1016/j.biopsych.2007.03.001.

[34] Martel MM. Sexual selection and sex differences in the prevalence of childhood externalizing and adolescent internalizing disorders. *Psychol Bull* (2013) 139:1221–59. doi:10.1037/a0032247.

[35] Canonaco M, Giusi G, Madeo A, Facciolo RM, Lappano R, Canonaco A, Maggiolini M. A sexually dimorphic distribution pattern of the novel estrogen receptor G-protein-coupled receptor 30 in some brain areas of the hamster. *J Endocrinol* (2008) 196:131–138. doi:10.1677/JOE-07-0392.

[36] Hart D, Nilges M, Pollard K, Lynn T, Patsos O, Shiel C, Clark SM, Vasudevan N. Activation of the G-protein coupled receptor 30 (GPR30) has different effects on anxiety in male and female mice. *Steroids* (2014) 81:49–56. doi:10.1016/j.steroids.2013.11.004.

[37] Cover KK, Maeng LY, Lebrón-Milad K, Milad MR. Mechanisms of estradiol in fear circuitry: implications for sex differences in psychopathology. *Transl Psychiatry* (2014) 4:e422. doi:10.1038/tp. 2014.67.

[38] Kessler RC, Sonnega A, Bromet E, Hughes M, Nelson CB. Posttraumatic stress disorder in the National Comorbidity Survey. *Arch Gen Psychiatry* (1995) 52:1048–1060. doi:10.1001/archpsyc.1995. 03950240066012.

[39] Eaton NR, Keyes KM, Krueger RF, Balsis S, Skodol AE, Markon KE, Grant BF, Hasin DS. An invariant dimensional liability model of gender differences in mental disorder prevalence: Evidence from a national sample. *J Abnorm Psychol* (2012) 121:282–288. doi:10.1037/a0024780.

[40] Marcell A V, Ford CA, Pleck JH, Sonenstein FL. Masculine beliefs, parental communication, and male adolescents' health care use. *Pediatrics* (2007) 119:e966–75. doi:10.1542/peds.2006-1683.

[41] Ochsner KN, Gross JJ. "The neural bases of emotion and emotion regulation: A valuation perspective," in *Handbook of Emotion Regulation, Second Edition* (New York, NY: Guilford Press), 23–41.

[42] Martel MM. Research review: a new perspective on attention-deficit/hyperactivity disorder: emotion dysregulation and trait models. *J Child Psychol Psychiatry* (2009) 50:1042–51. doi:10.1111/j.1469-7610.2009.02105.x.

[43] Casey B, Jones RM, Somerville LH. Braking and Accelerating of the Adolescent Brain. *J Res Adolesc* (2011) 21:21–33. doi:10.1111/j.1532-7795.2010.00712.x.

In: Neurobiology of Men's Mental Health
Editors: Leo Sher and Timothy R. Rice
ISBN: 978-1-63483-191-8
© 2015 Nova Science Publishers, Inc.

Chapter 2

TESTOSTERONE, EMOTION REGULATION, AND ENDOPHENOTYPES

Timothy R. Rice**, MD
Department of Psychiatry, Icahn School of Medicine at Mount Sinai, New York, NY, US

ABSTRACT

The concept of the endophenotype is presented as of eminent value to the neurobiology of men's mental health. The provision of a stable biological phenotype as an intermediary between the more basic levels of genetics and the high-level symptom presentations and clusters which effect men. The definition and concept of the endophenotype is presented as the background to our transition into a focus on the emotion regulation system in this chapter. This system and its history is reviewed. Recent studies examining the structural foundation of the emotion regulation system are cited as a means of demonstrating the value of emotion regulation in understanding brain processes. The correlation between emotion regulation deficits and psychiatric disorders of childhood, adolescence, and adulthood are put forth with an emphasis, based on the data presented in the preceding chapter, upon the externalizing behaviors of childhood as a distinctively "male" set of disorders. The impairment beset by attention deficit hyperactivity disorder (ADHD) and oppositional defiant disorder (ODD) can be perhaps best understood as products of this emotion regulation system. The value of this construct of men's mental health is reviewed with a return to the topic in light of therapeutics developed further in chapter 9.

Keywords: endophenotype, emotion regulation, men's mental health

* Corresponding author: Timothy R. Rice, M.D. Icahn School of Medicine at Mount Sinai, Box 1230, One Gustave L. Levy Place, New York, NY 10029, USA, Email: timothy.rice@mssm.edu

INTRODUCTION

A strong presentation has been made elsewhere for the use of the emotion regulation (ER) system construct as a neurobiologically-based transdiagnostic organizing principle in which to understand both childhood and adult psychopathology [1]. This chapter will extend this perspective of the importance of the ER construct by presenting the ER system as of eminent utility to the advancement of the field of men's mental health.

The ER construct will first be defined. It will then be presented as an optimal endophenotype for distinctly male mental health problems. The endophenotype concept will be defined and explored in relation to the ER construct. In the former chapter we have referenced the emerging data on the role of the sex hormones on central nervous system (CNS) development; in this chapter an emphasis will be placed on the role of these male sex hormones within this endophenotype construct. The utility of providing of a stable, inherited, biological phenotype as an intermediary between the more basic levels of genetics and the high-level symptom presentations and clusters which affect men will be emphasized and organize the discussion.

It is the perspective of this chapter that, broadly, the most crucial and pressing problems of men's mental health relate to difficulties of reactive aggression. Be it the externalizing disorders of boys or the violence of men, physical aggression, suicide, and homicide are highly significant untoward outcomes and all occur much more frequently in men than in women. These behavioral outcomes will be regarded as occurring the majority of the time as products of reactive aggression. The concepts of reactive as compared to proactive aggression will be described in relation to developmental deficits within the ER system. Models of ER development and the implications of developmental failures will be described.

With definitions developed and an argument consolidated for the value of the ER construct in men's mental health, a foundation will have been set for the development of a brain-based therapeutic approach to men's mental health by focusing on bolstering ER development. This discussion will proceed in chapter 9. The focus of this chapter will be to further place in context the field with its neurobiological underpinnings as well as the possible applications which can derive from such efforts.

EMOTION REGULATION AND THE INDIVIDUAL

The ER construct is a recent development within the field of the affective neurosciences. It has gained an almost exponentially-increasing academic interest as evidenced by the numebr of citations in the literature by year [2]. ER may be defined as the capacity to shape which emotions one has, when one has them, and how one experiences or expresses them [2, 3]. The neural correlates of the ER system are defined and, in the process of reducing unmanageable negative affect, broadly consist of prefrontal region inhibition of limbic over-arousal [4]. Interest in ER has expanded dramatically in the last decade [2] as psychiatry has shifted towards a brain-based dimensional approach to healthy functioning and psychopathology [5]. It additionally overlaps with some traditional models of intrapsychic processes which are utilized in traditional psychotherapeutic approaches [6], enabling the reconstruction of traditional psychotherapy within a contemporary brain-based model.

In this system, self-regulative processes may be divided into implicit and explicit processes [7]. Implicit regulation, or implicit environmental and self-evaluations with immediate response tendencies, is broadly dependent upon the ventral pre- frontal cortex (PFC), which includes the orbitofrontal cortex, ventromedial PFC and ventral anterior cingulate cortex. Explicit regulation, or deliberate or effortful conscious cognitive manipulations that monitor, adjust and select responses from a range of options, is broadly dependent upon the dorsal anterior cingulate cortex and dorsolateral PFC. These two divisions of the cerebral cortex both modulate lower brain structures, including the amygdala, hypothalamus and brainstem nuclei, through their respective regulative pathways. The successful modulation of these visceromotor centers through the PFC regulators leads to increased vagal tone and decreased sympathetic arousal. This affects a measurable neurochemical state that promotes calmness and inhibits behavioral dysregulation.

A variety of ER processes have been described. Two processes, effortful suppression [8, 9] and cognitive reappraisal [8, 10, 11], have been extensively studied. Both result in decreased sympathetic arousal and increased vagal tone promoting behavioral regulation via prefrontal-mediated attenuation of the amygdala, hypothalamus, and brainstem nuclei. Interestingly, effortful suppression is less effective than reappraisal in effecting these neurophysiologic changes [8].

ER and Development

Emotion regulation develops early in infancy [12]. Prior to age three months children either turn towards or turn away from stimuli in their attempts to self-regulate. By age three months children begin to self-soothe through thumb sucking, starting to move away, or by reflexive social signaling via crying or social smiling. This culminates by age six months in the ability to self-distract through focusing attention on neutral objects in lieu of distressing stimuli. Notably, this work mirrors that of Phebe Cramer who postulated the importance of denial in early childhood [13] in her studies of defense mechanisms.

Successful regulation requires extrinsic influence through flexible and supportive parental interactions . The toddler years mark the initiation of the organization of neural connectivity required for emotional regulation [15, 16]. Thereafter, toddlers learn a variety of specific strategies to manage affective states. Clinically, a diversity of defense mechanisms emerges. Suboptimal organization of this development may be hypothesized to result in deficient emotional regulation. Data suggests that trauma and other adverse life experiences do indeed result in ER deficits [17].

Neuroanatomic and functional correlates involve the concept of sensitive periods of development. Both progressive (myelination) and regressive (synaptic pruning) brain changes which occur in childhood and adolescence enable the clinician with a focus on men's mental health in children to intervene at a moment of cognitive, behavioral, and emotional malleability afforded by neural plasticity. The creation and implementation of therapeutic interventions designed to profit from these time-sensitive periods of neural plasticity are in alignment with the goals of the National Institute of Health Blueprint for Neuroscience Research Workshop [19], as has been noted by authors of alternate emergent RDoC-oriented psychotherapeutic approaches to child and adolescent psychopathology [20].

During these sensitive periods of childhood, subcortical areas including brainstem and limbic areas develop earlier than higher order cortical levels [21]. This may be particularly relevant to self-regulation and emotion regulation in particular [22]. As amygdalo-cortical connections develop . Shortcomings in neural network development, through constitution or environmental stress, may result in an underdeveloped implicit emotion regulation system. This may manifest in the affected child or adolescent's preponderance of use of maladaptive defenses. Its crystallization in adulthood into defined patterns of regulation may be a significant signature of adult psychopathology.

ER Deficits in the Common Presentations of Male Children, Adolescents, and Adults

ER deficits are very relevant to male child and adolescent populations. The ER system in infancy is present almost at birth: early infants can turn away from distressing stimuli, by three months of age infants can crawl away and self-sooth through thumb sucking, and by six months of age can focus attention on neutral objects in lieu of distressing stimuli. Reliance upon these orienting activities yield to more executive control when then prefrontal-limbic connectivity required for more advanced ER develops in the early childhood years [16].

Deficits in these skills, which may arise either through constitution, suboptimal caregiving experiences [14], or both, may result in the major neuropsychiatric conditions of childhood and adolescence. Whereas connections between adolescent mood disorders and ER deficits [26] have been previously understood and appreciated [27], it is increasingly clear that even disorders long characterized as behavioral or disruptive have significant ER deficits – and may even be characterized by them. These may be the most significant clinical problems for men's mental health as it is these disorders which present in males at a much higher frequency than in females.

The quintessential childhood behavioral disorder, oppositional defiant disorder (ODD), has in fact been proposed to be better conceptualized as a disorder of emotion regulation [28]. This reformulation has occurred in the context of several studies which found distinct affective dimensions within ODD [29–33]. ODD has reorganized within the DSM5 to emphasize this finding. The ER deficits in attention deficit hyperactivity disorder (ADHD) have now been (re)emphasized [34], and a new disorder, disruptive mood dysregulation disorder (DMDD) [35], has been added to the DSM5 partly to emphasize the importance of severe emotional dysregulation as a distinct entity. As many male children and adolescents either have these pathologies or have the direct products of these disorders [32], clinical evidence incorporating developmental models suggests that ER deficits are of direct relevance to adolescents.

This reorganization of pathology from behavioral models to those of affect reflects a broader transition towards understanding the role of emotions in behavioral disorders and linking models these with defined neural systems. This approach originated in the birth of affective neuroscience in the last decade of the twentieth century through the work of Damasio, Panksepp, and LeDoux and has driven advances in identifying the neurophysiologic biomarkers of affect. This premise applies to many childhood disorders manifesting with disruptive and externalizing behaviors as well as those with aggressive behaviors.

SEX HORMONES, ER, AND AN ENDOPHENOTYPE OF MEN'S MENTAL HEALTH

As we have presented in the fist chapter, it appears that testosterone-mediated developmental inhibition the ER system occurs. This may result in aggression. emotion regulation capacities in childhood, which may result in dysregulated aggression. A useful, if imperfect, emerging research heuristic by which to test this claim has been the division of aggression into reactive and proactive subtypes [39]. Whereas the latter has been described as calculated, callous and effected to achieve some personal gain, the former has been termed 'hot' aggression. Although these types of aggression likely exist on a continuum, the construction of the reactive aggression pole allows for the study of a homogenous and common form of male aggression [40].

Although earlier findings were often contradictory and inconclusive, more recent studies seem to support a positive association between undifferentiated aggression and the organizational effects of testosterone in young children [41, 42]. We have described in Chapter 1 some ways in which testosterone affects the emotion regulation pathways that are related to reactive aggression. We may further this suggestion by considering that one specific area inhibited in preadolescents in a study of the effects of fetal testosterone was the orbitofrontal PFC [43]. This suggests that self-regulative processes, and especially those that are applied effortlessly (a concept we will expand in Chapter 9) may be particularly affected by testosterone. We know that this area of the PFC is particularly important to background feelings or emotional states that make up conscious experience but are not noticed as such unless they are attended to [7]. Males may, therefore, be at a particular disadvantage relative to their female peers to appropriately modulate backgrounded affective states.

While Beck has emphasized hopelessness [44] and suicide "modes" deriving from cognitive distortions and depressed mood [45], hopelessness and depression may be less important to suicidal states in adolescent males than in females [46, 47]. Clinical experience suggests that this is the case for children, young adults, and adults as well. Homicide among men with psychiatric disorders is an additional highly clinically relevant behavior which is related to issues of reactive aggression in adolescents, young men, and many adults [49].

Thus, the ER system is an appropriate endophenotype for men's mental health. An endophenotype is defined as a stable biological phenotype as an intermediary between the more basic levels of genetics and the high-level symptom presentations and clusters. Given that we have demonstrated that men's mental health issues are often products of reactive aggression and include externalizing behaviors, suicide, and homicide in individuals with psychiatric disorders, the ER system appears a good candidate for further study of the neurobiology of men's mental health. The preliminary data has demonstrated that the male sex hormones directly influence the development of the ER system, a development which has been detailed in this chapter.

CONCLUSION

This chapter has presented the concept of ER as an endophenotype of eminent value to men's mental health. It provides a stable biological phenotype as an intermediary between the more basic levels of genetics and the high-level symptom presentations and clusters which effect men. Its development and problems which may arise in its development have been referenced. The correlation between emotion regulation deficits and psychiatric disorders of childhood, adolescence, and adulthood have been shown. This has been performed to further the data presented in Chapter 1 which demonstrate a direct influence of the male sex hormones on the development and function of the ER system.

This work paves a way for us to explore a brain-based psychotherapeutic approach targeted to the common issues in men's mental health. Such an approach would target the specific deficits in the ER system to which males are biologically predisposed. This topic will be returned to in light of therapeutics developed in Chapter 9.

REFERENCES

[1] Etkin A, Cuthbert B. Beyond the DSM: development of a transdiagnostic psychiatric neuroscience course. *Acad Psychiatry* (2014) 38:145–50. doi:10.1007/s40596-013-0032-4

[2] Gross JJ. Emotion regulation: taking stock and moving forward. *Emotion* (2013) 13:359–365. doi:10.1037/a0032135

[3] Gross JJ. "Emotion Regulation: Conceptual and Empirical Foundations," in Handbook of Emotion Regulation, Second Edition (New York, NY: *Guilford Press*), 3–20.

[4] Ochsner KN, Gross JJ. "The neural bases of emotion and emotion regulation: A valuation perspective," in Handbook of Emotion Regulation, Second Edition (New York, NY: *Guilford Press*), 23–41.

[5] Insel T, Cuthbert B, Garvey M, Heinssen R, Pine DS, Quinn K, Sanislow C, Wang P. Research domain criteria (RDoC): toward a new classification framework for research on mental disorders. *Am J Psychiatry* (2010) 167:748–751. doi:10.1176/appi.ajp.2010.09091379

[6] Rice TR, Hoffman L. Defense mechanisms and implicit emotion regulation: a comparison of a psychodynamic construct with one from contemporary neuroscience. *J Am Psychoanal Assoc* (2014) 62:693–708. doi:10.1177/0003065114546746

[7] Gyurak A, Gross JJ, Etkin A. Explicit and implicit emotion regulation: a dual-process framework. *Cogn Emot* (2011) 25:400–412. doi:10.1080/02699931.2010.544160

[8] Goldin PR, McRae K, Ramel W, Gross JJ. The neural bases of emotion regulation: reappraisal and suppression of negative emotion. *Biol Psychiatry* (2008) 63:577–86. doi:10.1016/j.biopsych.2007.05.031

[9] Dunn BD, Billotti D, Murphy V, Dalgleish T. The consequences of effortful emotion regulation when processing distressing material: a comparison of suppression and acceptance. *Behav Res Ther* (2009) 47:761–63. doi:10.1016/j.brat.2009.05.007

[10] McRae K, Hughes B, Chopra S, Gabrieli JDE, Gross JJ, Ochsner KN. The neural bases of distraction and reappraisal. *J Cogn Neurosci* (2010) 22:248–262. doi:10.1162/jocn.2009.21243

[11] Ochsner KN, Bunge SA, Gross JJ, Gabrieli JDE. Rethinking feelings: an FMRI study of the cognitive regulation of emotion. *J Cogn Neurosci* (2002) 14:1215–1229. doi:10.1162/089892902760807212

[12] Kopp C, Neufeld S. "Emotional development during infancy," in *Handbook of affective sciences* (Oxford, UK: Oxford University Press), 347–374.

[13] Cramer P. Protecting the Self: Defense Mechanisms in Action. New York, NY: *Guilford Press* (2006).

[14] Calkins SD, Hill A. "Caregiver influences on emerging emotional regulation: biological and environmental transactions in early development," in Handbook of emotion regulation, ed. J. J. Gross (New York, NY: *Guilford Press*), 229–248.

[15] Lewis MD, Lamm C, Segalowitz SJ, Stieben J, Zelazo PD. Neurophysiological correlates of emotion regulation in children and adolescents. *J Cogn Neurosci* (2006) 18:430–43. doi:10.1162/08989290 6775990633.

[16] Rothbart MK, Sheese BE, Rueda MR, Posner MI. Developing Mechanisms of Self-Regulation in Early Life. *Emot Rev* (2011) 3:207–213. doi:10.1177 /17540739 10387943.

[17] Marusak HA, Martin KR, Etkin A, Thomason ME. Childhood Trauma Exposure Disrupts the Automatic Regulation of Emotional Processing. N*europsychopharmacology* (2014) 40:1250–1258. doi:10.1038/npp. 2014.311.

[18] Lourenco F, Casey BJ. Adjusting behavior to changing environmental demands with development. *Neurosci Biobehav Rev* (2013) 37:2233–42. doi:10.1016 /j.neubiorev.2013.03.003.

[19] Cramer SC, Sur M, Dobkin BH, O'Brien C, Sanger TD, Trojanowski JQ, Rumsey JM, Hicks R, Cameron J, Chen D, et al. Harnessing neuroplasticity for clinical applications. *Brain* (2011) 134:1591–609. doi:10.1093/brain/awr039.

[20] Blom EH, Duncan L, Ho TC, Connolly CG, LeWinn KZ, Chesney M, Hecht F, Yang T. The development of an RDoC-based treatment program for adolescent depression: "Training for Awareness, Resilience, and Action" (TARA). Front Hum ... (2014) 8:1–19. doi:10.3389/fnhum. 2014.00630.

[21] Gogtay N, Giedd JN, Lusk L, Hayashi KM, Greenstein D, Vaituzis AC, Nugent TF, Herman DH, Clasen LS, Toga AW, et al. Dynamic mapping of human cortical development during childhood through early adulthood. *Proc Natl Acad Sci U S A* (2004) 101:8174–9. doi:10.1073/pnas.0402680101.

[22] Fjell AM, Walhovd KB, Brown TT, Kuperman JM, Chung Y, Hagler DJ, Venkatraman V, Roddey JC, Erhart M, McCabe C, et al. Multimodal imaging of the self-regulating developing brain. *Proc Natl Acad Sci U S A* (2012) 109:19620–5. doi:10.1073/ pnas.1208243109.

[23] Cunningham MG, Bhattacharyya S, Benes FM. Amygdalo-cortical sprouting continues into early adulthood: implications for the development of normal and abnormal function during adolescence. *J Comp Neuro*l (2002) 453:116–30. doi:10.1002/cne.10376.

[24] Arnsten AFT, Rubia K. Neurobiological circuits regulating attention, cognitive control, motivation, and emotion: disruptions in neurodevelopmental psychiatric disorders. *J Am Acad Child Adolesc Psychiatry* (2012) 51:356–67. doi:10.1016/j.jaac.2012.01.008.

[25] Rubia K. Functional brain imaging across development. *Eur Child Adolesc Psychiatry* (2013) 22:719–31. doi:10.1007/s00787-012-0291-8.

[26] Perlman G, Simmons AN, Wu J, Hahn KS, Tapert SF, Max JE, Paulus MP, Brown GG, Frank GK, Campbell-Sills L, et al. Amygdala response and functional connectivity during emotion regulation: a study of 14 depressed adolescents. *J Affect Disord* (2012) 139:75–84. doi:10.1016/j.jad.2012.01.044.

[27] Silk JS, Steinberg L, Morris AS. Adolescents' emotion regulation in daily life: links to depressive symptoms and problem behavior. *Child Dev* (2003) 74:1869–80.

[28] Cavanagh M, Quinn D, Duncan D, Graham T, Balbuena L. Oppositional Defiant Disorder Is Better Conceptualized as a Disorder of Emotional Regulation. *J Atten Disord* (2014)1–9. doi:10.1177/1087054713520221.

[29] Stringaris A, Goodman R. Longitudinal outcome of youth oppositionality: irritable, headstrong, and hurtful behaviors have distinctive predictions. *J Am Acad Child Adolesc Psychiatry* (2009) 48:404–12. doi:10.1097/CHI.0b013e3181984f30.

[30] Burke JD. An affective dimension within oppositional defiant disorder symptoms among boys: personality and psychopathology outcomes into early adulthood. *J Child Psychol Psychiatry* (2012) 53:1176–1183. doi:10.1111/j.1469-7610.2012.02598.x.

[31] Gadow KD, Drabick DAG. Anger and irritability symptoms among youth with ODD: cross-informant versus source-exclusive syndromes. *J Abnorm Child Psychol* (2012) 40:1073–85. doi:10.1007/s10802-012-9637-4.

[32] Whelan YM, Stringaris A, Maughan B, Barker ED. Developmental continuity of oppositional defiant disorder subdimensions at ages 8, 10, and 13 years and their distinct psychiatric outcomes at age 16 years. *J Am Acad Child Adolesc Psychiatry* (2013) 52:1–2. doi:10.1016/j.jaac. 2013.06.013.

[33] Rowe R, Costello EJ, Angold A, Copeland WE, Maughan B. Developmental pathways in oppositional defiant disorder and conduct disorder. *J Abnorm Psychol* (2010) 119:726–38. doi:10.1037/a0020798.

[34] Shaw P, Stringaris A, Nigg J, Leibenluft E. Emotion dysregulation in attention deficit hyperactivity disorder. *Am J Psychiatry* (2014) 171:276–93. doi:10.1176/appi.ajp. 2013.13070966.

[35] Dougherty LR, Smith VC, Bufferd SJ, Carlson G a, Stringaris a, Leibenluft E, Klein DN. DSM-5 disruptive mood dysregulation disorder: correlates and predictors in young children. *Psychol Med* (2014)1–12. doi:10.1017/S0033291713003115.

[36] Damasio A. Descartes' error: Emotion, rason, and the human brain. New York, NY: *Grosset/Putnam* (1994).

[37] Panksepp J. Affective neuroscience: The foundations of human and animal emotions. New York, NY: *Oxford University Press* (1998).

[38] LeDoux J. The emotional brain: The mysterious underpinnings of emotional life. New York, NY: *Simon and Schuster* (1996).

[39] Hubbard JA, McAuliffe MD, Morrow MT, Romano LJ. Reactive and proactive aggression in childhood and adolescence: precursors, outcomes, processes, experiences, and measurement. *J Pers* (2010) 78:95–118. doi:10.1111/j.1467-6494.2009.00610.x.

[40] Mullin BC, Hinshaw SP. "Emotion regulation and externalizing disorders in children and adolescents," in *Handbook of emotion regulation* (New York, NY: Guilford Press).

[41] Sánchez-Martín JR, Azurmendi A, Pascual-Sagastizabal E, Cardas J, Braza F, Braza P, Carreras MR, Muñoz JM. Androgen levels and anger and impulsivity measures as predictors of physical, verbal and indirect aggression in boys and girls. *Psychoneuroendocrinology* (2011) 36:750–60. doi:10.1016/j.psyneuen.2010.10.011.

[42] Liu J, Portnoy J, Raine A. Association between a marker for prenatal testosterone exposure and externalizing behavior problems in children. *Dev Psychopathol* (2012) 24:1–2. doi:10.1017/S0954579412000363.

[43] Lombardo M V, Ashwin E, Auyeung B, Chakrabarti B, Taylor K, Hackett G, Bullmore ET, Baron-Cohen S. Fetal testosterone influences sexually dimorphic gray matter in the human brain. *J Neurosci* (2012) 32:674–80. doi:10.1523/JNEUROSCI.4389-11.2012.

[44] Beck AT, Brown G, Berchick RJ, Stewart BL, Steer RA. Relationship between hopelessness and ultimate suicide: a replication with psychiatric outpatients. *Am J Psychiatry* (1990) 147:190–5.

[45] Beck AT. "Beyond belief: A theory of modes, personality, and psychopathology.," in Frontiers of cognitive therapy, ed. P. Salkovskis (New York, NY: *Guilford Press*), 1–2.

[46] Spirito A, Overholser J, Hart K. Cognitive characteristics of adolescent suicide attempters. *J Am Acad Child Adolesc Psychiatry* (1991) 30:604–8. doi:10.1097/00004583-199107000-00012.

[47] Cole DA. Psychopathology of adolescent suicide: hopelessness, coping beliefs, and depression. *J Abnorm Psychol* (1989) 98:248–55. doi:10.1037/0021-843X.98.3.248.

[48] Sher L. Teaching medical professionals and trainees about adolescent suicide prevention: five key problems. *Int J Adolesc Med Health* (2012) 24:121–3. doi:10.1515/ijamh.2012.018.

[49] Sher L, Rice TTi, on behalf of the World Federation of Biological Psychiatry (WFSBP) Task Force on Men's Mental Health. Prevention of homicidal behavior in men with psychiatric disorders. *World J Biol Psychiatry* (2015) .

In: Neurobiology of Men's Mental Health
Editors: Leo Sher and Timothy R. Rice
ISBN: 978-1-63483-191-8
© 2015 Nova Science Publishers, Inc.

Chapter 3

FUTURE DIRECTIONS CONCERNING THE IMPACT OF CHILDHOOD AND ADOLESCENT ADVERSITIES IN THE FIELD OF MEN'S MENTAL HEALTH: THE NEW YORK DECLARATION

Timothy R. Rice[1,*], MD, Julia Golier[1, 2], MD, Zoltan Rihmer[3], MD, PhD and Leo Sher[1,2], MD
[1] Department of Psychiatry, Icahn School of Medicine at Mount Sinai, New York, NY, US
[2] James J. Peters Veterans' Administration Medical Center, Bronx, New York, US
3Semmelweis University, Budapest, Hungary

ABSTRACT

Boys, male adolescents, and young men are more often the victims of physical abuse than their female counterparts. Boys may be particularly vulnerable to the psychological effects of physical abuse owing to androgen effects upon the developing central nervous system. Boys and male youth, unlike their female counterparts, tend to bear the stigma of abuse through externalizing behaviors. These disruptive behaviors often engender involvement with the legal system and negative longitudinal outcomes. To address these deficits in men's mental health, this chapter proposes further study and emphasis upon lay and professional education and trauma-focused treatments. Psychotherapeutic approaches which target brain-based vulnerabilities may be of increasing value as our knowledge of the neurobiology of Men's Mental Health increases and expands.

Keywords: men's mental health, childhood trauma, externalizing behaviors, trauma-focused treatment

---

* Corresponding author: Timothy R. Rice, M.D. Icahn School of Medicine at Mount Sinai, Box 1230, One Gustave L. Levy Place.

Members of the World Federation of Biological Psychiatry's Task Force on Men's Mental Health met in New York in November 2014 to discuss future directions concerning the impact of childhood and adolescent adversities in the field of Men's Mental Health. Timothy Rice M.D. (USA), Julia Golier M.D. (USA), Leo Sher, M.D. (USA), and Zoltan Rihmer, M.D., Ph.D. (Hungary) participated in the meeting. The following consensus assessments regarding the current state of knowledge in the field were made.

1. Targets of Abuse and Victims of Trauma

Whereas girls are most likely to be the victims of sexual abuse, boys are much more likely to experience physical abuse and trauma [1]. Boys are much more likely to be admitted to hospitals for serious physical abuse than girls ($P<0.0001$) [2]. Male youth are more likely to witness urban violence than females [3]. Young men are exposed more frequently and to more significant community trauma than young women [4]. Low income children and adolescents are at greatest risk [5].

2. Neural Vulnerability

Abuse disrupts development across multiple functional domains, including the physiological, cognitive, emotional, and social [6]. These central nervous system processes have defined neural correlates, and there is now evidence of direct effects of abuse on the developing brain [7]. For example, there are specific deleterious consequences on the emotion regulation system [8], a system which may be fundamental to a large swath of childhood disorders [9]. Boys relative to girls may be more vulnerable to these negative effects owing to the developmentally-delaying effects of male sex hormones, particularly in the early post-natal periods [10]. The developmental impact of early child abuse on boys may be more significant than on girls. Further studies are needed.

3. Endocrinological Vulnerability

Trauma in children affects the hypothalamic pituitary adrenal (HPA) axis [11]. Elevated basal cortisol levels and decreased HPA axis responsivity to stress have deleterious consequences throughout the life cycle. Boys may be more vulnerable to these trauma-induced alternations of normal physiology than girls. A study of youth exposed to the trauma of an environmental disaster [12] suggest that boys experience higher basal cortisol and increased distress following trauma. This gender-significant finding in children has been replicated [13] but not observed in a third trial [14]. To increase sensitivity, future studies will need to distinguish between the types of trauma.

4. Externalizing Behavior

Following trauma, girls present as more adjusted [15]. Though girls may experience more internalizing symptoms (depression, anxiety, and post-traumatic stress disorder symptoms) than their male counterparts, they generally receive less attention owing to the phenomenology of these symptoms. The attention which externalizing symptoms and behaviors grab towards boys is significant. Too often, these children can be labeled "bad" and either triaged to the justice system or to behavioral management programs to reduce target behaviors [16]. Often, these children are placed on antipsychotics [17]. The serious adverse effects of these agents are more pronounced in children and adolescents than in adults [18, 19].

5. The Justice System

The juvenile justice system is overwhelmingly a male-populated system [20]. These youth have a much higher frequency of trauma histories throughout the world, including in sub-Saharan Africa [21]. A history of trauma places children at a much greater risk for placement in the juvenile justice system [20]. In America, as many as 95% of youth in the juvenile justice system had experienced trauma, and there was a direct correlation between traumatic exposure and the degree of delinquency [22]. Involvement in the juvenile justice system may in itself be traumatic, and opportunities for meaningful therapeutics targeting traumatic histories may be suboptimal.

Given these findings and assessments, the following consensus recommendations were made:

1. Education

Public health and health care worker education interventions are needed to raise recognition in the importance of adverse childhood experiences in young men. Parents commonly underestimate their children's exposure to violence and trauma, with negative consequences [23]. Externalizing symptoms to which children and young men are particularly prone may too often be understood as simply "bad behavior." This can invoke harsh punishments in an escalating cycle of punishment and bad behavior which system theorists correctly note end up victimizing the child [24]. The result may be entry into the legal system, where the chances of rehabilitation decrease. Physicians, nurses, and health care workers should understand the role of trauma in boys and young men and be able to educate parents on the importance of alternate approaches to reaching their sons. This may be of particular importance in lower socioeconomic classes and in the underserved.

2. Trauma-focused treatments

When behavioral treatments, antipsychotics, and the juvenile justice system fail boys and young men curb what may be symptoms of trauma, trials of evidenced-based systematic treatments may be of merit. Boys with a history of interpersonal violence may be particularly response to trauma-focused components in treatments for externalizing behaviors [25]. A systematic review suggests that trauma-focused cognitive behavioral therapy (TF-CBT) has the greatest evidence for efficacy among abused children [26]. Other interventions including attachment-based interventions for parenting may also be of use. The finding that boys with a history of trauma may be particularly vulnerable to environmentally-induced deficits in the emotion regulation system [8] suggests that interventions targeting this neural system may be of particular use. Regulation Focused Psychotherapy for Children with Externalizing Behaviors [RFP-C; 27] is one such manualized, time-limited intervention developed to be used explicitly with this population. The integration of traditional methods of child psychotherapy with a brain-based model of dimensional neuropsychiatric pathology that is compatible with the Research Domain Criteria project [RDoC; 28, 29] of the National Institute of Mental Health (NIMH) permits a targeted intervention for this children that is capable of hypothesis testing within a classification system based on observable behaviors with neural correlates. A gradual transition to the exploration of further biomarkers in these population and the development of targeting therapeutic strategies will be of help.

Men's mental health in children and adolescents needs more attention from clinicians, researchers, and health policy makers. This declaration serves as a preliminary communication of select members of the World Federation of Biological Psychiatry's Task Force on Men's Mental Health concerning future directions for research, clinical care, and public health policy.

Conflict of Interest Statement

The authors declare that the research was conducted in the absence of any commercial or financial relationships that could be construed as a potential conflict of interest.

References

[1] Tolin DF, Foa EB. Sex differences in trauma and posttraumatic stress disorder: a quantitative review of 25 years of research. *Psychol. Bull.* (2006) 132:959–92.

[2] Leventhal JM, Martin KD, Gaither JR. Using US data to estimate the incidence of serious physical abuse in children. *Pediatrics* (2012) 129:458–64.

[3] Zona K, Milan S. Gender differences in the longitudinal impact of exposure to violence on mental health in urban youth. *J. Youth Adolesc.* (2011) 40:1674–90.

[4] Stein BD, Jaycox LH, Kataoka S, Rhodes HJ, Vestal KD. Prevalence of child and adolescent exposure to community violence. *Clin. Child Fam. Psychol. Rev.* (2003) 6:247–64.

[5] Wade R, Shea JA, Rubin D, Wood J. Adverse childhood experiences of low-income urban youth. *Pediatrics* (2014) 134:e13–20.

[6] Toth SL, Cicchetti D. A developmental psychopathology perspective on child maltreatment. Introduction. *Child Maltreat.* (2013) 18:135–9.

[7] Glaser D. The effects of child maltreatment on the developing brain. *Med. Leg. J.* (2014) 82:97–111.

[8] Kim J, Cicchetti D. Longitudinal pathways linking child maltreatment, emotion regulation, peer relations, and psychopathology. *J. Child Psychol. Psychiatry* (2010) 51:706–16.

[9] Rice TR, Hoffman L. Defense mechanisms and implicit emotion regulation: a comparison of a psychodynamic construct with one from contemporary neuroscience. *J. Am. Psychoanal Assoc.* (2014) 62:693–708.

[10] Martel MM, Klump K, Nigg JT, Breedlove SM, Sisk CL. Potential hormonal mechanisms of attention-deficit/hyperactivity disorder and major depressive disorder: a new perspective. *Horm. Behav.*(2009) 55:465–479.

[11] Tarullo AR, Gunnar MR. Child maltreatment and the developing HPA axis. *Horm. Behav.* (2006) 50:632–9.

[12] Vigil JM, Geary DC, Granger DA, Flinn M V. Sex differences in salivary cortisol, alpha-amylase, and psychological functioning following Hurricane Katrina. *Child Dev* (2010) 81:1228–40.

[13] Trickett PK, Gordis E, Peckins MK, Susman EJ. Stress reactivity in maltreated and comparison male and female young adolescents. *Child Maltreat.* (2014) 19:27–37.

[14] Susman EJ, Dockray S, Granger DA, Blades KT, Randazzo W, Heaton JA, Dorn LD. Cortisol and alpha amylase reactivity and timing of puberty: vulnerabilities for antisocial behaviour in young adolescents. *Psychoneuroendocrinology* (2010) 35:557–69.

[15] Villodas MT, Litrownik AJ, Thompson R, Jones D, Roesch SC, Hussey JM, Block S, English DJ, Dubowitz H. Developmental transitions in presentations of externalizing problems among boys and girls at risk for child maltreatment. *Dev. Psychopathol.* (2014)1–15.

[16] Kazdin AE, Rotella C. *The Kazdin Method for parenting the defiant child: With no pills, no therapy, no contest of wills*. Boston, MA: Houghton Mifflin (2008).

[17] Olfson M, Blanco C, Liu S-M, Wang S, Correll CU. National Trends in the Office-Based Treatment of Children, Adolescents, and Adults With Antipsychotics. *Arch. Gen. Psychiatry* (2012) 71:81–90.

[18] Correll CU. Metabolic side effects of second-generation antipsychotics in children and adolescents: a different story? *J. Clin. Psychiatry* (2005) 66:1–2.

[19] Maayan L, Correll CU. Weight gain and metabolic risks associated with antipsychotic medications in children and adolescents. *J. Child Adolesc. Psychopharmacol.* (2011) 21:1–2.

[20] Espinosa EM, Sorensen JR, Lopez MA. Youth pathways to placement: the influence of gender, mental health need and trauma on confinement in the juvenile justice system. *J. Youth Adolesc.* (2013) 42:1824–36.

[21] Atilola O, Omigbodun O, Bella-Awusah T. Lifetime exposure to traumatic events among adolescents in contact with the Nigerian juvenile justice systems compared with a comparison group of secondary school students. *Paediatr. Int. Child Health* (2014) 34:92–100.

[22] Becker SP, Kerig PK. Posttraumatic stress symptoms are associated with the frequency and severity of delinquency among detained boys. *J. Clin. Child Adolesc. Psychol.* (2011) 40:765–71.

[23] Zimmerman GM, Farrell AS. Gender differences in the effects of parental underestimation of youths' secondary exposure to community violence. *J. Youth Adolesc.* (2013) 42:1512–27.

[24] Patterson GR. "The aggressive child: Victim and architect of a coercive system," in *Behavior modification and families: Theory and research*, eds. L. Hamerlynck, L. Handy, E. Mash (New York, NY: Brunner/Mazel.), 1–2.

[25] Shenk CE, Dorn LD, Kolko DJ, Rausch JR, Insana SP. Prior exposure to interpersonal violence and long-term treatment response for boys with a disruptive behavior disorder. *J. Trauma Stress* (2014) 27:585–92.

[26] Leenarts LEW, Diehle J, Doreleijers TAH, Jansma EP, Lindauer RJL. Evidence-based treatments for children with trauma-related psychopathology as a result of childhood maltreatment: a systematic review. *Eur. Child Adolesc. Psychiatry* (2012) 22:269–283.

[27] Hoffman L, Rice T. *Regulation-Focused Psychotherapy for Children with Externalizing Behaviors (RFP-C).* New York, NY: Routledge.

[28] Insel TR. The NIMH Research Domain Criteria (RDoC) Project: Precision Medicine for Psychiatry. *Am. J. Psychiatry* (2014) 171:395–397.

[29] Insel T, Cuthbert B, Garvey M, Heinssen R, Pine DS, Quinn K, Sanislow C, Wang P. Research domain criteria (RDoC): toward a new classification framework for research on mental disorders. *Am. J. Psychiatry* (2010) 167:748–751.

In: Neurobiology of Men's Mental Health
Editors: Leo Sher and Timothy R. Rice

ISBN: 978-1-63483-191-8
© 2015 Nova Science Publishers, Inc.

Chapter 4

NEUROBIOLOGICAL BASIS OF ATTENTION DEFICIT HYPERACTIVITY DISORDER (ADHD). ARE MALES DIFFERENT?

Jose Martinez-Raga[1]*, MD, PhD, Carlos Knecht[2], MD, Raquel de Alvaro[3], MD, Amparo Ferreros[4], MD, and Nestor Szerman[5], MD

[1] Teaching Unit of Psychiatry and Psychological Medicine, Hospital Universitario Doctor Peset, University of Valencia, and University Cardenal Herrera CEU, Valencia, Spain

[2] Área de Salud Mental, Hospital Padre Jofré, Valencia, Spain

[3] Hospital General de Castellón, Consorcio Hospitalario Provincial, Castellón, Spain

[4] Teaching Unit of Psychiatry and Psychological Medicine, Hospital Universitario Doctor Peset, Valencia, Spain

[5] Department of Psychiatry, Hospital Universitario Gregorio Marañón, Madrid, Spain

ABSTRACT

Attention-deficit/hyperactivity disorder (ADHD) is a multifactorial and heterogeneous neurodevelopmental disorder characterized by a persistent pattern of age-inappropriate levels of inattention, hyperactivity, and/or impulsivity. Evidence from epidemiological studies indicate that the prevalence of ADHD in males and females may be more similar than once thought, with a ratio of approximately 2:1 in children and 1.6:1 in adults. In the present chapter it is aimed to review the neurobiological mechanisms that underlie gender differences in attention, hyperactivity and impulsivity, focusing on the differential aspects of ADHD in males. There are gender-related differences in comorbid

* Corresponding author: Jose Martinez-Raga MD, PhD, Unidad Docente de Psiquiatría y Psicologia Medica, Hospital Universitario Dr. Peset, Avda. Gaspar Aguilar 90 , 46017 Valencia, Spain, Tel./Fax: +34 96 1622328, E-mail: martinez_josrag@gva.es

psychiatric conditions among subjects with ADHD along the life-span, with males having higher rates of "externalizing" disorders, while females may have more frequently "internalizing" disorders. ADHD is a polygenic disorder and various evidence suggest that genetic effects may be more pronounced in boys, and boys may have a lower threshold for the expression of neurodevelopmental and conduct disorders. A growing literature reveals ADHD-related dysfunctions in multiple neuronal networks involved in higher-level cognitive and sensorimotor functions, including the visual system, as well as in the default network. However, the majority of neuroimaging studies of ADHD have used largely or exclusively male individuals. Scattered reports have suggested gender differences in patterns of neural activation consistent with hypotheses and evidences suggesting right frontal abnormalities, as well as altered cerebellar-prefrontal-striatal networks in ADHD, more prominent in males. Future research, using large enough samples should expand our understanding about gender effects in ADHD, as this will improve diagnosis, prognosis and treatment.

Keywords: ADHD, neurobiology, gender differences, males

INTRODUCTION

Attention-deficit/hyperactivity disorder (ADHD) is a multifactorial and heterogeneous neurodevelopmental disorder characterized by a persistent pattern of age-inappropriate levels of inattention, hyperactivity, and/or impulsivity [1, 2]. ADHD is the most common neurobiological disorder of childhood onset, with a prevalence of approximately 5-8% in children and 2.5-4% in adults worldwide [1, 3, 4], more common than other more researched psychiatric disorders like schizophrenia or bipolar disorder. The prevalence of ADHD is even higher in clinical samples with patients with other primary psychiatric disorders [5-7], as well as in studies conducted with prison populations [8, 9]. In addition, a substantial body of research suggests cross-cultural stability of the ADHD phenotype. Studies have revealed that ADHD is not exclusive of western societies; there are no significant cross-cultural differences in prevalence or symptomatic expression across countries and cultures [3, 4].

ADHD has been traditionally described as being more common in males than in females, particularly among children and adolescents, with male-female ratios ranging from 2:1 to 9:1 [10, 11]. However, evidence has emerged indicating that prevalence rates of ADHD in males and females may be more similar than once thought, although it is more frequent in males than in females in the general population, with a ratio of approximately 2:1 in children and 1.6:1 in adults [1,3]. In addition, research indicates that there are subtle, but significant gender differences in ADHD symptom profiles and clinical characteristics. The predominantly inattentive ADHD presentation is reportedly more prevalent amongst girls, whereas the hyperactive-impulsive and combined presentations have been described to be more common in boys [1, 12-15]. Gender differences in the clinical presentation of the disorder and in comorbid neurodevelopmental and other psychiatric disorders may reasonably account for gender specific differences in referrals and diagnosis and consequently may account for the disparity in gender differences in prevalence rates [14, 16].

Although once conceptualized solely as a childhood disorder, the persistence of ADHD from childhood throughout adolescence into adulthood is now fully acknowledged [1]. For a significant proportion of affected children and adolescents ADHD persists into adulthood, resulting in ongoing psychological and functional impairment that often persists through adult life, so that core symptoms of ADHD, as well as problems like low self-esteem or symptoms of impaired mood regulation may continue to be present into adult life in up to two thirds of affected children and adolescents [17, 18]. Along the lifespan, ADHD is associated with significant personal, familial, medical and social complications, important functional impairment, and high rates of comorbid psychiatric disorders. Among children and adolescents with ADHD, conduct disorders, anxiety disorders, and mood disorders are the most common concurrent psychiatric disorders, whilst mood disorders, anxiety disorders, substance use disorders, and personality disorders are the most frequent comorbid mental disorders in adult patients with ADHD [1, 5, 7, 19-23]. However, there does not appear to exist solid evidences that gender moderates the association between ADHD and other psychiatric disorders [24].

It is now generally agreed that there are inherent sex and gender differences in healthy individuals, as well in neuropsychiatric populations, in many species. These differences exist in regard to neurodevelopment, ongoing brain function and structure, behavior and cognition, and in rates of incidence, clinical features and progression [15, 25, 26]. Whilst advances in neuroscience, including recent neuroimaging and genetic studies have provided a much better understanding on the neurobiological substrates, the etiology and the clinical characteristics of ADHD, there is limited research regarding the neurobiological underpinnings of gender differences in ADHD. In the present chapter it is aimed to explore gender differences in the clinical presentation and in the commonly comorbid psychiatric disorders and to review the neurobiological mechanisms that underlie gender differences in attention, hyperactivity and impulsivity, focusing on the differential aspects of ADHD in males.

Gender Differences in ADHD Symptoms and Comorbid Neurodevelopmental and Other Psychiatric Disorders

Children, adolescents and adults with ADHD show various degrees of impairment in measures of neuropsychological functioning compared to their age-matched controls. These impairments include deficits in executive functions, such as planning, task flexibility, reasoning or decision-making, working memory deficits, failures to inhibit or delay behavioral responses, reading disabilities, or difficulties in cognitive processing, also known as sluggish cognitive tempo [27-31]. Females with ADHD are more likely to exhibit attention difficulties whilst symptoms of hyperactivity-impulsivity are typically more common among boys with the disorder [2, 12, 13]. In addition, earlier reports suggested gender differences in neuropsychological dysfunction on a variety of neurocognitive domains, particularly in executive functions, reporting more severe impairments in boys than in girls with ADHD [32]. However, growing evidence [33-36], including two meta-analyses [28, 37] suggests that males show similar degree of cognitive impairment as females, or even the latter may show poorer intellectual functioning than ADHD males [28, 37].

Along the lifespan, ADHD is associated with high rates of comorbid psychiatric disorders [5, 20, 22, 38-40]. During childhood and adolescence conduct disorder, oppositional defiant disorder (ODD), anxiety disorders, mood disorders, and specific learning disorders are the most common comorbid psychiatric disorders. In adult individuals, anxiety and mood disorders, as well as personality disorders and substance use disorders are amongst the most common comorbid psychiatric disorders. It is estimated that at least 60-80% of children, adolescents and adults with ADHD have another neurodevelopmental or psychiatric disorder. In a longitudinal, case-control study with large sample of males and females with and without ADHD assessed prospectively from childhood into adolescence and young adulthood with the aim to establish whether sex moderated course of ADHD and the course and pattern of covariance of comorbid psychopathology, Monuteaux and colleagues [41] found that in both genders, age exhibited a similar effect on the decline of ADHD symptoms, but among females there was a greater stability in comorbid psychopathology from childhood into adolescence. Moreover, while in females, the stability of comorbid psychopathology remained significant after accounting for the correlation between adolescent comorbid psychopathology and ADHD, in males, childhood and adolescent comorbid psychopathology were not correlated with adolescent ADHD [41].

It is generally accepted that there are gender-related differences in comorbid psychiatric conditions among subjects with ADHD along the life-span, these differences are commonly described in clusters, so that boys have been found to have higher rates of "externalizing" disorders, while girls have been reported to have more frequently "internalizing" disorders [15]. Compared to girls, boys with ADHD are also more likely to have more learning problems [16]. In addition, boys with ADHD may be more prone to present with comorbid conduct disorder, with ODD, and with externalizing symptoms, including stress intolerance, poor social skills and awkwardness [12, 16, 42]. Males with ADHD have also higher rates of criminal behaviors than their female counterparts [43, 44]. In contrast, girls and adult women with ADHD have been reported to exhibit higher levels of inattentive symptoms, greater intellectual impairments, and more internalizing symptoms, including greater emotional instability and a more negative self-image, and higher rates of separation anxiety disorder, eating disorders and somatization disorders, than males with ADHD [18, 33, 43, 45]. Similarly, adult women diagnosed with ADHD are more likely to report depressive symptoms, to be more stressed and anxious, to have more external locus of control and lower self-esteem than women who did not meet diagnostic criteria for ADHD [46]. Interestingly, in a naturalistic 10-year prospective case-control study in males and females with and without ADHD conducted to assess the effects of ADHD and its treatment on height and weight, whilst there was no no evidence that ADHD or that stimulant treatment was associated with differences at follow-up in any growth outcomes; among subjects with ADHD, major depression was associated with significantly larger weight in females and smaller height in males [47].

Pathophysiology of ADHD in the Context of Gender Differences

The precise etiological and pathophysiological substrates of ADHD remain to be established, but like the majority of neurodevelopmental disorders, ADHD is considered to have a multifactorial origin, resulting from the complex interaction of biological and environmental risk factors, including multiple genetic factors in conjunction with prenatal events and infant complications [48, 49]. Nonetheless, genetic predisposition appears to be the most important etiological factor for developing ADHD. Furthermore, multiple lines of evidence have suggested a strong genetic component of inattentive and impulsivity phenotypes.

Genetic Substrates of ADHD

In humans and experimental animals, many adult brain functions, such as cognitive abilities, attention, mood, reward and motivation, are influenced by the activational influences of gonadal factors in both sexes [50-52]. In addition, growing body of evidence indicates that neurodevelopmental and neuropsychiatric disorders, including ADHD, have significant gender differences with regards to vulnerability, prevalence, presenting symptoms, the neurobiological substrates, evolution and treatment outcomes [53-55], partly as a consequence of the interaction of hormonal and genetic factors throughout the lifespan to promote sex dimorphisms in normal brain circuitries regulating behavior [51]. Moreover, whilst sociocultural factors may contribute to underlying differences in neurological development, as well as in cognitive functions, even when sociocultural differences have been accounted for, sex differences in brain function may persist, possibly through the interaction of genetic and hormonal mechanisms [56].

Family, adoption and twin studies have evidenced a high heritability of ADHD. Indeed, it has been estimated that approximately 75-80% of the origin of ADHD is attributed to genetic factors [39, 57, 58]. In recent years, numerous molecular genetic studies, including linkage studies, candidate-gene association studies, genome-wide association studies and genome-wide copy number variation studies, have been published to investigate susceptibility loci for ADHD, and thus have identified several candidate genes involved in neurotransmission. Therefore, for the most part, ADHD is not caused by single common genetic variants, but is rather a polygenic disorder with at least 50 genes involved [57, 59, 60]. This may help to explain the clinical heterogeneity of ADHD symptom presentations and the high comorbidity with other psychiatric disorders [61,62]. Despite the importance of genetic factors, 20% of the variance of ADHD symptoms is due to environmental factors, such as prenatal exposure to tobacco or lead [62].

Various evidence suggest that genetic effects may be more pronounced in boys and, according to the multiple threshold theory, boys may have a lower threshold for the expression of neurodevelopmental and conduct disorders, and consequently may explain the higher prevalence of these disorders in males compared to females [45, 63]. Emerging evidence from clinical and from animal studies suggest the existence of multiple gene expression differences between regions of male and female brains relevant to attention and impulsivity, including the anterior cingulate cortex, the dorsolateral prefrontal cortex, and the cerebellum [64]. These expression differences are likely to be the result of a complex interaction between sex chromosomes and hormonal factors [15].

Neuroimaging Studies of ADHD

Over the recent two decades, neuroimaging studies have provided evidence of the brain areas involved and the nature of the disturbance in brain development that underpins ADHD. Convergent data have identified brain areas that most strongly contribute to challenges with of attention, impulsivity and activity control, consistently emphasizing the role of the frontostriatal network, that involves structures like the lateral prefrontal cortex, the dorsal anterior cingulate cortex, the caudate nucleus, and putamen [48, 65]. Other brain regions, all of them interconnected, such as the cerebellum, particularly the midline vermis have also been identified as being crucial in understanding the pathophysiology of ADHD [65, 66]. Nonetheless, a growing body of literature reveals ADHD-related dysfunctions in multiple neuronal networks involved in higher-level cognitive and sensorimotor functions, including the visual system, as well as in the default network [67]. In addition, a pivotal longitudinal study using serial anatomic magnetic resonance imaging (MRI) and computational neuroanatomic techniques exploring the longitudinal trajectories of brain maturation showed a delay in cortical thickness and surface area [68-70]. This delay in cortical maturation was most prominent in prefrontal regions important in the control of cognitive processes and complex executive functions [69, 70]. Furthermore, studies have revealed that in children with ADHD the grey matter peaks occurred about three years later than in healthy controls [69].

The total cerebral volume is slightly reduced in individuals with ADHD, compared to those without the disorder [68]. In addition, several studies have evidenced that that the dorsolateral prefrontal cortex, the dorsal anterior cingulate cortex, and the corpus callosum are slightly thinner in individuals with ADHD; the striatum has also been reported to be smaller in ADHD, more pronouncedly on the right side of the putamen and the anterior parts of the caudate [65, 66, 71]. Similarly, decreased volume of the cerebellum has been found in children with ADHD, compared to children without [72, 73]. These interconnected yet distinct brain areas are involved attention-control, motor-planning skills, temporal information processing, and complex executive functions, such as allocation of attention, planning, and decision-making. Indeed, low prefrontal cortex and right frontal-striatal-cerebellar activation while trying to modulate attention, impulsive behavior, and activity have been reported [74, 75]. Additionally, some authors have reported hyporeactivity of the autonomic nervous system among ADHD individuals, especially when comorbidity with conduct disorder or ODD is present [76, 77]. Dopamine and noradrenaline imbalances at dopamine D1 and noradrenergic α2a receptors are key neurotransmitters in differentiating signals from noise in the prefrontal cortex and is likely related to attention regulation in ADHD patients [78].

Relatively few neuroimaging studies have adequately examined gender differences in the neuroanatomical characteristics underlying symptoms of inattention, hyperactivity and impulsivity [79]. Indeed, the majority of neuroimaging studies exploring the functional neuroanatomical substrates of ADHD have used largely or exclusively male individuals, as evidenced in a meta-analysis of structural imaging studies of ADHD that revealed that most studies included either only male or mostly male samples [80]. In addition, separate results for male and female participants, as is also the case for ADHD subtypes, are not usually reported in the literature [67].

The scattered studies published exploring gender differences in the neuroanatomy of ADHD have not reported major differences in brain structure or brain function between males and females diagnosed of ADHD [81, 82]. This is thus consistent with the relatively subtle differences reported in attention, hyperactivity and impulsivity symptoms, and neuropsychological tests between male and female patients with ADHD [15]. However, it is also possible that true gender differences in the neuroanatomical characteristics of individuals with ADHD may be at least partly explained by confounding factors, such as the age of the patients being scanned, the small study sizes, the various treatment regimes and of the variability inherent in the diagnosis of the disorder. Some of these confounding factors were addressed in a study assessing functional MRI and a working memory task to examine gender differences in neural activation functioning of adults with ADHD during performance of a verbal working memory task [79].

In their study, using functional MRI, Valera and colleagues [79] reported that whilst adult ADHD males and females did not differ in their behavioral performance on a working memory task, male ADHD patients exhibited less neural activity in a network of brain regions that was lateralized to right frontal and subcortical regions and left occipital and cerebellar regions relative to control males; in contrast, adult female individuals with ADHD did not show a similar attenuation in activity relative to the female comparison subjects. This pattern of neural activation differences is consistent with hypotheses and data which implicate right frontal abnormalities, as well as altered cerebellar-prefrontal-striatal networks, in individuals with ADHD, at least in males [48, 65, 67]. It is important to note that in this study female individuals had levels of ADHD symptoms and behavioral performance similar to those observed in the male cohort with ADHD [79].

A positron emission tomography (PET) study with a small sample of adolescents with and without ADHD provides additional evidence on gender differences in the neuroanatomy of ADHD [83]. There were no statistically significant differences in global or regional cerebral glucose metabolism between ADHD and healthy adolescents, but girls with ADHD had significantly lower glucose metabolism rates than control girls and male adolescents with ADHD. These findings were not replicated in a subsequent PET study of adolescent girls with ADHD conducted by the same research group [84]. Likewise, in a functional MRI study exploring prefrontal cortex functioning during working memory tasks in a sample of adolescent girls with and without ADHD found no between-group differences in neural activation [85]. In addition, Castellanos and colleagues [81] reported significantly smaller volumes in the posterior-inferior cerebellar vermis, although total brain volume was not significantly smaller in a quantitative brain magnetic resonance imaging in a larger sample of girls with ADHD compared with matched control girls. These findings are consistent with the different neuroimaging studies reporting hypoactivation in the frontoparietal executive control network, putamen, and ventral attention network, as shown in a meta-Analysis of 55 fMRI studies which included mostly male study groups [67]. This may suggest that the effects of ADHD in prefrontal and possibly other brain regions could be different for females and males. Indeed, further insight on the putative gender differences in brain development comes from a key longitudinal study applying surface-based morphometry in >1250 longitudinally acquired brain scans from 647 healthy individuals aged 3–30 years, that revealed how biologically distinct determinants of cortical volume contribute to cortical volume change in a manner that varies dramatically as a function of sex and developmental stage [26]. Raznahan and colleagues [26] showed that not all cortical volume change is created equal, because cortical volume change is so differently composed of changes in cortical thickness, convex hull area, and gyrification index, depending on the sex and developmental phase under consideration. Furthermore, sex differences in cortical volume change with age and are largely driven by the delayed peak and slower subsequent decline of convex hull area in males compared with females [26].

Conclusion

ADHD is well–accepted to be more common among males than among females along the life-span. There are noticeable gender differences in predominant symptom dimensions, so that the hyperactive-impulsive and combined ADHD presentations have been reported to be more common in boys, whereas the inattentive ADHD presentation is more frequent in girls [1, 2, 14]. ADHD is the most researched neurodevelopmental disorder. Indeed, over the recent two decades, neuroimaging studies have provided solid evidence on the brain networks that most strongly contribute to attention, impulsivity and activity control [65-67, 71-73]. Likewise, evidence has emerged on the longitudinal trajectories of brain maturation, showing a delay in cortical maturation in children with ADHD compared to healthy age-matched controls [68-70]. However, relatively few neuroimaging studies have adequately examined gender differences in the neuroanatomical substrates of ADHD, partly as they largely or exclusively used male individuals; furthermore, many studies suffered from small sample sizes, referral biases, differences in diagnostic procedures and possible rater influences. Nonetheless, male ADHD patients have been found to have significantly less neural activity in a network of brain regions that was lateralized to right frontal and subcortical regions and left occipital and cerebellar regions [79]. Future research, including neuroimaging and longitudinal studies, using large enough sample sizes should expand these findings with large enough study groups to assess gender differences in specific cerebral networks and brain developmental trajectories. Understanding gender effects in ADHD is important in order improve diagnosis, prognosis and treatment.

References

[1] American Psychiatric Association. Diagnostic and statistical manual of mental disorders (5th ed.). Arlington, VA, *American Psychiatric Publishing,* 2013.

[2] Feldman HM, Reiff MI: Clinical practice. Attention deficit-hyperactivity disorder in children and adolescents. *N Engl J Med* 2014; 370:838-846.

[3] Fayyad J, De Graaf R, Kessler R, Alonso J, Angermeyer M, Demyttenaere K, De Girolamo G, Haro JM, Karam EG, Lara C, Lépine JP, Ormel J, Posada-Villa J, Zaslavsky AM, Jin R: Cross-national prevalence and correlates of adult attention-deficit hyperactivity disorder. *Br J Psychiatry* 2007; 190:402-409.

[4] Polanczyk GV, Willcutt EG, Salum GA, Kieling C, Rohde LA: ADHD prevalence estimates across three decades: an updated systematic review and meta-regression analysis. *Int J Epidemiol* 2014; 43:434-842.

[5] Kooij JJ, Huss M, Asherson P, Akehurst R, Beusterien K, French A, Sasané R, Hodgkins P: Distinguishing comorbidity and successful management of adult ADHD. *J Atten Disord* 2012; 16(5 Suppl):3S-19S.

[6] van Emmerik-van Oortmerssen K, van de Glind G, van den Brink W, Smit F, Crunelle CL, Swets M, Schoevers RA: Prevalence of attention-deficit hyperactivity disorder in substance use disorder patients: A meta-analysis and meta-regression analysis. *Drug Alcohol Depend* 2012; 122:11-19.

[7] Yoshimasu K, Barbaresi WJ, Colligan RC, Voigt RG, Killian JM, Weaver AL, Katusic SK: Childhood ADHD is strongly associated with a broad range of psychiatric disorders during adolescence: a population-based birth cohort study. *J Child Psychol Psychiatry* 2012; 53:1036-1043.

[8] Colins O, Vermeiren R, Vreugdenhil C, van den Brink W, Doreleijers T, Broekaert E: Psychiatric disorders in detained male adolescents: a systematic literature review. *Can J Psychiatry* 2010; 55:255-263.

[9] Gudjonsson GH, Sigurdsson JF, Sigfusdottir ID, Young S: A National Epidemiological study of offending and its relationship with ADHD symptoms and associated risk factors. *J Atten Disord* 2014; 18:3-13.

[10] American Psychiatric Association. Diagnostic and statistical manual of mental disorders. Fourth Edition, Text Revision (DSM-IV TR). Washington DC, *American Psychiatric Publishing,* 2000.

[11] Willcutt EG: The prevalence of DSM-IV attention-deficit/hyperactivity disorder: a meta-analytic review. *Neurotherapeutics* 2012; 9:490–499.

[12] Abikoff HB, Jensen PS, Arnold LLE, Hoza B, Hechtman L, Pollack S, Wigal T: Observed classroom behavior of children with ADHD: Relationship to gender and comorbidity. *J Abn Child Psychol* 2002; 30:349–359.

[13] Biederman J, Faraone SV: The Massachusetts General Hospital studies of gender influences on attention-deficit/hyperactivity disorder in youth and relatives. *Psychiatr Clin North Am* 2004; 27:225-232.

[14] Lahey BB, Applegate B, McBurnett K, Biederman J, Greenhill L, Hynd GW, Richters J: DSM-IV field trials for ADHD in children and adolescents. *Am J Psychiatry* 1994; 151:1673–1685.

[15] Trent S, Davies W: The influence of sex-linked genetic mechanisms on attention and impulsivity. *Biol Psychol* 2011; 89:1–13.

[16] Biederman J, Mick E, Faraone SV, Braaten E, Doyle A, Spencer T, Wilens TE, Frazier E, Johnson MA: Influence of gender on attention deficit hyperactivity disorder in children referred to a psychiatric clinic. *Am J Psychiatry* 2002; 159:36-42.

[17] Faraone SV, Biederman J, Mick E: The age-dependent decline of attention deficit hyperactivity disorder: a meta-analysis of follow-up studies. *Psychol Med* 2006; 36:159–165.

[18] Turgay A, Goodman DW, Asherson P, Lasser RA, Babcock TF, Pucci ML, Barkley R: Lifespan Persistence of ADHD: *J Clin Psychiatry* 2012; 73:192–201.

[19] Charach A, Yeung E, Climans T, Lillie E: Childhood attention-deficit/hyperactivity disorder and future substance use disorders: comparative meta-analyses. *J Am Acad Child Adolesc Psychiatry* 2011; 50:9-21.

[20] Martinez-Raga J, Knecht C, de Alvaro R, Szerman N, Ruiz P: Addressing dual diagnosis patients suffering from attention deficit hyperactivity disorders and comorbid

substance use disorders: A review of treatment considerations. *Addict Dis Treat* 2013; 12:213-230.

[21] Miller CJ, Flory JD, Miller SR, Harty SC, Newcorn JH, Halperin JM: Childhood ADHD and the emergence of personality disorders in adolescence: A prospective follow-up study. *J Clin Psychiatry* 2008; 69:1477-1484.

[22] National Collaborating Centre for Mental Health. Attention Deficit Hyperactivity Disorder. The NICE Guideline on diagnosis and management of ADHD in children, young people and adults. National Clinical Practice Guideline Number 72. London: *Alden Press;* 2009.

[23] Piñeiro-Dieguez B, Balanzá-Martínez V, García-García P, Soler-López B, CAT Study Group: Psychiatric comorbidity at the time of diagnosis in adults with attention deficit hyperactivity disorder (ADHD) – the CAT study. *J Atten Disord* 2014 Jan 24. [Epub ahead of print]

[24] Biederman J, Faraone SV, Monuteaux MC, Bober M, Cadogen E: Gender effects on attention-deficit/hyperactivity disorder in adults, revisited. *Biol Psychiatry* 2004; 55:692-700.

[25] Holden C: Sex and the suffering brain. *Science.* 2005; 308:1574.

[26] Raznahan A, Shaw P, Lalonde F, Stockman M, Wallace GL, Greenstein D, Clasen L, Gogtay N, Giedd JN: How does your cortex grow? *J Neurosci* 2011; 31:7174-7177.

[27] Alderson RM, Kasper LJ, Hudec KL, Patros CH: Attention-deficit/hyperactivity disorder (ADHD) and working memory in adults: a meta-analytic review. *Neuropsychology* 2013; 27:287-302.

[28] Gershon J: A metaanalytic review of gender differences in ADHD. *J Atten Disord* 2002; 5:143–154.

[29] Saxbe C, Barkley RA: The second attention disorder? Sluggish cognitive tempo vs. attention-deficit/hyperactivity disorder: update for clinicians. *J Psychiatr Pract* 2014; 20:38-49.

[30] Seidman LJ: Neuropsychological functioning in people with ADHD across the lifespan. *Clin Psychol Rev* 2006; 26:466-85.

[31] Yoshimasu K, Barbaresi WJ, Colligan RC, Killian JM, Voigt RG, Weaver AL, Katusic SK: Gender, attention-deficit/hyperactivity disorder, and reading disability in a population-based birth cohort. *Pediatrics* 2010; 126:e788-795.

[32] Seidman LJ, Biederman J, Faraone SV, Weber W, Mennin D, Jones J: A pilot study of neuropsychological function in girls with ADHD. *J Am Acad Child Adolesc Psychiatry* 1997; 36:366-373.

[33] Arcia E, Conners CK: Gender differences in ADHD? *J Dev Behav Pediatr* 1998; 19:77–83.

[34] Seidman LJ, Biederman J, Monuteaux MC, Valera E, Doyle AE, Faraone SV: Impact of gender and age on executive functioning: do girls and boys with and without attention deficit hyperactivity disorder differ neuropsychologically in preteen and teenage years? *Dev Neuropsychol* 2005; 27:79-105.

[35] Rucklidge JJ: Gender differences in ADHD: implications for psychosocial treatments. *Expert Rev Neurother* 2008; 8:643-655.

[36] Skogli EW, Teicher MH, Andersen PN, Hovik KT, Øie M: ADHD in girls and boys--gender differences in co-existing symptoms and executive function measures. *BMC Psychiatry* 2013; 13:298.

[37] Gaub M, Carlson CL: Gender differences in ADHD: a meta-analysis and critical review. *J Am Acad Child Adolesc Psychiatry* 1997; 36:1036-1045.
[38] Brown TE: ADHD comorbidities: handbook for ADHD complications in children and adults. *American Psychiatric Publishing*, 2009.
[39] Faraone SV, Biederman J, Mick E, Doyle AE, Wilens T, Spencer T, Frazier E, Mullen K: A family study of psychiatric comorbidity in girls and boys with attention-deficit/hyperactivity disorder. *Biol Psychiatry* 2001; 50:586-592.
[40] Schubiner H: Substance abuse in patients with attention-deficit hyperactivity disorder : therapeutic implications. *CNS Drugs* 2005; 19:643–655.
[41] Monuteaux MC, Mick E, Faraone SV, Biederman J: The influence of sex on the course and psychiatric correlates of ADHD from childhood to adolescence: a longitudinal study. *J Child Psychol Psychiatry* 2010; 51:233-241.
[42] Stein MA, Sandoval R, Szumowski E, Roizen N, Reinecke M, Blondis T, Klein A: Psychometric characteristics of the Wender Utah Rating Scale (WURS): reliability and factor structure for men and women. *Psychopharmacol Bull* 1995; 31:423–431.
[43] Levy F, Hay DA, Bennett KS, McStephen M: Gender differences in ADHD subtype comorbidity. *J Am Acad Child Adolesc Psychiatry* 2005; 44:368-376.
[44] Quinn PO: Treating adolescent girls and women with ADHD: Gender specific issues. *J Clin Psychol* 2005; 61: 579–587.
[45] Kerekes N, Lundström S, Chang Z, Tajnia A, Jern P, Lichtenstein P, Nilsson T, Anckarsätter H: Oppositional defiant-and conduct disorder-like problems: neurodevelopmental predictors and genetic background in boys and girls, in a nationwide twin study. *Peer* J 2014; 2:e359.
[46] Rucklidge JJ, Kaplan BJ: Psychological functioning in women identified in adulthood with attention deficit/hyperactivity disorder. *J Atten Disord* 1997; 2:167–76.
[47] Biederman J, Spencer TJ, Monuteaux MC, Faraone SV: A naturalistic 10-year prospective study of height and weight in children with attention-deficit hyperactivity disorder grown up: sex and treatment effects. *J Pediatr* 2010; 157:635-40.
[48] Curatolo P, D'agati E, Moavero R: The neurobiological basis of ADHD. *Ital J Pediatr* 2010; 36:79.
[49] Pennington BF, Mc Grath LM, Rosenberg J, Barnard H, Smith SD, Willcut EG, Friend A, Defries JC, Olson RK: Gene X environment interactions in reading disability and attention-deficit/hyperactivity disorder. *Dev Psychol* 2009; 45:77-89.
[50] Daniel JM: Effects of oestrogen on cognition: what have we learned from basic research? *J Neuroendocrinol* 2006; 18:787–795.
[51] Gillies GE, Virdee K, McArthur S, Dalley JW: Sex-dependent diversity in ventral tegmental dopaminergic neurons and developmental programing: A molecular, cellular and behavioral analysis. *Neuroscience* 2014; 282C:69-85.
[52] McEwen BS, Gianaros PJ: Central role of the brain in stress and adaptation: links to socioeconomic status, health, and disease. *Ann N Y Acad Sci* 2010; 1186:190–222.
[53] Cahill L: Why sex matters for neuroscience. *Nat Rev Neurosci* 2006; 7:477–484.
[54] Cosgrove KP, Mazure CM, Staley JK: Evolving knowledge of sex differences in brain structure, function, and chemistry. *Biol Psychiatry* 2007; 62:847–855.
[55] Gillies GE, McArthur S: Estrogen actions in the brain and the basis for differential action in men and women: a case for sex-specific medicines. *Pharmacol Rev* 2010; 62:155–198.

[56] Guiso L, Monte F, Sapienza P, Zingales L: Diversity. Culture, gender, and math. *Science* 2008; 320:1164–1165.

[57] Franke B, Faraone SV, Asherson P, Buitelaar J, Bau CH, Ramos-Quiroga JA, Mick E, Grevet EH, Johansson S, Haavik J, Lesch KP, Cormand B, Reif A; International Multicentre persistent ADHD Collaboration: The genetics of attention deficit/hyperactivity disorder in adults, a review. *Mol Psychiatry* 2012; 17:960-987.

[58] Smith AK, Mick E, Faraone SV. Advances in genetic studies of attention-deficit/hyperactivity disorder. *Curr Psychiatry Rep* 2009; 11: 143-148.

[59] Akutagava-Martins GC, Salatino-Oliveira A, Kieling CC, Rohde LA, Hutz MH: Genetics of attention-deficit/hyperactivity disorder: current findings and future directions. *Expert Rev Neurother* 2013; 13:435-445.

[60] Li Z, Chang SH, Zhang LY, Gao L, Wang J: Molecular genetic studies of ADHD and its candidate genes: a review. *Psychiatry Res* 2014; 219:10-24.

[61] Beaver KM, Nedelec JL, Rowland MW, Schwartz JA: Genetic risks and ADHD symptomatology: exploring the effects of parental antisocial behaviors in an adoption-based study. *Child Psychiatry Hum Dev* 2012; 43:293-305.

[62] Schilling CM, Walsh A, Yun I: ADHD and criminality: a primer on the genetic, neurobiological, evolutionary and treatment literature for criminologists. *J Criminal Justice* 2011; 39:3-11.

[63] Rhee SH,Waldman ID, Hay DA, Levy F: Sex differences in genetic and environmental influences on DSM-III-R attention-deficit/hyperactivity disorder. *J Abn Psychol* 1999; 108:24–41.

[64] Vawter MP, Evans S, Choudary P, Tomita H, Meador-Woodruff J, Molnar M: Gender-specific gene expression in postmortem human brain: localization to sex chromosomes. *Neuropsychopharmacology* 2004; 29:373–384.

[65] Shaw P: ADHD: 10 Years Later. *Cerebrum* 2013; 2013:11.

[66] Cherkasova MV, Hechtman L: Neuroimaging in attention-deficit hyperactivity disorder: beyond the frontostriatal circuitry. *Can J Psychiatry* 2009; 54:651-664.

[67] Cortese S, Kelly C, Chabernaud C, Proal E, Di Martino A, Milham MP, Castellanos FX: Toward systems neuroscience of ADHD: a meta-analysis of 55 fMRI studies. *Am J Psychiatry* 2012; 169:1038-1055.

[68] Castellanos FX, Lee PP, Sharp W, Jeffries NO, Greenstein DK, Clasen LS, Blumenthal JD, James RS, Ebens CL, Walter JM, Zijdenbos A, Evans AC, Giedd JN, Rapoport JL: Developmental trajectories of brain volume abnormalities in children and adolescents with attention-deficit/hyperactivity disorder. *JAMA* 2002; 288:1740-1748.

[69] Shaw P, Eckstrand K, Sharp W, Blumenthal J, Lerch JP, Greenstein D, Clasen L, Evans A, Giedd J, Rapoport JL: Attention-deficit/hyperactivity disorder is characterized by a delay in cortical maturation. *Proc Natl Acad Sci U S A* 2007; 104:19649-19654.

[70] Shaw P, Malek M, Watson B, Greenstein D, de Rossi P, Sharp W: Trajectories of cerebral cortical development in childhood and adolescence and adult attention-deficit/hyperactivity disorder. *Biol Psychiatry* 2013; 74:599-606.

[71] Frodl T, Skokauskas N: Meta-analysis of structural MRI studies in children and adults with attention deficit hyperactivity disorder indicates treatment effects. *Acta Psychiatr Scand* 2012; 125:114-126.

[72] Bledsoe JC, Semrud-Clikeman M, Pliszka SR: Neuroanatomical and neuropsychological correlates of the cerebellum in children with attention-

deficit/hyperactivity disorder--combined type. *J Am Acad Child Adolesc Psychiatry* 2011; 50:593-601.

[73] Mackie S, Shaw P, Lenroot R, Pierson R, Greenstein DK, Nugent TF 3rd, Sharp WS, Giedd JN, Rapoport JL: Cerebellar development and clinical outcome in attention deficit hyperactivity disorder. *Am J Psychiatry* 2007; 164:647-655.

[74] Arnsten AF: The emerging neurobiology of attention deficit hyperactivity disorder: the key role of the prefrontal association cortex. *J Pediatr* 2009; 154:I-S43.

[75] Giedd JN, Blumenthal J, Molloy E, Castellanos FX: Brain imaging of attention deficit/hyperactivity disorder. *Ann NY Acad Sci* 2001; 931:33–49.

[76] Crowell SE, Beauchaine TP, Gatzke-Kopp L, Sylvers P, Mead H, Chipman-Chacon J: Autonomic correlates of attention-deficit/hyperactivity disorder and oppositional defiant disorder in preschool children. *J Abnorm Psychol* 2006; 115:174-178.

[77] Posthumus JA, Böcker KB, Raaijmakers MA, Van Engeland H, Matthys W: Heart rate and skin conductance in four-year-old children with aggressive behavior. *Biol Psychol* 2009; 82:164-168.

[78] Stahl SM: The Prefrontal Cortex Is Out of Tune in Attention-Deficit/Hyperactivity Disorder. *J Clin Psychiatry* 2009; 70:950–951.

[79] Valera EM, Brown A, Biederman J, Faraone SV, Makris N, Monuteaux MC, Whitfield-Gabrieli S, Vitulano M, Schiller M, Seidman LJ: Sex differences in the functional neuroanatomy of working memory in adults with ADHD. *Am J Psychiatry* 2010; 167:86-94.

[80] Valera EM, Faraone SV, Murray KE, Seidman LJ: Meta-analysis of structural imaging findings in attention-deficit/hyperactivity disorder. *Biol Psychiatry* 2007; 61:1361–1369.

[81] Castellanos FX, Giedd JN, Berquin PC, Walter JM, Sharp W, Tran T: Quantitative brain magnetic resonance imaging in girls with attentiondeficit/hyperactivity disorder. *Arch Gen Psychiatry* 2001; 58:289–295.

[82] Rubia K, Alegria AA, Brinson H: Brain abnormalities in attention-deficit hyperactivity disorder: a review. *Rev Neurol* 2014; 58(S01):S3-S18.

[83] Ernst M, Liebenauer LL, King AC , Fitzgerald GA, Cohen RM, Zametkin AJ: Reduced brain metabolism in hyperactive girls. *J Am Acad Child Adolesc Psychiatry* 1994; 33:858–868.

[84] Ernst M, Cohen RM, Liebenauer LL, Jons PH, Zametkin AJ: Cerebral glucose metabolism in adolescent girls with attention-deficit/hyperactivity disorder. *J Am Acad Child Adolesc Psychiatry* 1997; 36:1399-1406.

[85] Sheridan MA, Hinshaw S, D'Esposito M: Efficiency of the prefrontal cortex during working memory in attention-deficit/hyperactivity disorder. *J Am Acad Child Adolesc Psychiatry* 2007; 46:1357–1366.

In: Neurobiology of Men's Mental Health
Editors: Leo Sher and Timothy R. Rice
ISBN: 978-1-63483-191-8
© 2015 Nova Science Publishers, Inc.

Chapter 5

Mental Health Service Use by Suicidal Boys and Girls

Anne E. Rhodes, [1*] ***PhD and Jennifer Bethell,*** [2] ***PhD***

[1]Associate Professor,
Department of Pyschiatry and the Dalla Lana School of Public Health,
University of Toronto, Toronto, ON, Canada
Adjunct Scientist, Institute for Clinical Evaluative Sciences
[2]Dalla Lana School of Public Health, University of Toronto, Toronto, ON, Canada

Abstract

Youth suicide is a significant public health problem globally, necessitating the systematic planning and delivery of resources. In this chapter, we review what is known about when and where suicidal boys and girls receive mental health services and why over time, drawing from community-based studies. We identify noteworthy age-sex/gender differences which may contribute, in part, to the age-dependent 'gender paradox' of higher suicide attempt rates in girls but higher suicide rates in boys. While more up to date information is needed, the receipt of mental health services close in time to a suicide attempt or death seems precarious, particularly in boys. Why this peril occurs requires a much stronger, developmentally sensitive and well-coordinated response.

Keywords: suicide, attempted suicide, sex, gender, child, adolescent, mental health service utilization, review

Introduction

Suicide is the second leading cause of death among young people worldwide with profound social and economic costs [1]. While youth suicide rates vary widely [2, 3], they are

[*] Corresponding Author address: Email: annie.rhodes@utoronto.ca. Supported by The Canadian Institutes of Health Research, No. 319379.

at least 2 to 3 times higher in boys than girls [4, 5] in developed countries with good-quality vital registration data [1]. While it is still unclear how best to intervene to reduce suicidal behavior in youth [6], the problem may be resolved, in part, by addressing the "gender paradox" of higher suicide attempt rates in females alongside higher suicide rates in males [7].

Of particular importance in working with children and adolescents, this paradox is age-dependent. For suicide attempt rates, the sex/gender difference (F > M) increases with age peaking in mid adolescence [8-10], whereas for suicide rates, the sex/gender difference (M > F) steadily climbs until early adulthood [1]. Could these sex/gender differences be explained, in part, by differences in the ways boys and girls interact with mental health services over time, impacting the lethality of their behaviors? In this chapter, developmental perspectives are built on [11] examining where, when and why suicidal youth receive care. Such knowledge is necessary, not only to enable policy makers to direct resources [12], but also to pilot interventions sensitive to the sex/gender developmental needs of suicidal youth.

Terminology

"Boys" and "girls" refer to youth ages 10 to 24 years [13]. "Sex/gender" is used rather than "sex" or "gender" to signify the complex interplay of social and biological determinants [14].

Findings are limited by dichotomous measures of sex/gender [15] and to Western cultures [16]. We define suicidal ideation, attempts and suicide according to standard nomenclature [17]. Hereafter, hospital presentation data on self-inflicted injuries and poisonings are referred to as "hospital presentations" or when admitted, "hospital admissions." For both, unless otherwise noted, suicidal intent was unspecified.

Mental Health Service Use and the Continuum of Suicidal Behavior in Boys and Girls

In this section, the epidemiologic evidence on differences in mental health service use in boys and girls is synthesized with respect to the continuum of suicidal ideation to behaviors. Community-based samples (household or school-based surveys) of youth tend to better represent more common, less lethal behaviors, most prevalent in adolescent girls. In contrast, studies of more lethal behaviors or suicide may slanted towards older boys [11].

Mental Health Service Use and Suicide Attempts in the Community in Boys and Girls

A suicide attempt is the strongest known risk factor for suicide among youth [2] and almost all those who attempt suicide have met criteria for a mental disorder over their lifetime [18], typically anxiety and/or disruptive disorders, followed by mood and/or substance use disorders [11]. However, only a few community-based studies have examined past year mental health service use among youth who attempted suicide in that same year [19-22]. This is, in part, a statistical power problem: roughly 8% of high school students report a suicide attempt in the previous year [16], with only a portion using mental health services in that same year. Thus, details regarding the care received (e.g., specific medications and/or therapy) are usually lacking [18]. The ability to test whether suicidal youth are more likely than their peers to use mental health services requires an adequate number of controls who used mental health services in the past year. Herein, the power to test for effect modification (e.g., by age and/or sex/gender) and mediation becomes problematic given small cell sizes.

To circumvent these problems, some have combined youth who have suicidal thoughts with those who attempt suicide [20, 21]. Still, combining these outcomes is debatable given youth who ideate and go onto attempt suicide likely differ from those who do not [18]. For example, suicidal ideation has been found to be a better predictor of a future suicide attempt in girls than boys [11]. Depression may be a strong predictor of suicidal thoughts, whereas disorders characterized by anxiety, agitation and poor behavioral control seem to better characterize those who attempt suicide [18]. Not many have examined the temporal sequence between suicidal ideation and/or behaviors, mental disorders and mental health service use [18, 23]. Nevertheless, retrospective recall is inconsistent, particularly when attempts occur at earlier ages [24]. Finally, most studies were conducted over a decade ago; thus, it is uncertain how applicable the findings are now. With these caveats in mind, the findings from these studies are reviewed.

Among youth who attempted suicide in the past year, 40-45% reported receiving mental health services in that same year [19, 22], almost identical to youth with a past year mental disorder [25]. Two studies [21, 22] found that suicidal boys and girls differed in where they received services. In one [22], among 12-17 year-olds who attempted suicide, boys were less likely than girls to contact outpatient mental health services (OR: 0.60) but only after adjustments and no sex/gender differences were found in mental health inpatient or school-based service use. However, inpatient mental health excluded stays caused by alcohol or drugs. In the other study [21], among 15-24 year-olds with suicidal ideation and/or behaviors, girls were more likely than boys to report contact with GPs (in both 15-18 and 19-24 year-olds). Still, neither study examined what factors may have explained these sex/gender differences.

Differences in where suicidal boys and girls receive care is likely influenced, in part, by their age. It is thought that younger youth still living at home and/or attending school are more likely to secure mental health services through family members and/or teachers. For example, among US youth with suicidal ideation and/or behaviors, those younger (aged 13 or less) were found more likely than those older (age 18 or more) to receive counseling in the past year [20]. However, it is unclear whether these findings are more applicable to boys or girls given that, in the general population, boys have a higher prevalence of disruptive disorders than girls [26] and among children (ages 4 to 7 years), behavioral (but not emotional) disorders were found to predict mental health service use over time. Further, while mental health service use was higher in boys than girls, this difference was attenuated after adjustments for whether their problems were perceived as burdensome by the parents as well as whether parents, day care teachers and nurses perceived the child needed professional help [27].

Sex/gender differences in youth [25] and adults [28] exist in terms of where healthcare is received and why (i.e., mediated by sex/gender differences in mental health problems).

Although these studies did not specifically examine suicidal youth, it seems that, in the general population, boys and girls differ in where and when they receive mental health services as they age. In particular, when they are young, boys are more likely than girls to receive mental health services because of their greater prevalence of disruptive disorders, which prompt adults to act. However, during the transition from adolescence to adulthood, girls surpass boys. This may be explained, in part, by their greater prevalence of internalizing (i.e., mood and/or anxiety) disorders. If these same patterns also apply to suicidal boys and girls, then compared to suicidal boys, suicidal girls may be more likely to receive care closer in time to their attempts, given they have internalizing disorders, and as noted earlier, in outpatient settings.

The two community-based studies that examined the temporal sequence between mental health service use and youth suicidal behavior [18, 23] offer insights on suicidal boys and girls as they age. Among American 13-17 year-olds who reported a lifetime suicide attempt, over 94% reported receiving some form of mental health services in their lifetime. It is striking that the most common form (82%) was specialty mental health (e.g., psychiatrist or psychologist); however, about 25% of this use occurred after the attempt. Despite an elevated Odds Ratio (OR 1.69), a suicide attempt in the past year was not significantly associated with specialty mental health service use during the past year, nor with any other type of mental health services [19]. With respect to sex/gender differences among youth with suicidal ideation (but no plan or attempt) in the past year, boys were less likely than girls to report receiving any form of mental health services [19]. This sex/gender difference is troubling given the transition from ideation to an attempt was usually reported within a year, for both boys and girls [18]. Why does this happen?

In a prospective study [23], when youth with suicidal ideation and/or attempts before age 18 were compared to their peers, suicidal youth were more likely to receive treatment from a mental health professional (i.e., GP, psychiatrist, psychologist, counsellor) for depression between the ages 18 and 25. This was also true for suicidal youth and later substance-related problems, but only among girls. In other words, when boys had substance use disorders, being suicidal did not influence whether they later received treatment for substance-related problems. While not examined, adjustment variables (such as disruptive disorders between ages 14 to 18) may have mediated this effect.

Taken together, these studies suggest that where, when and why suicidal boys and girls receive mental health services differs. However, such differences and the causal pathways that lead to these differences, which may impact the age-dependent gender paradox, have not been well explained in prospective studies, much less in intervention trials. In particular, if girls are better than boys at securing mental health services (e.g., for internalizing disorders) closer in time to an attempt, the lethality of their attempt(s) may be lessened, possibly partly accounting for the gender-paradox. In contrast, boys may not receive mental health services close in time to their attempt(s) possibly due to a history of disruptive disorders and later substance use problems as they transition away from home/high school.

The transition to adulthood is a vulnerable time given youths' brains are more 'plastic,' encouraging them to explore and master changing environments requiring greater autonomy with some engaging in 'risky' behaviors.

Further, the onset of mental illness may alter how youth respond to sex/gender cultural 'adulthood' expectations and how they cope with perceived threats, interpersonal loss and related-pain. Masculine norms of personal autonomy and attainment may prevent boys from disclosing their pain to others. If alcohol is misused, it may proximally reduce anxiety but increase impulsiveness, and depression [11]. Even if suicidal youth perceive they need help, they may not know where or how to access it given changes in their living situations. With age, youth become ineligible for children's services and need to meet new criteria for adult mental health services, endangering continuity of care [12]. Accordingly, if youth seek help close in time to a suicide attempt, the where may be somewhat haphazard and possibly, crisis-based, underscoring the importance of ongoing integration of care.

Findings from the broader health services literature indicate that during the transition between adolescence and adulthood, boys and girls do differ in their use of ambulatory care services. In particular, among girls, an increase in visits to primary care providers occurs between adolescence (ages 13 to 18) and adulthood (ages 19 to 24), largely for reproductive reasons; whereas, a decrease in such visits occurs in boys, but an increase in visits to the emergency department (ED) [29].

In theory, such visits could improve girls' chances of having their mental health problems detected and treated, in parallel with, and potentially impacting the age-dependent gender paradox. Nevertheless, any such improvements in care for girls may not extend to suicide attempts. A suicide attempt has not been found associated with General Practitioner (GP) visits in boys or girls [30], implying that if a suicide attempt occurs, if care is received, it may not be from GPs. This could reflect differences in the underlying populations of those who ideate *vs.* attempt and/or provider recommendations/referrals. Thus, in the next sections, we review what is known about what happens to suicidal boys and girls when they present to hospital emergency departments (EDs).

Mental Health Service Use and Hospital Presentations in Boys and Girls

There is some indirect evidence that when boys present to the ED after a suicide attempt, they are less likely than girls to have their mental health problems detected and treated. In a large school-based survey of 15-16 year-olds, among those who reported on their most recent episode of harming themselves (irrespective of intent), boys were more likely than girls to present to hospital (18.0% vs. 10.7%) [31]. In both boys and girls, suicidal thoughts are the strongest predictor of help-seeking in relation to the self-harm [32]. Yet, the incidence and prevalence of hospital presentations for self-inflicted injuries and poisonings are known to be much higher in girls than boys [33-36]. If suicidal intent is less likely to be detected in boys than girls, boys may be more likely to be discharged from the ED without follow-up arrangements. One prospective study of hospital presentations among Medicaid insured youth (aged 10 to 19), found boys were less likely to be admitted to hospital afterwards than girls [37]. Another study of hospital presentations [38], among 20-24 year-olds, found boys were more likely than girls to be without treatment contact in the 30 days afterwards (43% vs. 22%).

Given that suicidal boys are less likely to receive outpatient care than suicidal girls, and some may present to the ED but not have their intent detected/acted on, when boys present to the ED, a suicide attempt may only be detected/acted on when it is medically serious. Note that the sex/gender difference in youth suicide attempts (F > M) diminishes among hospital presentations in medically serious suicide attempts (largely self-poisonings) [39] and reverses (M > F) with increased lethality of methods (*e.g.*, hanging and firearms) [40, 41]. These suppositions need confirmation, but are troubling, given that prospective hospital presentation data demonstrate that these youth have a higher suicide risk (about 10 times) than their peers [42], particularly in the year after the presentation [43]. It is possible this level of risk would be lower, if some of these boys had presented to hospital beforehand, and had their intent detected/acted on then.

There has been some study of youth making medically serious suicide attempts. In one study, while youth who made a medically serious suicide attempt were much more likely than their population-based peers to have a current mental disorder, for the most part, they did not differ from those who died by suicide [40]. What distinguished youth who died by suicide from those who made a medically serious suicide was, respectively, male sex/gender (81.7 *vs* 45.6%) and a lower prevalence of a current mood disorder (30.0 *vs* 71.2%). Differential reporting of mood symptoms by the informants seemed less likely given the severity of the events being compared. Supplementary analyses revealed that the higher proportion of boys among youth who died by suicide was explained by their lower prevalence of a current mood disorder and greater prevalence of a highly lethal method. While not explicitly tested, care for mental health problems (50.0% *vs* 68.8%) over their lifetime appeared less prevalent in youth who died by suicide than among those with a medically serious suicide attempt. Thus, if these boys had a mood disorder, it may not have been detected and/or treated close in time to their deaths. Dependent on their age, symptoms may have been subthreshold and/or masked by other comorbid conditions, such as disruptive disorders.

Mental Health Service Use and Suicide in Boys and Girls

Retrospective 'psychological autopsy' studies have found that nearly all youth who died by suicide met diagnostic criteria for a mental disorder [2]. Compared to girls, boys' suicides more often include prior disruptive and substance use disorders. In contrast, girls' suicides are more likely to include prior mood or eating disorders [2, 44-47]. Schizophrenia, though rare, is also a strong risk factor and may be more common in boys than girls, due to an earlier onset in boys [48, 49]. Still, the presence of these disorders does not mean their illness was ever diagnosed and/or treated. In fact, studies among those who died by suicide under age 35, the proportion who were seen by a mental health professional was low about 25% [50], less than noted earlier for youth who attempt suicide or have a mental disorder in the past year. Others have found that girls who died by suicide were more likely than boys who died by suicide to receive some form of mental health care [51-54] and see a physician (other than a psychiatrist) closer in time to their death [55, 56]. Few have examined whether these youth were seen in the ED (and not by sex/gender) [57-61].

In one of the largest studies among youth who died by suicide, the use of past year insured health services (outpatient medical, ED and inpatient) were compared in boys and girls aged 10 to 25 years [55]. Most were seen by an outpatient physician (72.0% of boys and 81.8% of girls) and about half were seen in the ED (47.7% of boys and 58.3% of girls). However, only about half of boys and two-thirds of girls received mental health services in these settings. At a time when mental disorders would have become more prevalent (i.e., among older youth, aged 16 to 25 years), sex/gender differences in mental health care became more pronounced, suggesting additional under recognition (and treatment) of mental disorders in older boys. Among those seen by an outpatient physician, in younger boys and girls (10-15 year-olds) a similar proportion was seen for mental health reasons (50.0 *vs.* 52.0%). Yet, older girls were more likely than older boys to receive mental health services (70.5% *vs.* 56.7%). Girls were also more likely than boys to be seen in the ED for mental health reasons in both age groups, but boys were more likely than girls to be seen in the ED for 'other' problems. These sex/gender differences by age are reminiscent of those in the broader ambulatory care literature [29], underscoring the importance of the early recognition and treatment of mental disorders in outpatient medical and ED settings, especially in boys as they age.

CONCLUSION

Studying suicidal behavior in boys and girls and their interactions with the health care system over time is challenging, with only a few prospective studies. Further, the findings may now be out of date. Nevertheless, the weight of the evidence implies that receiving mental health services close in time to suicide attempt and suicide may be at best 50:50, particularly for boys. Whether and where suicidal boys and girls receive care differs with age. Still, the reasons for these differences have received much less attention, of great import to the design of effective interventions and the appropriate planning of service delivery and resources. A number of shortfalls in the health care system [62] along with promising, developmentally sensitive youth suicide prevention approaches have been identified, including youth and their families, across multiple settings over time [6, 63-65]. Such approaches will likely succeed if they are sensitive to sex/gender differences in the antecedents of suicidal behavior [11], which may also determine whether, when and where suicidal youth receive care and the lethality of their behaviors.

REFERENCES

[1] World Health Organization. *Preventing suicide. A global imperative*. Geneva, Switzerland, 2014, pp 1-89.

[2] Bridge J, Goldstein T, Brent D. Adolescent suicide and suicidal behavior. *J. Child. Psychol. Psychiatry*. 2006;47:372-94.

[3] Hawton K, Saunders K, O'Connor R. Self-harm and suicide in adolescents. *Lancet*. 2012;379:2373-82.

[4] Wasserman D, Cheng Q, Jiang G. Global suicide rates among young people aged 15-19. World Psychiatry. 2005;4:114-20.

[5] Pitman A, Krysinka K, Osborn D, King M. Suicide in young men. *Lancet*. 2012;379:2383-92.

[6] De Silva S, Parker A, Purcell R, Callahan P, Liu P, Hetrick S. Mapping the evidence of prevention and intervention studies for suicidal and self-harming behaviors in young people. *Crisis*. 2013;34(4):223-32.

[7] Canetto S. Women and suicidal behavior: A cultural analysis. *Am. J. Orthopsychiatry*. 2008;78(2):259-66.

[8] Lewinsohn P, Rohde P, Seeley J, Baldwin C. Gender differences in suicide attempts from adolescence to young adulthood. *J. Am. Acad. Child Adol. Psychiatry.* 2001;40(4):427-34.

[9] Nkansah-Amankra S. Adolescent suicidal trajectories through young adulthood: Prospective assessment of religiosity and psychosocial factors among a population-based sample in the United States. *Suicide Life Threat Behav.* 2013;43(4):439-59.

[10] Thompson M, Light L. Examining gender differences in risk factors for suicide attempts made 1 and 7 years later in a nationally representative sample. *J. Adol Health.* 2011;48:391-7.

[11] Rhodes A, Boyle M, Bridge J, Sinyor M, Links P, Tonmyr L, et al. Antecedents and sex/gender differences in youth suicidal behavior. *World Journal of Psychiatry.* 2014;In press.

[12] Pottick K, Bilder S, Vander Stoep A, Warner L, Alvarez M. US patterns of mental health service utilization for transition-age youth and young adults. *J. Beh. Health Serv. Res.* 2008;35(4):373-89.

[13] Sawyer S, Afifi R, Bearinger L, Blakemore S-J, Ezeh A, Patton G. Adolescence: a foundation for future health. *Lancet.* 2012;379:1630-40.

[14] Canadian Institutes of Gender and Health. Introduction. What a difference sex and gender make - A gender, sex and health research casebook. Ottawa, *Canadian Institutes of Health Research*, 2012, pp ix-xiii.

[15] Johnson J, Repta R. Sex and gender: Beyond the binaries, in Designing and conducting gender, sex and health research. Edited by Oliffe J, Greaves L. Los Angeles, California, Sage; 2012, pp 17-37.

[16] Kann L, Kinchen S, Shanklin S, Flint K, Hawkins J, Harriss W, Lowry R, Olsen EO, McManus T, Chyen D, Whittle L, Taylor E, Demissie Z, Brener N, Thornton J, Moore J, Zaza S.et al. Youth risk behavior surveillance — United States, 2013. MMWR. 2014;63(4):1-168.

[17] Silverman M, Berman A, Sanddal N, O'Carroll PW, Joiner TE. Rebuilding the Tower of Babel: A revised nomenclature for the study of suicide and suicidal behaviors. Part 2: Suicide-related ideations, communications and behaviors. *Suicide Life Threat Behav.* 2007;37(3):264-77.

[18] Nock M, Green J, Hwang I, McLaughlin K, Sampson N, Zaslavsky A, et al. Prevalence, Correlates, and Treatment of Lifetime Suicidal Behavior Among Adolescents. Results From the National Comorbidity Survey Replication Adolescent Supplement. *JAMA Psychiatry.* 2013.

[19] Husky M, Olfson M, He P, Nock M, Swanson S, Merikangas K. Twelve-Month Suicidal Symptoms and Use of Services Among Adolescents: Results From the National Comorbidity Survey. *Psychiatr. Serv.* 2012;63(10):989-96.

[20] Pirkis J, Irwin C, Brindis C, Sawyer M, Friestad C, Biehl M, et al. Receipt of psychological or emotional counseling by suicidal adolescents. *Pediatrics.* 2003;111(4):388-93.

[21] Cheung A, Dewa C. Mental Health Service Use Among Adolescents and Young Adults With Major Depressive Disorder and Suicidality. *Can. J. Psychiatry.* 2007;52:228-32.

[22] Wu P, Katic B, Liu X, Fan B, Fuller C. Mental Health Service Use Among Suicidal Adolescents: Findings From a U.S. National Community Survey. *Psychiatr. Serv.* 2010;61(1):17-24.

[23] Fergusson D, Horwood L, Ridder E, Beautrais A. Suicidal behaviour in adolescence and subsequent mental health outcomes in young adulthood. *Psychol. Med.* 2005;35:983-93.

[24] Hart S, Musci R, Ialongo N, Ballard E, Wilcox H. Demographic and clinical characteristics of consistent and inconsistent longitudinal reporters of lifetime suicide attempts in adolescence through childhood. Depress Anxiety. 2013;30(10):1-14.

[25] Costello E, He J, Sampson N, Kessler R, Merikangas K. Services for adolescents with psychiatric disorders: 12-month data from the National Comorbidity Survey-Adolescent. *Psychiatr. Serv*. 2014;65(3):359-66.

[26] Kessler R, Avenevoli S, Costello E, Georgiades K, Green J, Gruber M, et al. Prevalence, persistence, and sociodemographic correlates of DSM-IV disorders in the National Comorbidity Survey Replication Adolescent Supplement. *Arch. Gen. Psychiatry*. 2012;69(372-380).

[27] Wichstrom L, Belsky J, Jozefiak T, Sourander A, Berg-Nielsen T. Predicting service use for mental health problems among young children. *Pediatrics*. 2014;133:1054-60.

[28] Rhodes A, Goering P, To T, Williams J. Gender and outpatient mental health service use. *Soc. Sci. Med*. 2002;54:1-10.

[29] Callahan P, Cooper W. Changes in ambulatory health care use during the transition to young adulthood. *J. Adol Health*. 2010;46:407-13.

[30] Haller D, Michaud P, Suris J-C, Jeannin A, Narring F. Opportunities for prevention in primary care in a country with universal insurance coverage. *J. Adol Health*. 2008;43:517-9.

[31] Madge N, Hewitt A, Hawton K, de Wilde E, Corcoran P, Fekete S, van Heeringen K

[32] De Leo D, Ystgaard M. Deliberate self-harm within an international community sample of young people: comparative findings from the Child & Adolescent Self-Harm in Europe (CASE) study. *J. Child Psychol. Psychiatry*. 2008;49(6):667-77.

[33] Ystgaard M, Arensman E, Hawton K, Madge N, van Heeringen K, Hewitt A, de Wilde EJ, De Leo D, Fekete S. Deliberate self-harm in adolescents: Comparison between those who receive help and those who do not. *Journal of Adolescence*. 2009;32:875-91.

[34] Colman I, Yiannakoulias N, Schopflocher D, Svenson L, Rosychuk R, Rowe B. Population-based study of medically treated self-inflicted injuries. *CJEM*. 2004;6(5):313-20.

[35] Rhodes A, Bethell J, Spence J, Links P, Streiner D, Jaakkimainen L. Age-sex differences in medicinal self-poisonings. *Soc Psychiatry Psychiatric Epidemiol*. 2008;43(8):642-52.

[36] Bethell J, Rhodes A. Identifying deliberate self-harm in emergency department data. *Health Rep*. 2009;20(2):1-9.

[37] Bethell J, Bondy S, Lou W, Guttman A. Emergency department presentations for self-harm among Ontario youth. *Can. J. Public Health*. 2013;104:124-30.

[38] Bridge J, Marcus S, Olfson M. Outpatient care of young people after emergency treatment of deliberate self-harm. *J. Am. Acad. Child Adol* Psychiatry. 2012;51(2):213-22.

[39] Suominen K, Isometsä E, Martunnen A, Ostamo A, Lonnqvist J. Health care contacts before and after attempted suicide among adolescent and young adult versus older suicide attempters. *Psychol. Med*. 2004;34:313-21.

[40] Beautrais A, Joyce P, Mulder R. Risk factors for serious suicide attempts among youths aged 13 through 24 years. *J. Am. Acad. Child. Adolesc. Psychiatry*. 1996;35(9):1174-82.

[41] Beautrais AL. Suicide and serious suicide attempts in youth: A multiple-group comparison study. *Am. J. Psychiatry*. 2003;160:1093-9.

[42] Rhodes A, Lu H, Skinner R. Time Trends in Medically Serious Suicide-Related Behaviours in Boys and Girls. *Can. J. Psychiatry*. 2014;In Press.

[43] Hawton K, Harriss L. Deliberate self-harm in young people: characteristics and subsequent mortality in a 20-year cohort of patients presenting to hospital. *J. Clin. Psychiatry*. 2007;68(10):1574-83.

[44] Hawton K, Bergen H, Kapur N, Cooper J, Steeg S, Ness J, et al. Repetition of self-harm and suicide following self-harm in children and adolescents: findings from the Multicentre Study of Self-harm in England. *J. Child. Psychol. Psychiatry*. 2012;53(12):1212-9.

[45] Shaffer D, Gould M, Fisher P, Trautman P, Moreau D, Kleinman M, et al. Psychiatric diagnosis in child and adolescent suicide. *Arch. Gen. Psychiatry*. 1996;53:339-48.

[46] Brent D, Baugher M, Bridge J, Chen T, Chiappetta L. Age-and sex-related risk factors for adolescent suicide. *J. Am. Acad. Child Adol. Psychiatry*. 1999;38:1497-505.

[47] Fleischman A, Beautrais A, Bertolote J, Belfer M. Completed suicide and psychiatric diagnoses in young people: A critical examination of the evidence. *Am. J. Orthopsychiatry*. 2005;75(4):676-83.

[48] Renaud J, Berlim M, McGirr A, Tousignant M, Turecki G. Current psychiatric morbidity, aggression/impulsivity, and personality dimensions in child and adolescent suicide: A case-control study. *J. Affect Disord*. 2008;105:221-8.

[49] Messias E, Chen C-Y, Eaton W. Epidemiology of schizophrenia: Review of findings and myths. *Psychiatr Clin. North Am*. 2007;30(3):323-38.

[50] Gould M, Greenberg T, Velting D, Shaffer D. Youth Suicide Risk and Preventive Interventions: A Review of the Past 10 Years. *J. Am. Acad. Child. Adol Psychiatry*. 2003;42(4):386-405.

[51] Luoma J, Martin C, Pearson J. Contact with mental health and primary care providers before suicide: A review of the evidence. *Am. J. Psychiatry*. 2002;159:909-16.

[52] Windfuhr K, White D, Hunt I, Turnbull P, Lowe R, Burns J, Swinson N, Shaw J, Appleby L, Kapur N. Suicides in juveniles and adolescents in the United Kingdom. *J. Child Psychol. Psychiatry*. 2008;49(11):1155-65.

[53] Renaud J, Berlim M, Séguin M, McGirr A, Tousignant M, Turecki G. Recent and lifetime utilization of health care services by children and adolescent suicide victims: A case-control study. *J. Affect Disord*. 2009;117:168-73.

[54] Vassilas C, Morgan H. Suicide in Avon. *Br. J. Psychiatry*. 1997;174:453-5.

[55] Marttunen M, Henriksson M, Aro H, Heikkinen M, Isometsä E, Lönnqvist J. Suicide among female adolescents: Characteristics and comparisons with males in the age group 13 to 22 years. *J. Am. Acad. Child. Adolesc. Psychiatry*. 2011;34(10):1297-307.

[56] Rhodes A, Khan S, Boyle M, Wekerle C, Goodman D, Tonmyr L, et al. Sex differences in suicides among children and youth – the potential impact of help-seeking behaviour. *Can. J. Psychiatry*. 2013;58(5):274-82.

[57] Appleby L, Amos T, Doyle U, Tomenson U, Woodman M. General practitioners and young suicides. A preventive role for primary care. *Br J. Psychiatry*. 1996;168:330-3.

[58] Morrison K, Laing L. Adults' use of health services in the year before death by suicide in Alberta. *Health Rep*. 2011;22(3):1-8.

[59] Gairin I, House A, Owens D. Attendance at the accident and emergency department in the year before suicide: retrospective study. *Br. J. Psychiatry*. 2003;183:28-33.

[60] Da Cruz D, Pearson A, Saini P, Miles C, While D, Swinson N, et al. Emergency department contact prior to suicide in mental health patients. *Emerg. Med. J.* 2011;28:467-71.

[61] Gunnarsdottir O, Rafnsson V. Risk of suicide and fatal drug poisoning after discharge from the emergency department: a nested case-control study. *Emerg. Med. J.* 2010;27:93-6.

[62] Weis M, Bradberry C, Carter L, Ferguson J, Kozareva D. An exploration of human services system contacts prior to suicide in South Carolina: An expansion of the South Carolina Violent Death Reporting System. *Inj. Prev*. 2006;12 (Suppl 2):17-21.

[63] Renaud J, Seguin M, Lesage A, Marquette C, Choo B, Turecki G. Service use and unmet needs in youth suicide: A study of trajectories. *Can. J. Psychiatry*. 2014;59(10):523-30.

[64] Brent D, McKakin D, Kennard B, Goldstein T, Mayes T, Douaihy A. Protecting adolescents from self-harm: A critical review of intervention studies. *J. Am. Acad. Child Adol. Psychiatry*. 2013;52(12):1260-71.

[65] Bridge J, Horowitz L, Fontanella C, Grupp-Phelan J, Campo J. Priortizing research to reduce youth suicide and suicidal behavior. *Am. J. Prev. Med.* 2014;47(3S2):229-34.

[66] Wyman P. Developmental approach to prevent adolescent suicides. Research pathways to effective upstream preventive interventions. *Am. J. Prev. Med.* 2014;47(3S2):251-6.

In: Neurobiology of Men's Mental Health
Editors: Leo Sher and Timothy R. Rice
ISBN: 978-1-63483-191-8
© 2015 Nova Science Publishers, Inc.

Chapter 6

ALL JUST FUN AND GAMES? MEN AND MEDIA VIOLENCE

Michele Knox, PhD [*] ***and Vipul Shukla, MS***
University of Toledo College of Medicine, Toledo, OH, US

ABSTRACT

Decades of research have documented associations between violent video game use and a host of negative outcomes including aggression, desensitization to violence, decreased empathy and prosocial behavior, and increased hostile attribution bias. Boys' and men's involvement in violent video gaming is pronounced and increasing. These factors, as well as the changing content of video games, present cause for concern regarding the potentially widespread impact of violent video games. The present chapter summarizes the hormonal and neurobiological aspects of aggression in males as well as the impact of violent video game use on these factors. Negative outcomes, including relational problems, impact on the children of gamers, and internet gaming disorder are considered, as are prevention and intervention efforts addressing these issues.

INTRODUCTION

Summary of the Research on the Effects of Violent Video Games

Video games are not what they used to be. Gone are the days when a player sits passively in a chair, moving a stick from side to side in order to make a cartoon figure move left or right on a screen. Today, the graphics in video games have improved to the point where the scenes depicted in video games are highly realistic. For example, in massively multiplayer online role-playing games (MMORPG), players role-play and interact with one another in real time within a virtual game world. Players take on the role of a character and control many

[*] Corresponding author: Michele Knox, PhD, Professor of Psychiatry, Director, ACT Great Lakes Regional Center, University of Toledo College of Medicine, Kobacker Center, 3130 Glendale Avenue, Toledo, OH 43614

of that character's actions. While playing video games, many players engage in the actions that are taking place on the screen, engaging actively in motoric behavior, verbal and nonverbal communication as if the events were happening in real life. Thanks to these advancements, for the video game player, the line between fantasy and reality is increasingly blurred.

Today's youth are often heavily involved in media and gaming in particular. A recent study of over 800 young adults demonstrated that over 50% of young men play video games weekly or more frequently and nearly 40% play *violent* video games weekly or more frequently [1]. Results of a U.S. national survey indicated that most (88%) American youths aged 8 and 18 years play video games at least occasionally [2]. A recent report by the Kaiser Family Foundation stated that the average youth experiences an average of 10 hours, 45 minutes of media content in an average day [3]. Typically, media use is not subject to parental limits [3].

Youth gaming involvement patterns vary by gender. Boys are exposed to an average of nearly one hour more of media each day than girls, primarily due to their greater use of console video games. Boys also tend to spend more time using computer games, and to continue their involvement in such games into their later teen years [3]. Another study found an even greater gap, with boys playing on average several more hours per week than girls [2]. Boys also are significantly more likely to endorse symptoms of pathological gaming [2]. Furthermore, boys have more exposure to violent video games [4]. In particular, boys with lower educational abilities prefer violent games, perceive video games to be more realistic, and wishfully identify more with video game characters than other boys do [5].

A great deal of research has been conducted examining the effects of video games. In a meta-analytic review of 136 studies which involved over 130,000 participants, video game violence was found to be positively associated with aggressive behavior, aggressive cognitions, and aggressive affect [6]. For example, in one study, sustained violent video game play was associated with increases in youth aggressive behavior, and more violent video game play predicted higher levels of aggression over time [7]. This finding held even after controlling for previous levels of aggression [7]. Results of meta-analyses also indicated that exposure to video game violence was associated with desensitization to violence/decreased empathy and decreased prosocial behavior [6]. Effects sizes were small to medium ($r+=.189$ for aggressive behavior, $r+=-.177$ for empathy/desensitization and $r+=-.101$ for prosocial behavior) and significant in both eastern and as western cultures. In yet another recent meta-analysis of 98 studies representing 36,965 participants, violent video games related to increased aggression and aggression-related variables and decreased prosocial outcomes ($r=.18$; 8).

Sexual violence in video games and their potential impact are also serious concerns. In several recent games, game players engage in violent or coercive virtual sexual behavior. This may encourage gamers to associate sex with violence. For example, players in one game may hire a virtual prostitute, engage in sex with her, then kill her. That these experiences might influence young minds is a serious concern. Some recent research is beginning to address this concern. For example, in a recent study, use of a video game depicting sexual objectification of women and violence against women related to a statistically significant increased acceptance of rape myths in the gamers [9].

If boys have a relatively higher degree of exposure to violent video games, then it stands to reason that they may be highly prone to the established negative effects of playing such games. For young males in particular, given the plasticity of the developing brain, it is possible that boys may be particularly susceptible to the influence of violence in video games. Much research over the past several years has converged upon the fact that the brain is far from fully developed at birth, and that early experiences shape brain architecture [10]. For example, early life stress seems to alter normal brain development, leading to deficits, particularly in the affective domain. Such deficits persist years after the stressors occur and may relate to an increased risk for later psychopathology [11]. Characteristics of modern video games, such as real-time interaction, motoric simulation of aggressive acts, and high-level graphics may make video game experiences similar to real-life experiences. Thus, it is possible that video game experiences may have impacts and effects approaching those of real life experiences. Given their marked presence in the daily lives of children, it stands to reason that experiences with video games may impact the developing brain. This may be of distinctive relevance for boys, who have greater exposure to games.

The Male Brain and Aggression

A great deal of research has been conducted addressing the human brain and aggression. The human brain, consisting of billions of neurons, controls our sensory, motor, and cognitive abilities. It also creates and controls aspects of emotion and behavior, including aggression and violence. It is divided into multiple regions, termed lobes, and each lobe consists of different sets of neural inputs and networks and related functions. The frontal, temporal and insular lobes and the thalamus of the brain have been shown to be most closely linked to human aggression and violence. Signals that travel within and between neurons in these regions of the brain influence whether a person will behave in aggressive and violent ways.

When confronted with or witnessing a violent event, many individuals would be able to control their feelings of aggression and violence voluntarily; they would be able to refrain from participating in the violent behavior. This type of control and restraint is due to the interaction between the frontal, temporal, and insular lobes of the brain. The temporal and insular lobes are the "generators of emotions" and the frontal lobe controls the amount of emotions generated by the temporal and insular lobes. In most individuals, the frontal, temporal, and insular lobes of the brain maintain a relatively constant level of activity and interact with each other in a manner that prevents violence.

The portions of the brain that are linked to aggression and violence start to form before birth, and some continue to develop well into adolescence and adulthood. Patterns of brain growth can be influenced by a variety of contextual forces, such as repeated experiences related to parenting and socialization. These factors in turn influence an individual's propensity for violence. The activity of key areas of the temporal, insular, and frontal lobes of the brain related to aggression also depends on the presence and amount of certain hormones. Various physiological and social factors and cues determine if hormones are expressed, the amount of hormone produced, and hormonal interaction with the brain.

It is established that males are more prone to violence compared to females. In fact, male gender is a key predictor of violence, more than doubling the odds for later violence [12]. Testosterone is a key hormone affecting aggression and violent behavior in males. Sex differences in physical aggression peak during the teenage years and these changes have been linked in part to changes in the hormone testosterone. Compared to their female counterparts, the levels of testosterone in males begin to increase at an exponential rate during puberty [13]. Some research indicates that testosterone decreases the amount of fear and increases the levels of aggression in males [14]. Increased levels of testosterone appear to activate the amygdala. Activation of the amygdala results in increased amounts of "cross-talk" and activity between the temporal, insular, and frontal lobes of the brain, influencing the likelihood of aggression and violent behavior.

In addition to testosterone, the hormones cortisol and vasopressin aid in controlling violent behavior in males. Recent research examined the effects and levels of vasopressin and cortisol in aggressive behavior. Decreased levels of cortisol and vasopressin are associated with increased levels of aggressive and violent behavior in males [15, 16].

Neurobiological and hormonal changes may occur relative to exposure to violent content. In a sample of boys aged 11 to 14 years, for example, violent television and video game exposure had greater effects on indicators of physiological stress than did media with nonviolent content [17].

In light of this knowledge about links between physiological changes and violent gaming, one may question whether the regions of the brain which control aggression and violence are altered in males who play video games, and whether any identified changes persist after video game use. The onset of the digital age has led to advancements in medical imaging and the creation of novel imaging techniques, like functional magnetic resonance imaging (fMRI), that have the potential to examine these questions. Researchers can utilize these imaging techniques to assess the activity of various regions of the brain. Many neuroimaging studies, including fMRI and electroencephalography (EEG), have now been conducted to assess the potential impact of violent video game use on the brain. This research shows that the regions of the brain related to aggressive behavior in fact may be altered in individuals involved in video game play. These changes have been demonstrated after short-term and long-term video game use using numerous imaging techniques. These neuroimaging studies demonstrate the behavioral, structural, and functional brain changes during, and in some cases, after, violent video game usage. For example, one study [18] examined the activity of the temporal lobe in thirteen males, ages 18-26 years, as they played violent video games. Upon failure of a violent video game task, males playing violent video games evidenced altered activity within the caudate nucleus and the orbitofrontal cortex compared to their baseline level of activity. Such changes may help players to maintain a stable mood despite the failure events, and correspond with other research indicating decreased levels of emotional reactivity to scenes of violence. Another research study monitored the electrical activity of the brain of men who played violent video games [19]. To assess the brain's electrical activity, the amount of maximal signal strength, termed "amplitude" was measured. This study indicated specific reduced amplitudes associated with violent, as compared to nonviolent video game play. This reduced brain response predicted increased aggressive behavior in a later task.

Although to date, numerous research studies have evidenced changes to the brain during active violent video game use, the question of whether these changes persist over time is just beginning to be addressed. In one study, for example, violent media exposure was associated with reduced frontoparietal white matter volume [20]. Since the frontal lobe controls emotions by interacting with regions of the temporal lobe, insular lobe, and the thalamus, decreased volume and activity in key areas may increase the risk for aggressive behavior. Decreased activity may also lead to diminished emotional responsiveness to aggressive behavior and violence following repetitive violent video game playing. This may, in essence, correlate with the observation that players become "numb" to violence and aggression during and after playing video games. A recent study used fMRI to examine whether changes in areas of the brain including the dorsal anterior cingulate cortex related to subsequent changes in aggression in violent video game players [21]. Findings indicated such changes are predictive of subsequent aggression but are dependent on baseline executive functioning. These findings were similar to those of a previous fMRI study which examined the anterior cingulate cortex in males during violent video game use [22]. These researchers found relative changes in activity in the rostral and dorsal portions of the anterior cingulate cortex associated with aggressive cognition and behavior. It is possible that the neural patterns of the anterior cingulate cortex represent a mechanism to suppress positive emotions (such as empathy) so that the individual can play the game successfully. Taken together, these studies suggest that interactions between changes to frontal lobe and other regions of the brain induced from violent video game play may lead to both desensitization towards violence and an increased propensity toward violent behavior.

The brain changes that have been observed to occur relative to video game exposure appear to have important similarities to characteristics of the brains of individuals known to be prone to aggression. For example, several fMRI imaging studies have examined the frontal lobe and aspects of aggression in patients with behavioral, psychiatric, and personality disorders. Among other findings, decreased frontal lobe activity has been evidenced in aggressive patients with dementia, substance abuse, schizophrenia, and attention deficit hyperactivity disorder. Less activation in the frontal lobes is also seen in males diagnosed with an aggressive/disruptive behavior disorder [23]. Antisocial behavior and impulsive aggression, characteristics present in violent criminals, also are linked to decreased activity of the frontal lobes [24, 25]. Although more research is needed, an important area of continued inquiry is whether the changes that have been observed to occur relative to video game exposure lead to lasting changes in the brain that are characteristic of antisocial and behavior disorders.

Impact of Gaming on Functioning and Interpersonal Relationships

Popular views of the nature of violence in the U.S. are often driven by media and other sources suggesting that violent acts are typically perpetrated by strangers, often in a public setting. However, the more typical scenario is that of violent acts being committed between people who know or are related to one another [26]. Thus, when investigating the impact of violence in men, it is critical to consider the qualities and contexts of men's relationships. The presence and quality of close relationships relates to important health and mental health outcomes. For example, positive paternal involvement in the lives of children and youths is highly beneficial [27]. Father involvement and various characteristics of fathers' behavior and relationships with their children have been shown to relate to better psychosocial adjustment, coping, cognitive and social competence, capacity for empathy, self-esteem, academic and maturity in children, and even better mental health and occupational achievement when children grow into adulthood [28, 29, 30]. In families with fathers, maintaining fathers' positive involvement in childrens' lives, and ensuring that parent-child interactions are marked by warmth, support, and positive discipline are increasingly viewed as beneficial to healthy child development and positive outcomes.

In contrast, when father-child relationships are marked by negative interactions, and/or no or inconsistent involvement, child outcomes may be less than optimal. Research has shown, for example, that parent conflict and antisocial behavior, as well as paternal hostility, are positively associated with child externalizing problems such as aggression and disruptive or oppositional behavior [31, 32].

In the same vein, men's involvement in intimate relationships can influence both partners' health and well-being for better or for worse. For example, numerous studies suggest that marital satisfaction is associated with psychological and physical health and well-being [33]. For gay, lesbian, and heterosexual couples alike, support provided by one's romantic partner appears to buffer individual well-being from the potentially negative impacts of stress [34].

In contrast, when hostility and aggression are present in interpersonal relationships, profound human suffering and other negative outcomes can ensue. When men have characteristics such as hostile masculinity and misperceptions of women's sexual cues, sexually aggressive acts are more likely [35]. Perpetrators of interpersonal and sexual violence have been shown to be more likely to have histories of exposure to domestic violence, parental physical punishment, and childhood sexual abuse. Furthermore, factors such as low empathy, traditional gender role beliefs, hostility toward women, and psychopathic and antisocial tendencies also predict male aggression [36, 37]

In addition to associations with aggressive behavior, violent video game use consistently has been shown to be associated with desensitization, decreased empathy and decreased prosocial behavior [6]. These factors potentially have enormous implications for the relational lives of gamers. As stated above, empathy, the degree to which one identifies or commiserates with a victim and feels emotional distress, is key to healthy relational functioning, and low empathy is a predictor of aggression. The flipside of empathy is desensitization, defined as a reduction in negative emotional response to scenes of violence. It stands to reason that these established effects of violent video game use may lead gamers to be impervious to the pain and suffering others, including their partners, spouses and children.

It is hard to imagine more important traits for parents than empathy and the ability to respond to childrens' distress. From birth, the responsiveness of parents to infants' distress is key not only to mental and physical development, but also to survival. Repeated, empathic, reciprocal parent-infant exchanges are critical to infants' ability to self-soothe and to regulate affect. During childhood, children rely on the warmth and nurturing of empathic parents to be able to learn social problem-solving, interpersonal skills, and other capacities. In fact, the security provided by a loving parent who will notice and respond if something goes wrong allows children to explore the world and learn. In contrast, when parents lack empathy and are desensitized to children's suffering, a host of negative outcomes can be expected. Although additional research is required, it is very possible that violent video game use may be linked to myriad problems in parent/child interactions and child outcomes in families where video game use is, or has been, a regular activity.

Violent video game use also relates to the development of hostile attribution bias [38], defined as the tendency to attribute hostile intentions to others' actions. People with hostile attribution bias have a tendency to assume that others are hostile toward them, even when this is not the case. Such people develop hostility-related schemas that filter information from the environment causing them to "see" others as potentially dangerous or threatening to them, even when this is not true. Parents with such biases are prone to see children in this light. In turn, parental hostile attributions and parental hostility relate to harsh and abusive behavior toward children in a cycle that then seems to perpetuate children's own aggressive and externalizing behavior [39, 40].

Lastly, violent video game use is also predictive of less prosocial behavior, including interpersonal actions such as giving to and helping others [6]. Prosocial behavior is a key aspect of positive relational functioning. If violent video game use diminishes such behaviors, it stands to reason that the close relationships of gamers would suffer.

In sum, if violent video game use makes individuals less empathic, desensitized to others' suffering, more likely to see others as hostile, and more aggressive, then partners and spouses of gamers may experience relationships that are less supportive and helpful, and potentially more aggressive and hostile. Perhaps even more concerning, the existence of these qualities in parents may be devastating for children.

Video Game Content

The content of today's video games also may have important implications for men's relational functioning. For example, while many adults may be familiar with content of video games in which players "shoot" at distant cars, much of the public may not realize that today's video games may involve, for example, acting as a school shooter and killing students, killing large groups of people in an airport line with a machine gun, killing faith leaders, killing homeless people, killing the elderly, and killing people of color while making racist remarks. As stated previously, many games now allow the player to engage in acts of sexual violence. It seems that no type of violence is off-limits. In fact, some modern video games now allow the player to engage in acts of child abuse including physical abuse, torture, using children as "shields" against gunfire, and child prostitution.

To date, there are no comprehensive published lists detailing the exact types of violent acts that players of certain video games may commit. The Entertainment Safety Review Board (ESRB) "assigns the age and content ratings for video games and mobile apps, enforces advertising and marketing guidelines for the video game industry, and helps companies implement responsible online privacy practices" [41]. While ESRB and other reviews of games do indicate violent content, specific types of violence, such as rape and child abuse, are not included in their descriptions. Thus, short of playing the full game, there is no way for an individual to know the full content of a given game.

The content of violent video games is problematic in part because it gives players new conceptions for violent acts. One of the ways that violent video games are thought to promote violent behavior is through the development and increased accessibility of cognitive scripts for violent behavior [42]. That is, players are more able to conceive of, imagine, or picture violent acts that they see in games. Playing the game "primes" those scripts, making them more cognitively available to the player both during and after game play. If games now include content depicting acts such as child abuse, torture, and rape, then players may be more likely to develop scripts for such behaviors, whereas if they had not played the game, they would not have such scripts. For example, children who had never conceived of necrophilia may become newly aware of this concept and form a necrophilia "script" after playing the video game with this content. Adults who had never conceived of committing acts of rape or child abuse may begin to consider such.

In addition to the content and established negative outcomes of violent video games use, the time spent playing video games has the potential to be harmful to families with children. For instance, cases of child neglect have been attributed to parents' excessive immersion in video games use. In one case, a couple who were avid players of an on-line game neglected their 30 month old daughter to the point where she starved to death [38].

Due to these and other reports of excessive and harmful gaming, a new mental health condition called "Internet Gaming Disorder" has been proposed. This disorder was included in the most recent version of the Diagnostic and Statistical Manual of Mental Disorders, Fifth Edition [43] as a "Condition for further study." According to the manual, the proposed condition involves persistent and recurrent use of games leading to clinically significant impairment or distress. To qualify for the disorder, it is proposed that the individual should have five of the following symptoms: preoccupation, withdrawal symptoms (irritability, anxiety, etc.), tolerance (need to spend more time), unsuccessful attempts to stop/cut down, loss of other interests/activities, continued use despite problems, history of deceiving others regarding game use, use of games to escape or relieve negative mood, experience with jeopardizing or losing a relationship, job, or education/career, and/or parental game addiction and child neglect. According to the DSM-5 [43], the proposed disorder is most prevalent among males aged 12 to 20 years of age. Also according to the DSM-5 [43], "internet gaming disorder may lead to school failure, job loss, or marriage failure. The compulsive gaming behavior tends to crowd our normal social, scholastic, and family activities. Students may show declining grades and eventually failure in school, and family responsibilities may be neglected," (p. 797). Given the functional impact described, if this disorder turns out to be legitimate, there may be devastating consequences for gamers and their families. Especially given the ubiquitous presence of internet and game systems in the lives of families, the threat of this proposed problem warrants serious concern.

Conclusion

In sum, research has solidly established a host of potential negative effects related to violent video game use. The status of the literature base on this issue is definitive. The insidious presence of violent video games in the U.S. and other cultures is a serious cause for concern for many reasons, and these issues may be particularly relevant for males. From a young age, males are more likely to use video games, and use them for longer periods of time. Males, already more prone to violence for a variety of contextual, physiological, and neurobiological reasons, may be highly prone to the negative effects of violent video games. These effects have the propensity to impair boys' and mens' functioning in a variety of life domains.

Whether these effects can be prevented or treated remains to be seen. Some evidence suggests that early childhood interventions, aimed at promoting media literacy and preventing violent media exposure, may be useful [44]. These methods involve the education and guidance of caregivers who are in the position of preventing, limiting, and monitoring violent media exposure. Another promising method involves school-based media literacy education

[45]. This method, aimed at children and adolescents, has been shown to bolster media literacy and, in particular, to improve students' knowledge about the negative effects of violent media.

Evidence-based methods for the assessment of pathological or addictive gaming are not yet well established. The use of a "Pathological Computer/Video Game Use Questionnaire" has been suggested (46). Furthermore, to date, few studies have addressed treatment for excessive gaming. Two studies have found that psychopharmacological interventions (e.g., Bupropion and Methylphenidate for comorbid ADHD) may be useful [47, 48]. One study [49] utilized a randomized, controlled design to investigate the efficacy of school-based group CBT for adolescents. The results indicated that this intervention may be effective for treatment of internet addiction in adolescents. Treatment services based on 12-step self-help treatment philosophies that traditionally have been used for other types of addiction are also available [50] but have not yet been sufficiently empirically evaluated.

Moving forward, research is needed to better understand more specific issues such as who is most at risk for these effects, and how these effects can be effectively prevented and treated. There also remains much to learn about the potential long-term impact of violent media on the brain and physiology of gamers. In the meantime, it will be imperative for the public to become much more educated about the potential effects of violent video game use, as well as the content of video games, especially those marketed to children and youths. Professionals dedicated to violence prevention in all domains, including dating violence, elder abuse, interpersonal violence, child maltreatment, sexual assault, and other areas, should be made aware of these themes in video games and their potential ramifications. In contrast to the traditionally discrete nature of prevention initiatives, this effort could, and should be, highly inter-professional and collaborative, involving diverse groups, because the virtual world of gaming now cuts across all sectors. Clearly, in an era when the vast majority of the households in the U.S. have gaming systems or devices containing violent games, and as public use of these games is increasing, the time is upon us to address this issue. The situation may not be, as they say, "all just fun and games."

REFERENCES

[1] Padilla-Walker LM, Nelson LJ, Carroll JS, Jensen AC: More than a just a game: video game and internet use during emerging adulthood. *Journal of Youth and Adolescence* 2010; 39:103-113.

[2] Gentile D: Pathological Video-Game Use Among Youth Ages 8 to 18 A National Study. *Psychol. Sci* 2009; 20:594-602.

[3] Rideout VJ, Foehr UG, Roberts DF: "Generation M 2—media in the lives of 8-to 18 year olds: A Kaiser Family Foundation Study. 2010. Retrieved from http://kff.org/other/event/generation-m2-media-in-the-lives-of/

[4] Funk JB, Buchman DD, Jenks J, Bechtoldt H: Playing violent video games, desensitization, and moral evaluation in children. Journal of *Applied Developmental Psychology* 2003; 2:413-436.

[5] Nije BM, Konijn EA, Bushman BJ: "We don't need no education": Video game preferences, video game motivations, and aggressiveness among adolescent boys of different educational ability levels. *J. Adolesc.* 2012; 35:153-162.

[6] Anderson CA, Shibuya A, Ihori N, Swing EL, Bushman BJ, Sakamoto A, Rothstein HR, Saleem M: Violent video game effects on aggression, empathy, and prosocial behavior in eastern and western countries: a meta-analytic review. *Psychol. Bull* 2010; 136:151.

[7] Willoughby T, Adachi PJ, Good M: A longitudinal study of the association between violent video game play and aggression among *adolescents. Dev. Psychol* 2012; 4:1044.

[8] Greitemeyer T, Mügge DO: Video games do affect social outcomes; a meta-analytic review of the effects of violent and prosocial video game play. *Personality and Social Psychology Bulletin* 2014; 0146167213520459.

[9] Beck VS, Boys S, Rose C, Beck E: Violence against women in video games: A prequel or sequel to rape myth acceptance? *Journal of Interpersonal Violence* 2012; 0886260512441078.

[10] National Scientific Council on the Developing Child: The Timing and Quality of Early Experiences Combine to Shape Brain Architecture 2007; Working Paper No. 5. Retrieved from www.developingchild.harvard.edu.

[11] Pechtel P, Pizzagalli DA: Effects of early life stress on cognitive and affective function: an integrated review of human literature. Psychopharmacology 2011: 214:55-70.

[12] Herrenkohl TI, Maguin E, Hill KG, Hawkins JD, Abbott RD, Catalano RF: Developmental risk factors for youth violence. *J. Adolesc. Health* 2000; 26:176-186.

[13] Archer J: Testosterone and human aggression: an evaluation of the challenge hypothesis. *Neurosci. Biobehav. Rev*. 2006; 30:319-45.

[14] Stanton SJ, Wirth MM, Waugh CE, Schultheiss OC: Endogenous testosterone levels are associated with amygdala and ventromedial prefrontal cortex responses to anger faces in men but not women. *Biol. Psychol.* 2009 81:118-22.

[15] Terburg D, Morgan B, van Honk J. The testosterone-cortisol ratio: A hormonal marker for proneness to social aggression. *Int. J. Law Psychiatry* 2009 Jul-Aug;32(4):216-23.

[16] Ferris CF, Stolberg T, Kulkarni P, Murugavel M, Blanchard R, Blanchard DC, Febo M, Brevard M, Simon NG. Imaging the neural circuitry and chemical control of aggressive motivation. *BMC Neurosci*. 2008; 9:111.

[17] Maass A, Lohaus A, Wolf OT: Media and Stress in Adolescent Boys in Germany: Psychophysiological effects of violent and nonviolent television programs and video games. *Journal of Children and Media* 2010; 4:18-38.

[18] Mathiak KA, Klasen M, Weber R, Ackermann H, Shergill SS, Mathiak K: Reward system and temporal pole contributions to affective evaluation during a first person shooter video game. *BMC Neurosci*. 2011; 12:66.

[19] Bartholow BD, Dushman BJ, Sestir, MA: Chronic violent video game exposure and desensitization to violence: Behavioral and event-related brain potential data. *Journal of Experimental Social Psychology* 2006; 42:532-539.

[20] Hummer TA, Kronenberger WG, Wang Y, Anderson CC, Mathews VP: Association of television violence exposure with executive functioning and white matter volume in young adult males. *Brain Cogn*. 2014; 88:26-34.

[21] Chester DS, Eisenberger NI, Pond RS Jr, Richman SB, Bushman BJ, Dewall CN: The interactive effect of social pain and executive functioning on aggression: an fMRI experiment. *Soc. Cogn. Affect Neurosci.* 2014; 9:699-704.

[22] Weber R, Ritterfeld U, Mathiak K: Does playing violent video games induce aggression? Empirical evidence of a functional magnetic resonance imaging study. *Media Psychology* 2006; 8:39-60.

[23] Mathews VP, Kronenberger WG, Wang Y, Lurito JT, Lowe MJ, Dunn D: Media violence exposure and frontal lobe activation measured by functional magnetic resonance imaging in aggressive and nonaggressive adolescents. *J. Comput. Assist. Tomogr.* 2005, 29:287–292.

[24] Raine A, Lencz T, Bihrle S, LaCasse L, Colletti P: (2000). Reduced prefrontal gray matter volume and reduced autonomic activity in antisocial personality disorder. *Arch. Gen. Psychiatry* 2000; 57:119-127.

[25] Raine A, Meloy J.R, Bihrle S, Stoddard J, LaCasse L, Buchsbaum, MS: (1998). Reduced prefrontal and increased subcortical brain functioning assessed using positron emission tomography in predatory and affective murderers. *Behav. Sci. Law* 1998; 16:319- 332.

[26] Payne BJ, Gainey RR: Family violence and criminal justice: A life-course approach. Routledge, New York, NY 2010.

[27] Wilson KR, Prior MR: Father involvement and child well-being. *J. Paediatr. Child Health* 2011; 47:405-407.

[28] Lamb ME, (Ed.): The role of the father in child development. Hoboken, NJ: Wiley, 2004.

[29] Lamb, Michael E., and Charlie Lewis. The Development and Significance of Father–Child Relationships in Two-Parent Families. The role of the father in child development 2010; 94-153. Hoboken, NJ, US: John Wiley & Sons Inc.

[30] Ehrenberg MF, Gearing-Small M, Hunter MA, Small BJ: Childcare task division and shared parenting attitudes in dual-earner families with young children. *Fam. Relat.* 2001; 50:143–.

[31] Harold GT, Elam KK, Lewis G, Rice F, Thapar A: Interparental conflict, parent psychopathology, hostile parenting, and child antisocial behavior: Examining the role of maternal versus paternal influences using a novel genetically sensitive research design. *Dev. Psychopathol* 2012; 24:1283-1295.

[32] Harold GT, Leve LD, Elam KK, Thapar A, Neiderhiser JM, Natsuaki MN, Shaw DS Reiss, D: The nature of nurture: Disentangling passive genotype–environment correlation from family relationship influences on children's externalizing problems. *Journal of Family Psychology* 2013; 27:12.

[33] Proulx C, Helms H, Buehler C: Marital quality and personal well-being: A meta-analysis. *Journal of Marriage and Family* 2007; 69:576-593.

[34] Graham JM, Barnow ZB: Stress and social support in gay, lesbian, and heterosexual couples: direct effects and buffering models. *Journal of Family Psychology* 2013; 27:569.

[35] Abbey A, Jacques-Tiura, AJ, LeBreton JM: Risk factors for sexual aggression in young men: An expansion of the confluence model. *Aggressive Behavior* 2011; 37:450-464.

[36] Eisenberg N, Kazdin AE (Eds.): Encyclopedia of psychology, Vol. 3 American Psychological Association, Washington, DC 2000: pp.179–182.

[37] Zhen S, Xie H, Zhang W, Wang S, Li D: Exposure to violent computer games and Chinese adolescents' physical aggression: The role of beliefs about aggression, hostile expectations, and empathy. Computers in Human Behavior 2011; 27:1675-1687.

[38] Bavelier D, Green CS, Han DH, Renshaw PF, Merzenich MM, Gentile DA: Brains on video games. *Nature Reviews Neuroscience* 2011; 12:763-768.

[39] Nix RL, Pinderhughes EE, Dodge KA, Bates JE, Pettit GS, McFadyen-Ketchum SA: The relation between mothers' hostile attribution tendencies and children's externalizing behavior problems: The mediating role of mothers' harsh discipline practices. *Child. Dev.* 1999; 70:896-909.

[40] Wilson SR., Rack JJ, Shi X, Norris AM: Comparing physically abusive, neglectful, and non-maltreating parents during interactions with their children: A meta-analysis of observational studies. *Child Abuse and Negl* 2008; 32:897–911.

[41] Entertainment Software Rating Board, http://www.esrb.org/index-js.jsp accessed 10/13/2014.

[42] Anderson C: American Psychological Association Psychological Science Agenda Science Briefs: Violent Video Games: Myths, Facts, and Unanswered Questions Studies provide converging evidence that exposure to media violence is a significant risk factor for aggressive and violent behavior October 2003 http://www.apa.org /science/about/psa/2003/10/anderson.aspx Access online 10/13/14.

[43] American Psychiatric Association: Diagnostic and statistical manual of mental disorders (5th Ed.). Washington, DC, 2013.

[44] Porter B, Howe, T: Pilot evaluation of the ACT Parents Raising Safe Kids violence prevention program. *Journal of Child. & Adolescent Trauma* 2008; 1:193-206.

[45] Webb T, Martin K: Evaluation of a US school-based media literacy violence prevention curriculum on changes in knowledge and critical thinking among adolescents. *Journal of Children and Media* 2012; 6:430-449.

[46] Sim T, Gentile DA, Bricolo F, Serpelloni G, Gulamoydeen F: (2012). A conceptual review of research on the pathological use of computers, video games, and the Internet. *International Journal of Mental Health and Addiction* 2012; 10:748-769.

[47] Han DH, Lee YS, Na C, Ahn JY, Chung US, Daniels MA, Haws CA, Renshaw PF: The effect of methylphenidate on Internet video game play in children with attention-deficit/hyperactivity disorder. *Compr Psychiatry* 2009; 50:251–256.

[48] Han, DH, Hwang JW, Renshaw PF: Bupropion sustained release treatment decreases craving for video games and cue-induced brain activity in patients with Internet video game addiction. *Exp. Clin. Psychopharmacol.* 2010, 18:297–304.

[49] Du YS, Jiang W., Vance A: Longer term effect of randomized, controlled group cognitive behavioural therapy for Internet addiction in adolescent students in Shanghai. *Aust. and N Z J Psychiatry* 2010; 44:129-134.

[50] King DL, Delfabbro PH, Griffiths MD, Gradisar M: Assessing clinical trials of internet addiction treatment: A systematic review and CONSORT evaluation. *Clin. Psychol. Rev.* 2011; 31:1110-1116.

In: Neurobiology of Men's Mental Health
Editors: Leo Sher and Timothy R. Rice
ISBN: 978-1-63483-191-8
© 2015 Nova Science Publishers, Inc.

Chapter 7

PHYSIOLOGICAL AND PSYCHOLOGICAL ASSOCIATIONS OF TESTOSTERONE IN SPORTS AND EXERCISE WITH DUE REGARD TO ADOLESCENTS

Mirko Wegner[1],PhD, Claudia Niemann[2], MS and Henning Budde[3,4],PhD
[1]University of Bern, Switzerland
[2]Jacobs University Bremen, Germany
[3]Medical School Hamburg, Germany
[4]Reykjavik University, Iceland

ABSTRACT

This chapter aims to review physiological mechanisms and psychological associations of testosterone in the field of sport and exercise science with due regard to an adolescent age group. The chapter will start with an introduction to the biosynthesis and regulation of testosterone. Following, we will have a look at the regulation of testosterone in areas related to sport and exercise, then discuss on the importance of testosterone for motor learning and describe how endurance and resistance training affects testosterone levels acutely and chronically.

Another focus will be on sport- and exercise-related psychological associations of testosterone. This includes a discussion of the relationship of testosterone and dominance behavior as well as aggression. We will also describe emotional consequences of changes in testosterone levels. Additionally, the relevance of testosterone for motivational processes involved in competitive behavior such as those found in sport and exercise will be discussed. Finally, we talk about effects of altered testosterone levels on cognitive functions.

Keywords: testosterone, sport, exercise, adolescents, children, cognition, dominance, aggression

Introduction

Testosterone (T) is best known as the principal male sex hormone (androgen) and an anabolic (protein synthesizing) steroid. It is the final product of the hypothalamic-pituitary-gonadal (HPG) axis. External stimuli like physical or mental stress as well as endogenous stimuli like biological rhythms or hormones themselves activate or inactivate the HPG axis. The HPG axis also shows interaction with the hypothalamic-pituitary-adrenal (HPA) axis (final product: cortisol).

Biosynthesis and Regulation of Testosterone

The HPG axis is responsible for testosterone regulation. As illustrated in Figure 1, the gonadotrophin-releasing hormone (GnRH) is pulsatilely delivered from the hypothalamus to the anterior pituitary via the hypophyseal portal system. The gonadotrophins follicle-stimulating hormone (FSH) and luteinizing hormone (LH) then cause the synthesis of oestrogen and testosterone in the gonads. The Leydig cells of the testes produce 5-10 mg of testosterone per day. Only 5% of testosterone is produced by the adrenal glands. Testosterone levels are then fed back to the pituitary and the hypothalamus to control further hormone production (inhibition). At the same time high levels of FSH and LH inhibit the secretion of GnRH from the hypothalamus and the gonadotrophin release in the anterior pituitary. The hypothalamus also responds with slowing down testosterone secretion with increased GnRH levels.

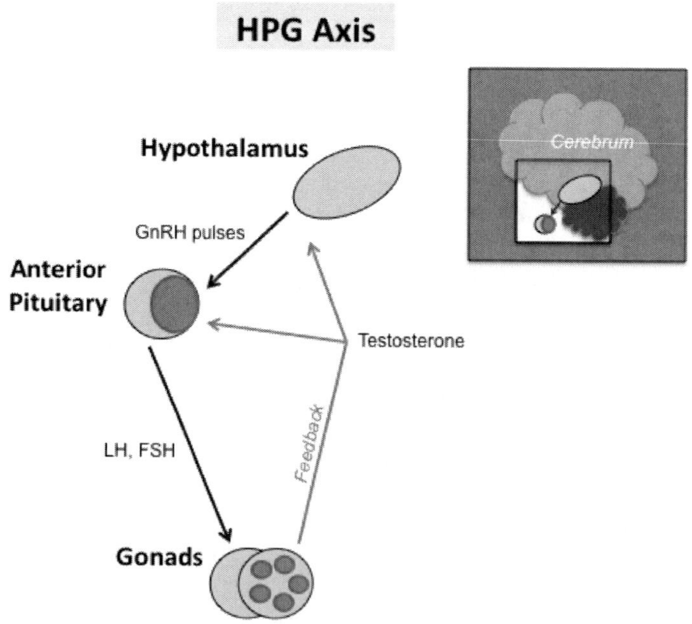

Figure 1. Functioning of the HPG axis.

The principal pathway of testosterone synthesis in humans starts with cholesterol that is converted to pregnenolone, 17α-hydroxypregnenolone, dehydroepiandrosterone (DHEA), androstendione, and finally testosterone. In the process approximately 7% of the testosterone is reduced to its more active form, 5α-dihydrotestosterone (DHT), which is the most important metabolite of testosterone being active in tissues expressing 5α-reductase, like in prostate tissue. DHT binds to the same androgen receptor even more strongly than testosterone, so that its androgenic potency is about five times that of testosterone.

Two main mechanisms contribute to the effectiveness of testosterone. First, testosterone becomes effective through the activation of the androgen receptor, and second, through aromatization to oestradiol and activation of oestrogen receptors.

Testosterone as well as DHT is transported into the cytoplasm of the target tissue cells and binds to the androgen receptor. A structural change then allows the testosterone-receptor complex to move into the cell nucleus where it binds to specific nucleotide sequences of the chromosomal DNA. The chromosomal DNA in turn triggers the transcriptional activity of genes leading to protein biosynthesis. Respective anabolic and androgenic effects of testosterone are effectively used in medical treatment, of e.g., malnourished or weak patients, but are also abused in sports by the use of doping substances, e.g., to improve muscle growth.

Testosterone Levels and Age

In early childhood (age 6-8) the zona reticularis in the adrenal cortex maturates which leads to an increase of androgen precursors like dehydroepiandrosterone (DHEA). These androgen precursors are then converted to T and DHT leading to the prepubertal growth spurt and the development of secondary sex characteristics.

Puberty is then initiated by an increase in the pulsatile secretion pattern of hypothalamic GnRH. During puberty, there are developmental changes in the release of GnRH, for example, the negative feedback sensitivity to steroid hormones in the anterior pituitary and the hypothalamus decreases. In addition, the concentration of gonadotrophin hormones and gonadal steroids increases. The decrease of feedback sensitivity can be made responsible for the increase in testosterone production by the beginning of puberty at about age 11-12 years [1]. Primary and secondary sex characteristics then grow until they reach adult levels. Testosterone is also responsible for the end of growth after puberty by closing the epiphyses [2]. Testosterone levels are highest at early adulthood between the ages of 20-30 years [1]. From 30 years of age on total and free levels of testosterone begin to decline. In elderly men the diurnal pattern and total amount of T is lower compared to adolescents and young men [3].

In aging healthy men the decrease in androgen production is progressing slowly also due to increased levels of sex hormone-binding globulin (SHBG) and reduced production of growth hormone and insulin-like growth factor 1 (IGF-1) [4]. Clinical significance of hypogonadism can be found in 20% of men aged 60 and older, and 40% of men aged 80 and older [5]. The drop in circulating levels of testosterone in hypogonadism can usually be attributed to failures of the hypothalamus and pituitary as well as disorders of testicular functions.

Testosterone and Sports and Exercise

The components of the neuroendocrine system that control and regulate the reproductive functions are very sensitive to physical exercise. Numerous studies reported the bidirectional effect of physical exercise and testosterone focusing mostly on adult men. In the following we give a short overview about possible underlying mechanisms of this bidirectional effect and report empirical findings on the exercise-testosterone relationship for acute as well as chronic physical exercise (training) in different kinds of sport.

Testosterone and Motor Performance

Studies on the link between exercise and motor performance often produce inconsistent findings mainly due to differences in the characteristics of the stress intervention and the kind of motor performance tested [6, 7]. From animal research it can be assumed that testosterone improves fine motor performance by its effects on the nigrostriatal dopaminergic (NSDA) system and by improving the expression of dopamine transporter (DAT) and tyrosine hydroxylase (TH) [8, 9]. In humans it could be shown that androgen deprivation causes decreases in fine motor performance of the hand [10].

Findings in adolescents. For adolescents, it has been shown that testosterone is involved in the development of perceptual motor skills especially during puberty [11-13]. Not surprisingly, Wegner and colleagues [7] were recently able to show that fine motor performance was directly linked to testosterone responses to different stressors in an adolescent sample. Participants with higher T increases committed fewer errors in the fine motor task.

Testosterone and Acute Bouts of Exercise

Arena and Maffulli [14] suggest various mechanisms as causes for the increase in hormonal concentration through acute exercise. Among these explanations is the increase in body core temperature. The rise in core temperature interferes with the binding of hormones to plasma proteins, which is accompanied by an increase in the metabolically active hormonal fraction.

Moreover, plasma volume reductions and a reduced hepatic clearance caused by a drop in hepatic blood flow or higher secretion of testosterone through the testes may result in acute elevation of plasma testosterone. Moreover, higher testosterone levels may also cause an interaction between the hormone and a receptor of the target cells, which might for example increase muscle protein synthesis. Other possible explanations for an acute increase of plasma testosterone (besides plasma volume reductions and a reduced hepatic clearance due to a drop in hepatic blood flow) are potential adaptations in testosterone synthesis or secretory capacity of the Leydig cells in the testes and adrenergic stimulation or lactate-stimulated secretion [15].

Testosterone and Acute Bouts of Endurance Exercise

The effect of acute bouts of exercise on testosterone concentration is dependent on various factors including the individual's training status and duration and intensity of the exercise [16, 17]. For example, testosterone levels show significant increases after exercise that reaches a certain level of the maximal aerobic power (VO_{2max}). A 10-minute run between 70 to 80% of VO_{2max} may lead to significant increases in total testosterone concentration of endurance-trained men directly after the physical stressor [18]. Further, Gray and colleagues showed that an intense interval exercise protocol may lead to a 38% rise of total testosterone in trained men [19]. Testosterone levels even dropped 16% under resting levels after six hours of recovery. Brownlee et al., showed a significant increase in free testosterone after 60-90 minutes of endurance exercise at an intensity of 65-75% of VO_{2max} [20]. Increased testosterone concentrations following exercise were also described [14]. With respect to the training level, Crewther et al., showed that the reactivity of the HPG axis in response to an acute bout of exercise is more pronounced in trained individuals when compared to untrained individuals [21]. These results suggest an adaptation to the physical stress situation.

There is a strong relationship between the HPA (cortisol) and the HPG (testosterone) axis. Daly and colleagues, for example, found that endurance-trained men showed elevated cortisol levels but no change in testosterone directly after a 75-minute exercise session [22]. Cortisol levels then remained high for at least 60 minutes during the recovery period and did not drop to baseline levels until 90 minutes past the exercise were reached. Total and free testosterone levels, in turn, were significantly lower than at baseline levels.

Findings in adolescents. Testosterone responses to exercise are subject to age differences [23, 24]. For example, it could be shown that salivary testosterone in male adolescents aged 10-16 years increased by 32% after a 90-minute football training session [25]. Eliakim et al., [26] also found testosterone increases in 13-18 year-old volleyball players after one hour of volleyball practice. Budde and colleagues observed significant increases in salivary free testosterone concentrations of normal high-school boys and girls aged 15-16 years after a 12-minute exercise at 70-85% of HR_{max} [27, 28].

The inverse relationship between free cortisol and testosterone, however, could not be observed in high-school students after 12 minutes of intense exercise [27, 28]. By contrast, both cortisol and testosterone rose significantly in these studies with 14-15 year old teenagers. Brownlee and colleagues, by contrast, illustrated a negative relationship between cortisol and testosterone only for total but not for free fractions of T [20].

Testosterone and Acute Bouts of Resistance Exercise

For resistance training acute elevations of anabolic hormones in circulating blood during and immediately after exercise are well documented. Most studies show a significant increase in T concentrations in young and old men [15, 29-31]. Only few studies found no significant testosterone changes in men [32].

Findings in adolescents. There are some findings on testosterone changes for adolescents involved in strength training. Kraemer et al., [33], for example, found testosterone increases following an acute weightlifting exercise only in adolescent weightlifters who had been practicing for more than two years. Pullinen and colleagues, however, found no testosterone increases in adolescents after five sets of ten knee extensions (40% of their one-repetition maximum) and two sets to exhaustion [34, 35]. This pattern of results was different compared to men sampled in the same experiment.

Testosterone and Chronic Exercise/ Training

Certain interactions between the HPA and the HPG axes could possibly explain long-term effects of exercise on testosterone concentrations [36]. Practicing as well as participating in sports competitions is always accompanied by physical stress. Stress also means that the HPG axis regulation might be disrupted or has a negative impact on T secretion through inhibition of the activity of the Leydig cells in the testes [37]. Thus, testosterone decreases in response to endurance training might be seen as a general response of the HPG to physical stress [21, 37]. Van Eckardstein and colleagues put forward that testosterone decreases may have a cardiovascular protective effect and may lead to positive health consequences for endurance athletes [38]. Van Eckardstein et al., were able to associate high-density lipoprotein (HDL) – a marker of cardiovascular health – with decreased testosterone concentrations.

Zitzmann and Nieschlag [2] argued that with dropping T levels gonadotropin levels (e.g., LH, FH) should rise, which however sometimes is not the case in studies comparing exercising vs. non-exercising men. They attribute the lack of this pattern of results to the suppression of the feedback regulation of the HPG axis. Hackney [39] exactly observed a lack of increasing basal LH concentration in correspondence with reduced levels of testosterone.

Testosterone and Endurance Training

For endurance training it has been suggested that intensive training bouts lead to prolonged effects on androgen levels. For example, Hackney et al., [40] were able to show that endurance athletes practicing 10-20 hours a week showed lower testosterone levels compared to a control group. This finding was also confirmed by a study showing that long-distance runners develop chronically low basal T levels, which are 20 to 40% below the pre-training resting levels [39]. Physical training conducted over weeks, months, and years therefore seems to lead to systemic changes in the HPG axis functioning. The majority of the athletes exhibit very low T levels, which are still in a normal range. However, Hackney [39] also reviewed studies that did not report significant changes in basal T concentrations following 2-12 months of endurance training. One explanation for Hackney for the different HPG axis adaptations found in research studies is that it is strongly dependent on the exercise dosage used as well as participants' physical fitness.

Findings in adolescents. Mero, Jaakkola, and Komi [41] found that one year of regular endurance training significantly increased testosterone levels of boys aged 11-12 years when compared to a control group. However, in a longitudinal study with soccer players no significant long-term changes in testosterone could be found when compared to a control group [11].

Testosterone and Resistance Training

In resistance sports, effects on testosterone levels are ambiguous. Kraemer et al., found higher testosterone responses to heavy resistance exercise in trained compared to untrained weightlifters [31]. These results could be confirmed by Izquierdo and colleagues [29] for a resistance exercise program lasting seven weeks. Typically, a single hypertrophic type of resistance exercise is able to increase serum T concentrations. However, it was previously argued that the duration and magnitude of the acute hormone response may change in response to the length of the resistance training program.

Increases in serum testosterone levels in response to strength training was shown after 14 weeks of a resistance training program [15] and in untrained men and women after 8 weeks of resistance training [31]. Crewther et al., [21] additionally pointed to the relation between exercise performance and testosterone, arguing that this relation can only be found in athletes but not non-athletes. However, the testosterone concentration of athletes participating in seven weeks of additional resistance training dropped when compared to before the training [15]. Ahtianinen et al., found that total training volume and participants' isometric strength may be an explanation for the observed decreases in testosterone. They could also show that untrained individuals showing high acute T responses showed more muscle gain compared to athletes with low responses. These results suggest that basal T concentrations may be one factor involved in strength development.

Findings in adolescents. Gorostiaga and colleagues investigated the effects of six weeks of heavy resistance training on serum testosterone in 14-16 year-old handball players [42]. They did not find a significant difference in testosterone levels of the experimental group when compared to the control group. However, after eleven weeks of strength training with soccer players basal serum concentrations of testosterone increased significantly when compared to a control group [43].

Testosterone and Psychological Associations

There is a bidirectional link between testosterone and behavior. Testosterone has been linked to several psychological processes including emotions like dominance and aggression, anxiety and depression, and motivation; also, it has been shown to affect cognitive abilities [16].

Testosterone has been shown to be intertwined with psychological variables regarding its support of approach behavior and its involvement in the response to social threats [44]. Moreover, testosterone is involved in the control of impulses. When a threat occurs or an impulse needs to be controlled, testosterone affects behavior especially when the activation of both the serotonergic systems (5-HTTLPR) and the HPA axis (cortisol) is low [36, 45, 46]. That means only when cortisol levels and serotonergic tone are low an individual's behavior will be testosterone-driven. In men, testosterone additionally benefits the identification of socially ambiguous situations especially when vasopressin levels are high [47].

Testosterone and Dominance and Aggression

Aggression may be defined as a specific emotional reaction to threat [44] whereas dominance refers to the attempt to gain or maintain high status within a social hierarchy [48]. Several studies have pointed to the positive link between testosterone and dominance and aggression, which is especially high between the age of 25-35 in both athletes and non-athletes [49-51]. Moreover, the link between dominance and testosterone has been shown to be pronounced when the individual's status is evaluated by others, e.g., by means of performance in a test [52]. Archer [49] puts it differently by saying that individuals high in T respond more strongly to tasks that challenge their status. In sports, higher testosterone levels were previously associated with more offensive play during judo compctition [53]. In particular, judo players who showed high testosterone levels before the competition attacked more often, put more effort into fighting to gain advantage, and posed more threats to their opponent [54-56]. Moreover, judoists with high levels of testosterone are described as being angrier, more prone to challenges, and more violent competitors by their coaches [49, 53]. Furthermore, athletes who used anabolic steroids have been shown to be more aggressive compared to non-users [57-59].

Testosterone, Anxiety, and Depression

Higher testosterone levels are usually associated with low levels of anxiety and depression [60]. For example, older men whose testosterone levels are likely to be at the low end of the physiological range usually show higher levels of depression compared to younger men. However, it could be shown that testosterone substitution may improve negative mood states [61]. For adolescent boys it could be shown that testosterone levels are lower in participants showing higher levels of depression or anxiety [62]. In sports like tennis it has been described that participants who had just won a tennis match showed elevated testosterone levels only when they were proud about their victory [63]. Similarly, Booth et al., [64] linked the pre-contest rise in testosterone to more positive emotional states. In judoists, Salvador and colleagues could show that testosterone was negatively related to negative emotions [65]. Accordingly, it was shown that increases in testosterone accompanied by low levels of cortisol were found in athletes who perceived their levels of anxiety before the competition as facilitative of performance [66]. However, other authors could not find mood differences depending on testosterone after winning or losing in basketball [67] or judo [68].

Testosterone and Motivation

The relationship between testosterone and sports performance has not always been described consistently. For example, no differences were found for judoists prior to a competition [65] or after winning or losing a judo fight [68]. Schultheiss and colleagues suggested that individual differences in the power motive may be responsible for such null-findings [69, 70]. According to Winter [71] and McClelland [72] the power motive is defined as the capacity to derive pleasure from the impact on other persons. It could be previously shown that men high in the power motive showed elevated testosterone levels after winning a competition [48, 72]. Schultheiss and colleagues argue that increases in testosterone levels reflect a reinforcement mechanism for the behavior that led to success [73, 74].

Researchers also looked at how changes in testosterone levels affected individuals' perseverance to subsequent competitions. It could, for example, be shown that losers of a puzzle game preferred to participate in another game over a non-competitive activity only when their testosterone levels were high after the first game. Accordingly, participants in a computer game who found it rewarding to react aggressively not only had higher testosterone levels but also preferred to engage in further competitive games [75].

Testosterone and Cognition

It has been previously shown that testosterone can be linked to cognitive performance [76]. This association has been found for men and women and is assumed to follow a curvilinear function with optimal cognitive performance at medium serum testosterone levels [77-79]. For example, optimal levels of spatial ability were associated with intermediate testosterone levels. Similarly, Wolf and colleagues reported that a high dose of testosterone negatively affected performance in a verbal fluency task in elderly men [80, 81]. However, less is known about testosterone-related changes in cognitive performance after varying bouts of acute exercise. In a study with adolescents Budde and colleagues were able to significantly raise testosterone levels following an intense 12-minute bout of exercise which negatively affected working memory performance in those participants with higher T levels after exercise compared to those with lower T levels [28]. This finding is in line with the assumption of a curvilinear association between testosterone and cognitive performance found in other studies [77]. The results of these studies suggest that testosterone might be involved in the modification of the functioning of the prefrontal cortex [82]. However, the exact biological mechanisms behind this phenomenon have not yet been adequately explained [83].

Many of the behavior effects of testosterone occur only after conversion to DHT and oestradiol through 5α-reductase or enzyme aromatase showing that T interacts with both oestradiol [84] and androgen receptors [85]. A possible explanation for the beneficial effects of testosterone on cognitive performance is that T acts via aromatization to oestradiol in the striatal dopamine system with accompanying effects on prefrontal functions [86]. Another possible explanation is that the structural changes of synapses and their associated dendritic spines after motor training can be modified over a period of even minutes to hours [87] and have been shown to be driven by testosterone [88]. At present it is not possible to explain whether testosterone is the critical hormone for the beneficial effects on cognition or is only involved through its conversion to its metabolic derivates (oestradiol, DHT).

Conclusion

Testosterone as a principal androgen reaches its highest levels between the age of 20-30 years. Influence on testosterone by acute and chronic exercise in adult men has been reported in the past. Acute exercise needs to be of a certain length and of at least moderate intensity in order to significantly increase T levels. This has been shown in adult men as well as in adolescent boys. Chronic endurance exercise has been shown to decrease testosterone levels in male adults, an effect which could not be observed in adolescents. Regarding motor and cognitive performance, it could be shown that T levels are associated with improved motor and cognitive performance in adolescents. Other psychological variables related to the androgen are dominance and aggressive behavior as well as well-being, low anxiety, and low depression.

REFERENCES

[1] Granger DA, Shirtcliff EA, Booth A, Kivlighan KT, Schwartz EB: The 'trouble' with salivary testosterone. *Psychoneuroendocrinology*. 2004;29(10):1229-40.

[2] Zitzmann M, Nieschlag E: Testosterone levels in healthy men and the relation to behavioural and physical characteristics: Facts and constructs. *Eur. J. Endocrinol.* 2001;144(3):183-97.

[3] Diver MJ, Imtiaz KE, Ahmad AM, Vora JP, Fraser WD: Diurnal rhythms of serum total, free and bioavailable testosterone and of SHBG in middle-aged men compared with those in young men. *Clin. Endocrinol.* 2003;58:710-7.

[4] Vermeulen A, Kaufmann JM, Giagulli VA: Influence of some biological indexes on sex hormone-binding globulin and androgen levels in aging or obese males. *J. Clin. Endocrinol. Metab*. 1996;81:1821-6.

[5] Morley JE, Kaiser FE, Perry HM, Patrick P, Morley PMK, Stauber PM: Longitudinal changes in testosterone luteinizing hormone and follicle-stimulating hormone in healthy older men. *Metabolism.* 1997;46(4):410-3.

[6] Waldron EM, Anton BS: Effects of exercise on dexterity. *Percept Mot.* Skills. 1995;80(3):883-9.

[7] Wegner M, Koedijker JM, Budde H: The effect of acute exercise and psychosocial stress on fine motor skills and testosterone concentration in the saliva of high school students. *PLoS ONE.* 2014;9(3):e92953.

[8] Patchev VK, Schroeder J, Goetz F, Rohde W, Patchev AV: Neurotropic action of androgens: Principles, mechanisms and novel targets. *Exp. Geront.* 2004;39:1651-60.

[9] Emborg ME, Ma SY, Mufson EJ, Levey AI, Taylor MD, Brown WD: Age-related declines in nigral neuronal function correlate with motor impairments in rhesus monkeys. *J. Comp. Neurol*. 1998;401:253-65.

[10] Soyupek F, Soyupek S, Perk H, Özorak A: Androgen deprivation therapy for prostate cancer: Effects on hand function. *Urol. Oncol.-Semin. Ori*. 2008;26:141-6.

[11] Vänttinen T, Blomqvist M, Häkkinen K: Development of body composition, hormone profile, physical fitness, general perceptual motor skills, soccer skills and on-the-ball performance in soccer- specific laboratory test among adolescent soccer players. *J. Sport Sci. Med*. 2010;9:547-56.

[12] Tan Ü: Testosterone and hand skill in right-handed men and women. *International J. Neurosci*. 1990;53(2-4):179-89.

[13] Fink B, Manning JT, Neave N, Tan U: Second to fourth digit ratio and hand skill in Austrian children. *Biol. Psychol*. 2004;67:375-84.

[14] Arena B, Maffulli N: Endocrinology changes in exercising women. *Sports Med. Arthrosc.* 2002;10:10-4.

[15] Ahtiainen JP, Pakarinen A, Alen M, Kraemer WJ, Häkkinen K: Muscle hypertrophy, hormonal adaptations and strength development during strength trainng in strength-trained and untrained men. *Eur. J. Appl.* Physiol. 2003;89:555-63.

[16] Windisch C, Wegner M, Budde H: Testosterone. In: Ehrlenspiel F, Strahler K, editors. Psychoneuroendocrinology of sport and exercise: Foundations, markers, trends. London: Routledge; 2012. p. 61-85.

[17] Guglielminin C, Paolini AR, Conconi F: Variations of serum testosterone concentrations after physical exercises of different duration. *Int. J. Sports Med.* 1984;5:246-9.

[18] Kenefick RW, Maresh CM, Armstrong LE, Castellani JW, Whittlesey M, Hoffman JR: Plasma testosterone and cortisol responses to training-intensity exercise in mild and hot environments. *Int. J. Sports Med.* 1998;19:177-81.

[19] Gray AB, Telford RD, Weidemann MJ: Endocrine response to intense interval exercise. *Eur. J. Appl. Physiol. Occup. Physiol.* 1993;66(4):366-71.

[20] Brownlee KK, Moore AW, Hackney AC: Relationship between circulating cortisol and testosterone: Influence of physical exercise. J. Sports Sci. Med. 2005;4:76-83.

[21] Crewther PYL, Cook C, Cardinale M, Weatherby RP, Lowe T: Two emerging concepts for elite athletes: The short-term effects of testosterone and cortisol on the neuromuscular system and dose-response training role of these endogenous hormones. *Sports Med.* 2011;41:103-23.

[22] Daly W, Seegers CA, Rubin DA, Dobridge JD, Hackney AC: Relationship between stress hormones and testosterone with prolonged endurance exercise. *Eur. J. Appl. Physiol.* 2005;93:375-80.

[23] Kraemer WJ, Staron RS, Hagermann FC, Hikida RS, Fry AC, Gordon SE: The effect of short-term resistance training on endocrine function in men and women. *Eur. J. Appl. Physiol.* 1998;78:69-76.

[24] Kraemer WJ, Ratamess NA: Hormonal responses and adaptations to resistance exercise and training. *Sports Med.* 2005;35(4):339-61.

[25] Di Luigi L, Baldari C, Gallotta MC, Perroni F, Romanelli F, Lenzi A: Salivary steroids at rest and after a training load in young male athletes: Relationship with chronological age and pubertal development. *Int. J. Sports Med.* 2006;27(9):709-17.

[26] Eliakim A, Portal S, Zadik Z, Rabinowitz J, Adler-Portal D, Cooper DM: The effect of a volleyball practice on anabolic hormones and inflammatory markers in elite male and female adolescent players. *J. Strength Cond.* 2009;23(5):1553-9.

[27] Budde H, Pietrassyk-Kendziorra S, Bohm S, Voelcker-Rehage C: Hormonal responses to physical and cognitive stress in a school setting. *Neurosci. Lett.* 2010;474(3):131-4.

[28] Budde H, Voelcker-Rehage C, Pietrassyk-Kendziorra S, Machado S, Ribeiro P, Arafat AM: Steroid hormones in the saliva of adolescents after different exercise intensities and their influence on working memory in a school setting. *Psychoneuroendocrinology.* 2010;35(3):382-91.

[29] Izquierdo M, Ibañez J, Calbet JAL, Navarro-Amezqueta I, González-Izal M, Idoate F: Cytokine and hormone response to resistance training. *Eur. J. Appl. Physiol.* 2009;107:397-409.

[30] McCaulley GO, McBride JM, Cormie P, Hudson MB, Nuzzo JL, Quindry JC: Acute hormonal and neuromuscular response to hypertrophy, strength and power type resistance exercise. *Eur. J. Appl.* Physiol. 2009;105:695-704.

[31] Kraemer WJ, Häkkinen K, Newton RU, McCormick M, Nindl BC, Volek JS: Acute hormonal response to heavy resistance exercise in younger and older men. *Eur. J. Appl. Physiol.* 1998;77:206-11.

[32] Smilios I, Pilianidis T, Karamouzis M, Tokmakidis SP: Hormonal responses after various resistance exercise protocols. *Med. Sci. Sports Exerc.* 2003;35(4):644-54.

[33] Kraemer WJ, Fry AC, Warren BJ, Stone MH, Fleck SJ, Kearney JT: Acute hormonal responses in elite junior weightlifters. *Int. J. Sports Med.* 1992;13(20):103-9.

[34] Pullinen T, Mero A, Huttunen P, Pakarinen A, Komi PV: Resistance exercise-induced hormonal responses in men, women, and pubescent boys. *Med. Sci. Sports Exerc.* 2002;34(5):806-13.

[35] Pullinen T, Mero A, Huttunen P, Pakarinen A, Komi PV: Resistance exercise-induced hormonal response under the influence of delayed onset muscle soreness in men and boys. *Scand. J. Med. Sci. Sports.* 2011;21(6):e184-e94.

[36] Viau V: Functional cross-talk between the hypothalamic-pituitary-gonadal and -adrenal axes. *J. Neuroendocrinol.* 2002;14(6):506-13.

[37] Granger DA, Weisz JR, Kauneckis D: Neuroendocrine reactivity, internalizing behaviour problems, and control-related cognitions in clinic-referred children and adolescents. *J. Abnorm. Psychol.* 1994;103:267-76.

[38] van Eckardstein A, Kliesch S, Nieschlag E, Chirazi A, Assmann G, Behre H: Suppression of endogenous testosterone in young men increases serum levels of high density lipoprotein sbclass lipoprotein A-I and lipoprotein(a). *J. Clin. Endocrinol. Metab.* 1997;82:3367-72.

[39] Hackney AC: Effects of endurance exercise on the reproductive system of men: the "exercise-hypogonadal male condition." *J. Endocrinol.* Invest. 2008;31(10):932-8.

[40] Hackney AC, Sinning WE, Bruot BC: Reproductive hormonal profiles of endurance-trained and untrained males. *Med. Sci. Sports Exerc.* 1988;20:60-5.

[41] Mero A, Jaakkola L, Komi PV: Serum hormones and physical performance capacity in young boy atheltes during a 1-year training period. *Eur. J. Appl. Physiol. Occup. Physiol.* 1990;60:32-7.

[42] Gorostiaga EM, Izquierdo M, Iturralde P, Ruesta M, Ibáñez J: Effects of heavy resistance training on maximal and explosive force production, endurance and serum hormones in adolescent handball players. *Eur. J. Appl. Physiol.* 1999;80:485-93.

[43] Gorostiaga EM, Izquierdo M, Ruesta M, Iribarren J, González-Badillo JJ: Strength training effects on physical performance and serum hormones in young soccer players. *Eur. J. Appl. Physiol.* 2004;91:698-707.

[44] Liening SH, Josephs RA: It is not just about testosterone: Physiological mediators and moderators of testosterone's behavioral effects. *Soc. Personal Psychol. Compass.* 2010;11(4):982-94.

[45] Dickerson SS, Kemeny ME: Acute stressors and cortisol responses: A theoretical integration and synthesis of laboratory research. *Psychol. Bull.* 2004;130(3):355-91.

[46] Helmich I, Latini AS, Sigwalt A, Carta MG, Machado S, Velasques B: Neurobiological alterations induced by exercise and their impact on depressive disorders. Clin. Pract. Epidemiol. Ment. Health. 2010;6:115-25.

[47] Thompson RR, Goerge K, Walton JC, Orr SP, Benson J: Sex-specific influences of vasopressin on human social communication. *Proc. Natl Acad. Sci. USA*. 2006;103(20):7889-94.

[48] Mazur A, Booth A: Testosterone and dominance in men. Behav Brain Sci. 1998;21(3):353-97.

[49] Archer J: Testosterone and human aggression: An evaluation of the challenge hypothesis. *Neurosci. Biobehav. Rev.* 2006;30(3):319-45.

[50] Archer J, Graham-Kevan N, Davies M: Testosterone and aggression: A reanalysis of Book, Starzyk, and Quinsey's (2001) study. *Aggress. Violent Behav.* 2005;10(2):241-61.

[51] Book AS, Starzyk KB, Quinsey VL: The relationship between testosterone and aggression: A meta-analysis. *Aggress. Violent Behav.* 2001;6(6):579-99.

[52] Josephs RA, Neman ML, Brown RP, Beer JM: Status, testosterone, and human intellectual performance: stereotype threat as status concern. *Psychol. Sci.* 2003;14(2):158-63.

[53] Salvador A, Simón V, Suay F, Llorens L: Testosterone and cortisol responses to competitive fighting in human males: A pilot study. *Aggress. Behav.* 1987;13(1):9-13.

[54] Salvador A, Suay F, Martinez-Sanchis S, Simon VM, Brain PF: Correlating testosterone and fighting in male participants in judo contests. *Physiol. Behav.* 1999;68(1-2):205-9.

[55] Parmigiani S, Bartolomucci A, Palanza P, Galli P, Rizzi N, Brain PF: In Judo, Randori (free fight) and Kata (highly ritualized fight) differentially change plasma cortisol, testosterone, and interleukin levels in male participants. *Aggress. Behav.* 2006;32(5):481-9.

[56] Bateup HS, Booth A, Shirtcliff EA, Granger DA: Testosterone, cortisol, and women's competition. *Evol. Hum. Behav.* 2002;23(3):181-92.

[57] Pope HG, Katz DL: Psychiatric and medical effects of anabolic-androgenic steroid use: A controlled study of 160 athletes. *Arch. Gen. Psychiatry.* 1994;51(5):375-82.

[58] Moss HB, Panzak GL, Tarter RE: Personality, Mood, and Psychiatric Symptoms Among Anabolic Steroid Users. *Am. J. Addiction*. 1992;1(4):315-24.

[59] Choi PYL, Parrott AC, Cowan D: High-dose anabolic steroids in strength athletes: Effects upon hostility and aggression. *Hum. Psychopharm. Clin.* 1990;5(4):349-56.

[60] Daley A: Exercise and depression: A review of reviews. *J. Clin. Psychol Med. Settings*. 2008;15(2):140-7.

[61] Barrett-Connor E, von Mühlen DG, Kritz-Silverstein D: Bioavailable testosterone and depressed mood in older men: the Rancho Bernardo Study. *J. Clin. Endocrinol. Metab.* 1999;84(2):573-7.

[62] Granger DA, Shirtcliff EA, Zahn-Waxler C, Usher B, Klimes-Dougan B, Hastings P: Salivary testosterone diurnal variation and psychopathology in adolescent males and females: Individual differences and developmental effects. *Dev. Psychopathol.* 2003;15(2):431-49.

[63] Mazur A, Lamb TA: Testosterone, status, and mood in human males. *Horm. Behav.* 1980;14(3):236-46.

[64] Booth A, Shelley G, Mazur A, Tharp G, Kittok R: Testosterone, and winning and losing in human competition. *Horm. Behav.* 1989;23(4):556-71.

[65] Salvador A, Suay F, González-Bono E, Serrano MA: Anticipatory cortisol, testosterone and psychological responses to judo competition in young men. *Psychoneuroendocrinology.* 2003;28(3):364-75.

[66] Eubank M, Collins D, Lovell G, Dorling D, Talbot S: Individual temporal differences in precompetition anxiety and hormonal concentration. *Pers. Indiv. Differ.* 1997;23(6):1031-9.

[67] Gonzalez-Bono E, Salvador A, Serrano MA, Ricarte J: Testosterone, cortisol, and mood in a sports team competition. *Horm. Behav.* 1999;35(1):55-62.

[68] Suay F, Salvador A, González-Bono E, Sanchís C, Martínez M, Martínez-Sanchis S: Effects of competition and its outcome on serum testosterone, cortisol and prolactin. *Psychoneuroendocrinology.* 1999;24(5):551-66.

[69] Schultheiss OC, Rhode W: Implicit power motivation predicts men's testosterone changes and implicit learning in a contest situation. *Horm. Behav.* 2002;41(2):195-202.

[70] Wirth MM, Welsh KM, Schultheiss OC: Salivary cortisol changes in humans after winning or losing a dominance contest depend on implicit power motivation. *Horm. Behav.* 2006;49(3):346-52.

[71] Winter DG: The power motive. New York: The Free Press; 1973.

[72] McClelland DC: Power: The inner experience. New York: Irvington Publishers; 1975.

[73] Schultheiss OC, Campbell KL, McClelland DC: Implicit power motivation moderates men's testosterone responses to imagined and real dominance success. *Horm. Behav.* 1999;36(3):234-41.

[74] Schultheiss OC, Wirth MM, Torges CM, Pang JS, Villacorta MA, Welsh KM: Effects of implicit power motivation and men's and women's implicit learning and testosterone change after social victory or defeat. *J. Pers. Soc. Psychol.* 2005;88(1):174-88.

[75] Carré JM, Gilchrist JD, Morrissey MD, McCormick CM: Motivational and situational factors and the relationship between testosterone dynamics and human aggression during competition. *Biol. Psychol.* 2010;84(2):346-53.

[76] Hampson E: Spatial cognition in humans: Possible modulation by androgens and estrogens. *J. Psychiatry Neurosci.* 1995;20(5):397-404.

[77] Moffat SD, Hampson E: A curvilinear relationship between testosterone and spatial cognition in humans: Possible influence of hand preference. Psychoneuroendocrinology. 1996;21(3):323-37.

[78] Gouchie C, Kimura D: The relationship between testosterone levels and cognitive ability patterns. *Psychoneuroendocrinology.* 1991;16(4):323-34.

[79] Muller M, Aleman A, Grobbee DE, de Haan EH, van der Schouw YT: Endogenous sex hormone levels and cognitive function in aging men: Is there an optimal level? *Neurology.* 2005;64(5):866-71.

[80] Wolf OT, Preut R, Hellhammer DH, Kudielka BM, Schurmeyer TH, Kirschbaum C: Testosterone and cognition in elderly men: A single testosterone injection blocks the practice effect in verbal fluency, but has no effect on spatial or verbal memory. *Biol. Psychiatry.* 2000;47(7):650-4.

[81] Wolf OT, Kirschbaum C: Endogenous estradiol and testosterone levels are associated with cognitive performance in older women and men. *Horm. Behav.* 2002;41(3):259-66.

[82] Gold JM, Carpenter C, Randolph C, Goldberg TE, Weinberger DR: Auditory working memory and Wisconsin Card Sorting test performance in schizophrenia. *Arch. Gen. Psychiatry*. 1997;54(2):159-65.

[83] Beauchet O: Testosterone and cognitive function: Current clinical evidence of a relationship. *Eur. J. Endocrinol*. 2006;155(6):773-81.

[84] Moffat SD: Effects of testosterone on cognitive and brain aging in elderly men. *Ann. NY Acad. Sci*. 2005;1055:80-92.

[85] Sternbach H: Age-associated testosterone decline in men: Clinical issues for psychiatry. *Am. J. Psychiatry*. 1998;155(10):1310-8.

[86] Janowsky JS, Chavez B, Orwoll E: Sex steroids modify working memory. *J. Cognitive Neuroscience*. 2000;12(3):407-14.

[87] Johansen-Berg H, Sampaio Baptista C, Thomas AG: Human structural plasticity at record speed. *Neuron*. 2012;73(22):1058-60.

[88] Devoogd TJ, Nixdorf B, Nottebohm F: Synaptogenesis and changes in synaptic morphology related to acquisition of a new behavior. *Brain Res*. 1985;329(1–2):304-8.

In: Neurobiology of Men's Mental Health
Editors: Leo Sher and Timothy R. Rice
ISBN: 978-1-63483-191-8
© 2015 Nova Science Publishers, Inc.

Chapter 8

THE EFFECTS OF HIGH-FAT DIET (WITH AND WITHOUT WEIGHT GAIN) ON TESTOSTERONE LEVELS (PUBERTAL PERSPECTIVE)

Leandro Freire dos Santos*[1*]*, PhD, Rosália Rubel*[1,2] *and Carlos Ricardo Soccol*[1]*,PhD
[1] Federal University of Paraná,
Department of Bioprocesses and Biotechnology,
Curitiba, PR, Brazil
[2] Pelé Pequeno Príncipe Institute, Curitiba, PR, Brazil

Previous studies have revealed a relationship between high-fat diets (HFDs) and precocious puberty (Ahmed, Ong, & Dunger, 2009; Dos Santos et al., 2012). HFDs promote weight gain and possibly higher leptin levels (Ahmed et al., 2009) in individuals who consume HFDs. Leptin is one of the most important adipokines in the control of energy homeostasis. It is secreted into the circulation in a pulsatile manner by subcutaneous differentiated adipocytes. Thus, leptin is considered to be an adiposity signal, as its levels are directly related to the amount of fat stored by the body (Moreno, Chowen, & Argente, 2010).

Yet in recent years, circulating leptin levels have been shown to be directly correlated with the initiation of pubertal development (Moreno et al., 2010). Regarding the role of leptin on the reproductive system, higher leptin levels are associated with higher testosterone levels (Cortes & Olivera, 2010). Thus, these higher leptin levels, which are presumably present in obese versus thin adolescent boys, could be correlated with the precocious puberty. In fact, rare human individuals with leptin deficiency or leptin receptor mutations have hypogonadotrophic hypogonadism (Clement et al., 1998; Strobel, Issad, Camoin, Ozata, & Strosberg, 1998).

However, there are a considerable number of children who consume HFDs without weight gain. How might pubertal development in these individuals be affected? According to the cases reported to the health agencies, HFDs and liver damage are routinely related. In addition, studies have shown that low testosterone levels are common in men with severe liver disease (Grossmann et al., 2012). The pathogenesis of low testosterone levels in men with chronic liver disease involves dysregulation of the hypothalamo-pituitary-gonadal axis at multiple levels (Grossmann et al., 2012). Thereby, the severity of chronic liver disease could become obviously a serious problem for male adolescents at puberty.

Insight into the relationship between HFDs (without weight gain) and lower testosterone levels may be provided by at least three reasons: (a) dysregulation of the hypothalamo-pituitary gonadal axis at multiple levels since liver damage and HFDs are routinely related (Grossmann et al., 2012), (b) liver disease could effectively decrease the insulin-like growth factor 1 (IGF-1) production, (c) the progression of chronic liver disease to cirrhosis results in a state of acquired growth hormone (GH) resistance with high GH and low IGF-1 levels. IGF-1 is a single chain polypeptide consisting of 70 amino acids and regulated by liver. The liver is the central organ of the endocrine growth hormone/insulin-like growth factor (GH/IGF) axis. The IGF system is involved in all aspects of male reproductive physiology and it increases at the onset of puberty (Caregaro et al., 1997; Donaghy, Delhanty, Ho, Williams, & Baxter, 2002; Flores, Sanchez, Gonzalez, & Pizarro, 1998; Lackey, Gray, & Hendricks, 2000; Yoon, Berger, & Roser, 2011). Localization of IGF-1 receptor was observed in Leydig cells suggesting IGF-1 may play a role in steroidogenesis (Yoon & Roser, 2011). Therefore, children who consume HFDs (chronic ingestion) without weight gain could precipitate liver damage, decrease IGF-1 levels and delay pubertal development.

REFERENCES

Ahmed, M. L., Ong, K. K., & Dunger, D. B. (2009). Childhood obesity and the timing of puberty. *Trends in Endocrinology and Metabolism*, *20*(5), 237–242.

Caregaro, L., Alberino, F., Amodio, P., Merkel, C., Angeli, P., Plebani, M., Bolognesi, M., et al. (1997). Nutritional and prognostic significance of insulin-like growth factor 1 in patients with liver cirrhosis. *Nutrition*, *13*(3), 185–190.

Clement, K., Vaisse, C., Lahlou, N., Cabrol, S., Pelloux, V., Cassuto, D., Gourmelen, M., et al. (1998). A mutation in the human leptin receptor gene causes obesity and pituitary dysfunction. *Nature*, *392*(6674), 398–401.

Cortes, Z. T. R., & Olivera, M. (2010). Association between leptin receptors in the testicle, leptin and testosterone levels in puber male calves. *Revista MVZ Cordoba*, *15*(3), 2204–2214.

Donaghy, A. J., Delhanty, P. J. D., Ho, K. K., Williams, R., & Baxter, R. C. (2002). Regulation of the growth hormone receptor/binding protein, insulin-like growth factor ternary complex system in human cirrhosis. *Journal of Hepatology*, *36*(6), 751–758.

Dos Santos, L. F., Rubel, R., Bonatto, S. J. R., Zanatta, A. L., Aikawa, J., Yamaguchi, A. A., Torres, M. F., et al. (2012). Cordyceps sinensis biomass produced by submerged fermentation in high-fat diet feed rats normalizes the blood lipid and the low testosterone induced by diet. *EXCLI Journal*, *11*, 767–775.

Flores, J. M., Sanchez, M. A., Gonzalez, M., & Pizarro, M. (1998). Caprine testicular hypoplasia associated with sexual reversion decreases the expression of insulin-like growth factor II (IGF-II) mRNA in testes. *Animal Reproduction Science*, *52*(4), 279–288.

Grossmann, M., Hoermann, R., Gani, L., Chan, I., Cheung, A., Gow, P. J., Li, A., et al. (2012). Low testosterone levels as an independent predictor of mortality in men with chronic liver disease. *Clinical Endocrinology*, *77*(2), 323–328.

Lackey, B. R., Gray, S. L. L., & Hendricks, D. M. (2000). Physiological basis for use of insulin-like growth factors in reproductive applications: A review. *Theriogenology*, *53*(5), 1147–1156.

Moreno, G. A. M., Chowen, J. A., & Argente, J. (2010). Metabolic signals in human puberty: Effects of over and undernutrition. *Molecular and Cellular Endocrinology*, *324*(1-2, SI), 70–81.

Strobel, A., Issad, T., Camoin, L., Ozata, M., & Strosberg, A. D. (1998). A leptin missense mutation associated with hypogonadism and morbid obesity. *Nature Genetics*, *18*(3), 213–215.

Yoon, M. J., Berger, T., & Roser, J. F. (2011). Localization of insulin-like growth factor-I (IGF-I) and IGF-I Receptor (IGF-IR) in equine testes. *Reproduction in Domestic Animals*, *46*(2), 221–228.

Yoon, M. J., & Roser, J. F. (2011). A synergistic effect of insulin-like growth factor (IGF-I) on equine luteinizing hormone (eLH)-induced testosterone production from cultured Leydig cells of horses. *Animal Reproduction Science*, *126*(3-4), 195–199.

In: Neurobiology of Men's Mental Health
Editors: Leo Sher and Timothy R. Rice
ISBN: 978-1-63483-191-8
© 2015 Nova Science Publishers, Inc.

Chapter 9

INTEGRATING CHILD AND ADOLESCENT PSYCHOTHERAPY WITH MALE NEUROBIOLOGY: REGULATION-FOCUSED PSYCHOTHERAPY FOR CHILDREN

Timothy R. Rice*, MD, and Leon Hoffman, MD
Department of Psychiatry, Icahn School of Medicine at Mount Sinai,
New York, NY, US

ABSTRACT

Regulation Focused Psychotherapy for Children (RFP-C) is presented as an intervention tailored to the underlying neurobiology of men's common mental health problems. To explore this statement, the concepts of implicit emotion regulation and its parallels with defense mechanisms are developed.

The procedure of the approach is sketched and a clinical vignette demonstrates its applicability to the common presentation of externalizing behaviors and reactive aggression in boys. The externalizing behaviors and aggressions of male children are understood as maladaptive products of an underdeveloped implicit emotion regulation system. Implications for the creation of a therapeutic approach developed in relation to the neurobiology of men is presented as a transition into the following two chapters.

Keywords: men's mental health, psychotherapy, implicit emotion regulation

INTRODUCTION

In chapters 1 and 2 the importance of sex hormones in male development and the impact of these hormones upon the emotion regulation system were reviewed. These data were used

* Corresponding author: Timothy R. Rice, M.D. Icahn School of Medicine at Mount Sinai, Box 1230, One Gustave L. Levy Place, New York, NY 10029, USA, Email: timothy.rice@mssm.edu

to support a perspective as expanded in chapter 2 that the emotion regulation (ER) system construct is of primary importance to men's mental health. Data concerning male development suggests that this system may be predisposed to deficits in inhibitory processes in men relative to women. The previous discussion of the male disorders of reactive aggression and poor emotion regulation, which included the externalizing behaviors of childhood and a large portion of the suicides and homicides in adolescents and young men, may be the product of these deficits. Thus, the ER system is a key brain-based model through which a developed understanding of the neurobiology of men's mental health may be achieved.

It will be the focus of this chapter to focus on the therapeutic implication of this finding. In chapter 2 it was briefly referenced that the ER system model overlaps with some traditional models of intrapsychic processes which are utilized in traditional psychotherapeutic approaches [1], enabling the reconstruction of traditional psychotherapy within a contemporary brain-based model. Following a review of the neurobiology of the implicit ER system, this statement will be expanded within this chapter in an exploration of current therapeutics targeted to the ER system. The emerging evidence that implicit ER processes may be even more important to health mental functioning than explicit ER will be used to develop the distinction between explicit and implicit ER and demonstrate that a psychotherapeutic approach which targets implicit ER may be of high value to the provision of care to men.

This precision targeting offers an advantage over alternative psychotherapeutic programs for male psychopathology where therapeutic interventions are dispersed to address multiple domains by virtue of heterogeneous Diagnostic and Statistical Manual of Mental Disorders (DSM) clinical entities. The organizational simplicity of a relevant brain-based approach via the ER system construct provides an unparalleled opportunity for the development of a Research Domain Criteria (RDoC) compatible approach.

We will thus present Regulation Focused Psychotherapy for Children (RFP-C), a novel short-term psychotherapeutic approach that has been constructed based on the implicit ER system construct. The application of this approach will be demonstrated in a brief vignette. The potential of this approach to address deficits in both medical student and adult psychiatry resident education which have traditionally diminished recruitment into the field of child and adolescent psychiatry will then be developed in the following two chapters, chapters 10 and 11. The conclusion of this chapter will identify future opportunities for hypothesis testing concerning the ability of this technique to both reduce male children's overt behavioral and affective symptomology as well as improve biomarkers of ER processes.

Implicit and Explicit ER

In chapter 2, we discussed that the self-regulative processes of the ER system may be divided into implicit and explicit processes [2]. Implicit emotion regulation has been defined as "any process that operates without the need for conscious supervision or explicit intentions, and which is aimed at modifying the quality, intensity, or duration of an emotional response. Implicit ER regulation can thus be instigated even when people do not realize that they are engaging in any form of emotion regulation and when people have no conscious intention of regulating their emotions" [3].

Implicit regulation, or implicit self and environmental evaluations with immediate response tendencies, is dependent upon the ventral prefrontal cortex (vPFC), which includes the orbitofrontal cortex (OFC), ventromedial PFC (vmPFC), and ventral anterior cingulate cortex (vACC). Explicit regulation, or deliberate or effortful conscious cognitive manipulations that monitor, adjust, and select responses from a range of options, is dependent upon the dorsal anterior cingulate cortex (dACC) and the dorsolateral PFC (dPFC). There is likely great overlap within the neurophysiology of both symptoms. These two divisions of the cerebral cortex both modulate lower brain structures, including the amygdala, hypothalamus and brainstem nuclei, through their respective regulative pathways.

The successful modulation of these visceromotor centers through the PFC regulators leads to increased vagal tone and decreased sympathetic arousal. This affects a measurable neurochemical state that promotes calmness and inhibits behavioral dysregulation. Indeed, the success of this process leads to decreased sympathetic arousal and increased parasympathetic/vagal tone [4]. Unsuccessful modulation as may occur through an underdeveloped implicit ER system affects the autonomic nervous system and predisposes children to disruptive behaviors; it may also influence immunological and hormonal systems implicated in psychosomatic pathology. Internal physiological effects include diminished metabolic and immunologic control as evidenced by increased fasting glucose and hemoglobin A1c levels, overnight urinary cortisol, and proinflammatory cytokines and acute-phase proteins [5]. Externally, we see the agitated, disruptive behaviors and reactive aggressions of male children with externalizing disorders as well as young adolescent males and young men who "in the heat of the moment," often when under significant psychosocial stress, go on to commit homicide and/or suicide.

Therapeutic Relevance for Men's Mental Health

One promising means of reaching these males, then, may be therapies, which directly target the ER system. Such an approach transitions from behavioral approaches to categorical diagnoses and directly addresses the neural structures, which testosterone may effect. For example, training in explicit mechanisms of emotional control within preadolescents with externalizing problems, has been found to directly increase electroencephalographic correlates (increased prefrontal neural activation signature) of explicit emotion regulation processes [6]. Anger management training and other therapies which include elements that are focused on explicit process of emotion regulation, including Emotion Regulation Therapy (ERT), Affect Regulation training (ART), and elements of Dialectical Behavioral Therapy (DBT) may be of help [7].

As we reviewed in chapter 2, two processes, effortful suppression [8, 9] and cognitive reappraisal [8, 10, 11], have been extensively studied through functional magnetic resonance imaging (fMRI) and other mediums of investigation. The fMRI work demonstrates quite clearly that while both supression and cognitive reappraisal result in decreased sympathetic arousal and increased vagal tone promoting behavioral regulation via prefrontal-mediated attenuation of the amygdala, hypothalamus, and brainstem nuclei, effortful suppression is less effective than reappraisal in effecting these neurophysiologic changes [8]. These brain-based findings then may be used to inform psychotherapeutic approaches which train the at-risk male to rely to a greater extent on reappraisal than on suppression. This concept is not a novel revelation; however, its reorientation within a brain-based model offers new elements and opportunities for development. Suicide safety plans and other elements to reduce risk, including the distress tolerance skills of DBT, already include elements of promoting this conscious shift when in crisis mode; increasingly, researchers will need to consider how to further improve psychotherapeutic approaches within this brain-based context.

One important means may be through further investigation of the role of psychotherapy in implicit ER processes. It is now established that implicit ER may be even more important to healthy mental functioning than explicit ER [2]. There is face validity to this claim as implicit ER is employed automatically and do not require conscious triggering, effort, or monitoring. Males under duress are less likely to have the resources to convert from unconscious reactive process towards the conscious and effortful application of self-regulation strategies. Models of care based on explicit processes male fall short in addressing psychopathology and its untoward outcomes. Indeed, empirical evidence now suggests that implicit ER deficits may be more tightly associated with psychopathology, including anxiety disorders and mood disorders, than explicit ER [12].

Regulation Focused Psychotherapy for Children

In light of this, we have proposed [13] a Manual for Regulation-Focused Psychotherapy for Children with Externalizing Behaviors (RFP-C). This approach has harnessed the realization of an overlap with implicit ER and the traditional concept of defense mechanisms to advocate for an approach to mental health that in theory promotes development in the implicit ER system. The approach is structured upon a persistent focus upon the child's defenses with the goal of replacing maladaptive defenses with those that are more functionally adaptive and theoretically more productive in reducing limbic hyperarousal. This may be particularly appropriate in the mental health care of men, who as we have reviewed may have biologically-determined deficits in their implicit ER systems and who indeed present with salient disorders of ineffective reactive aggression as products of ineffective ER (see Chapter 2).

This then is very much similar to the suppression-reappraisal difference in ER efficacy which connotes that there is a hierarchy of effectiveness within explicit ER process. In theory, defense mechanisms also have been historically understood as existing within an empirically-confirmed hierarchy of “maturity” [14], or, as we now propose within the language of the affective neurosciences, as effective in promoting effective implicit ER.

With Regulation Focused Psychotherapy for Externalizing Behaviors (RFP-C), the focus of the treatment is not to promote cognitive (intellectual) change in the child (classical insight) ("Oh, now I understand") but to promote the development of the implicit ER system by advancing an implicit awareness that:

- The emotional state that is feared or is seen as overwhelming by the child is not as overwhelming as he/she thinks it is and that it will not destroy him/her
- There are better ways to manage those emotions than to fight
- The clinician will not be hurt by those feelings and then the child will feel that those feelings will not destroy him or her
- The clinician, by understanding that the behavior has meaning, is always implicitly communicating that to the child and communicating that the child is not bad.

We have commented in our manual [13] that, in contrast to behavioral approaches the clinician tries to understand the meaning of the child's maladaptive disruptive behavior. The clinician does not try to teach the child how to behave properly nor tries to teach parents management techniques. Instead the clinician communicates to the child that there is meaning to the behavior and that it is safer to act up than feel painful emotions, such as sadness, shame, or other negative affect. Child treatments have to include an active parent component. In the manual we describe how to help parents understand the meaning of their children's behavior and how to approach their children in a more effective way, understanding the working of their child, his/her own individual needs, worries, and responses to stress.

With many dysregulated children, participating in a course of RFP-C will result in symptomatic improvement as a result of the development of an increased tolerance of painful emotions. The therapeutic process facilitates development of implicit ER skills. The male child reaches this implicit awareness within the relationship with the clinician that can then be expanded to his life situations in home and at school. In other words, painful emotions can be mastered more effectively, and the boy's use of aggression as his main coping device is diminished. As a result of Regulation Focused Psychotherapy the child can master painful emotions more effectively and has less need for the use of maladaptive protective devices such as fighting. The child reaches this implicit awareness within the relationship with the clinician that can then be expanded to his/her life situations in home and at school. In other words, painful emotions can be mastered more effectively and the child's use of aggression as his/her main coping device is diminished.

VIGNETTE

Pete, an 8-year-old boy came for treatment because of tantrums and nightmares. Within a short period of time he became aggressive in sessions, trying to cut the analyst's beard off, going into the closet, and making noises to scare him. Pete alluded to his parents' fierce arguments in their bedroom and said he protected himself by building a fort around his bed. In sessions he built a model fort with blocks and other toys, saying he wanted to play "hit the donkey on the butt." Pete presented his clothed butt to the clinician, asking him to throw a soft ball at it. Before the clinician could respond, Pete said he was tired, "Never mind, I don't want to play." The clinician said, "*Maybe you feel tired because you feel uncomfortable.*" Pete ignored this comment, but said his older half-brother could beat anyone because *he* was not afraid, and came close to the clinician's face with a menacing expression. The clinician suggested *"Pete, you're acting tough so you don't have to worry about getting hurt,"* and thought to himself that Pete seemed to be provoking him to hurt Pete. Pete's aggression subsided.

Later Pete missed a session because of illness. On returning, he threw lit matches toward the clinician and, in an off-hand barely audible whisper said, "I'm going to the orthodontist right after the session." Pete calmed down when the clinician said, *"Oh, I understand, it's easier to attack me with matches than to worry about getting hurt by the orthodontist."* Pete replied that he worried that *he* (Pete) would get hurt if he did not attack, and jokingly asked if the clinician thought words could hurt him. Pete took a dictionary and playfully threw it at the clinician, who said, *"You get angry and attack when you worry about being hurt."* Pete became friendlier and demonstrated karate moves saying that the clinician should learn how to defend himself if he were mugged. The clinician said *"You're teaching me how to protect myself when you attack me!"* Pete laughed: "I get you angry, don't I?" The clinician replied, *"You know how to do that well."*

The clinician pointed out to Pete that he (1) warded off uncomfortable feelings by becoming tired and (2) acted tough to avoid scary feelings. In the later session, the analyst interpreted Pete's utilization of the mechanism of turning passive into active in order to avoid painful feelings about the dentist, after which Pete behaved more adaptively.

Conclusion

Our current understanding of the neurobiology of male development suggests that the ER system may be predisposed to deficits in inhibitory processes in men relative to women. The consequences include the most relevant clinical presentations of men which occur to a far greater degree in men than in women, including the aggression of male children in childhood and the homicides and suicides of adults. Thus the ER system serves as an optimal endophenotype in the development of men's mental health in that it bridges clinically relevant presentations in men with the underlying specific male neurobiology.

The capability to ground men's mental health in a neurobiological framework enables the application of psychothereapeutic approaches that are targeted to the distinctively male problems of living. Already there is a wide literature in anger management, DBT, and other cognitive behavioral approaches which work to reduce reactive aggression and improve explicit emotion regulation processes through the application of coping strategies and other conscious mechanisms.

The data which suggests that implicit processes of ER may be even more related to psychopathology suggests a role for a therapeutic approach which addresses the implicit ER system. This system has a defined neural signature that is dissectible from the explicit ER system and may serve later as a biomarker indiciating the need for treatment in men or, alternatively, as a measure of therapeutic response to a targeted intervention.

RFP-C has been presented as one such targeted intervention to address deficits in the implicit ER system. There will be future opportunities for hypothesis testing concerning the ability of this technique to both reduce male children's overt behavioral and affective symptomology as well as improve biomarkers of ER processes. Planning for clinical trials is underway.

The potential of this approach to address deficits in both medical student and adult psychiatry resident education which have traditionally diminished recruitment into the field of child and adolescent psychiatry will be developed in the following two chapters, chapters 10 and 11.

REFERENCES

[1] Rice TR, Hoffman L. Defense mechanisms and implicit emotion regulation: a comparison of a psychodynamic construct with one from contemporary neuroscience. *J. Am. Psychoanal .Assoc.* (2014) 62:693–708. doi:10.1177/0003065114546746.

[2] Gyurak A, Gross JJ, Etkin A. Explicit and implicit emotion regulation: a dual-process framework. *Cogn. Emot.* (2011) 25:400–412. doi:10.1080/02699931.2010.544160.

[3] Koole SL, Rothermund K. "I feel better but I don't know why": the psychology of implicit emotion regulation. *Cogn. Emot.* (2011) 25:389–399. doi:10.1080/02699931.2010.550505.

[4] Lane RD, McRae K, Reiman EM, Chen K, Ahern GL, Thayer JF. Neural correlates of heart rate variability during emotion. *Neuroimage* (2009) 44:213–22. doi:10.1016/j.neuroimage.2008.07.056.

[5] Thayer JF, Sternberg E. Beyond heart rate variability: vagal regulation of allostatic systems. *Ann. N Y Acad. Sci.* (2006) 1088:361–72. doi:10.1196/annals.1366.014.

[6] Woltering S, Granic I, Lamm C, Lewis MD. Neural changes associated with treatment outcome in children with externalizing problems. *Biol. Psychiatry (*2011) 70:873–9. doi:10.1016/j.biopsych.2011.05.029.

[7] Gross JJ. Handbook of Emotion Reguation, Second Edition. , ed. J. Gross New York, NY: Guilford Press (2014).

[8] Goldin PR, McRae K, Ramel W, Gross JJ. The neural bases of emotion regulation: reappraisal and suppression of negative emotion. Biol. *Psychiatry* (2008) 63:577–86. doi:10.1016/j.biopsych.2007.05.031.

[9] Dunn BD, Billotti D, Murphy V, Dalgleish T. The consequences of effortful emotion regulation when processing distressing material: a comparison of suppression and acceptance. *Behav. Res. Ther.* (2009) 47:761–63. doi:10.1016/j.brat.2009.05.007.

[10] McRae K, Hughes B, Chopra S, Gabrieli JDE, Gross JJ, Ochsner KN. The neural bases of distraction and reappraisal. *J. Cogn. Neurosci.* (2010) 22:248–262. doi:10.1162/jocn.2009.21243.

[11] Ochsner KN, Bunge SA, Gross JJ, Gabrieli JDE. Rethinking feelings: an FMRI study of the cognitive regulation of emotion. *J. Cogn. Neurosci.* (2002) 14:1215–1229. doi:10.1162/089892902760807212.

[12] Ehring T, Tuschen-Caffier B, Schnülle J, Fischer S, Gross JJ. Emotion regulation and vulnerability to depression: spontaneous versus instructed use of emotion suppression and reappraisal. *Emotion* (2010) 10:563–72. doi:10.1037/a0019010.

[13] Hoffman L, Rice T. Manual for Regulation-Focused Psychotherapy for Children with Externalizing Behaviors (RFP-C): *A Psychodynamic Approach.* New York, NY: Routledge (2015).

[14] Vaillant GE, Bond M, Vaillant CO. An empirically validated hierarchy of defense mechanisms. *Arch. Gen. Psychiatry* (1986) 43:786–94.

In: Neurobiology of Men's Mental Health
Editors: Leo Sher and Timothy R. Rice
ISBN: 978-1-63483-191-8
© 2015 Nova Science Publishers, Inc.

Chapter 10

REGULATION-FOCUSED PSYCHOTHERAPY FOR CHILDREN IN MEDICAL EDUCATION: AN ACCOUNT FROM THE PERSPECTIVE OF MEN'S MENTAL HEALTH

Timothy R. Rice*[*,1], MD, BS, Fareedat Oluyadi[1], BA, Emma Gaines[2], MA Tracy Prout[2], PhD. and Leon Hoffman[1], MD

[1]Department of Psychiatry, Icahn School of Medicine at Mount Sinai, New York, NY, US

[2]Ferkauf Graduate School of Psychology, Yeshiva University, New York, NY, US

ABSTRACT

Male toddlers, preschool- and school-aged boys, male adolescents, and young male adults often react to trauma and other adverse life experiences with externalizing behaviors. This constellation of presentation and means of interpersonal relating are highly specific to Men's Mental Health. Often, the raw aggression in these presentations can be off-putting or even disturbing to mental health professionals, especially those who have yet to build confidence through gained experience in working with male youth. This may be particularly true with prospective mental health professionals, including medical students. There is a large literature documenting the difficulties of recruitment of medical students into child and adolescent psychiatry. These Men's Mental Health issues may be a contributing factor to this long-standing problem. This paper describes the value to the medical student experience of the perspective of Regulation Focused Psychotherapy for Children, a short-term psychotherapeutic approach targeting externalizing behaviors through a means of attempting to understand the meaning of the behaviors in lieu of coercive control or management. A vignette involving one medical student's experience

[*] Corresponding author: Timothy R. Rice, M.D. Icahn School of Medicine at Mount Sinai, Box 1230, One Gustave L. Levy Place, New York, NY 10029, USA, Email: timothy.rice@mssm.edu

in working with a male child through this framework is provided and areas for further study outlined.

Keywords: men's mental health, male children, externalizing behaviors, medical student education, child and adolescent psychiatry as a career choice

INTRODUCTION

In this section of this book, we have assembled a collection of chapters with the intent to demonstrate the unique issues to men's mental health in child and adolescent populations. We have aimed to demonstrate that male children have a unique set of behaviors and internal experiences that distinguish them from their female peers. This is a product of both biological and psychosocial determinants. Nowhere is it more clear perhaps than in the expression of disruptive behavior disorders as a consequence of exposure to childhood trauma and other adverse life experiences (Chapter 3). In "The impact of adversities during childhood and/or adolescence on men's mental health: The New York declaration," we have attempted to explore the unique set of factors that these male children experience and sketch the recommendations for a mental health plan adapted to the specific of men's mental health entail.

In this and the following chapters, we broaden our perspective to consider in extended detail the consequences of these factors among the mental health professionals interacting with these male children. Specifically, we turn to the well-established problem of our educational system's difficulty in recruiting physicians in training and general psychiatry residents into the field of child and adolescent psychiatry. This difficulty has been extensively and repeatedly stated in the literature (in fact an entire issue of *Academic Psychiatry* was dedicated to the problem in 2008 [1]), yet no clear consensus has seemed to emerge as to why recruitment into the field remains so low when demand is so high.

Whereas in the following two chapters we will explore the offered reasons and preliminary survey findings among medical students and general psychiatric residents as to why recruitment remains so problematic, we will offer an original hypothesis that it is the current underdevelopment and poor integration of the men's mental health perspective into the psychiatric educational system that may be one of the primary determinants of this poor recruitment. We have already described how the aggression and tendency to go on the offensive in children with disruptive behavior disorders can be off-putting and disturbing to mental health professionals, and that this reaction may play a major role among even experienced clinicians to avoid working with children in an individual setting. It is by no means a great leap to consider that this reaction may be especially pronounced among those who have yet to build confidence in working with such children through gained experience in working with these youth. Both medical students and trainees may have difficulty verbalizing this reaction owing to the multiple unconscious mechanisms at play in counter-transferential reactions that is complicated by a comparatively much less engagement in and exposure to dynamic principles. In fact, the emphasis in contemporary education away from psychodynamic principles makes it such that many medical students in particular may not have any exposure to the concept at all. Given that disruptive behavior disorders are so prevalent amongst boys and such a common response to the very common occurrence of childhood adversity (Chapter 3), educators must consider the value of developing a more nuanced perspective in men's mental health into educational models.

In this chapter, we will focus on the literature unique to medical student education. We will review a sub-section of the prior literature which demonstrates the difficulty recruiting medical students into child and adolescent psychiatry. Many of these papers have very preliminary survey data which explores medical student attitudes about child psychiatry. Often attempts to connect attitudes with the low recruitment rates are made. We will review some of these observations and demonstrate how attention to the above hypothesis may offer an improved explanatory model.

Next, we will review the smaller literature which explores the effect of targeted interventions in medical student education to improve medical student interest in and recruitment into child and adolescent psychiatry. We will review the indicators of effective components of these interventions and again consider how the perspective offered by a unique focus on men's mental health may be of further value in addressing the concerns set out in the literature.

We will utilize this reviewed work to describe the hypothesized value to the medical student experience of the perspective of Regulation Focused Psychotherapy for Children. We have previously described (Chapter 9) this short-term psychotherapeutic approach to externalizing behaviors that is targeted to the unique biological and psychosocial vulnerabilities of male children as well as its organizing principle of attempting to understand the meaning of the behaviors in lieu of coercive control or management. We will demonstrate how an integration of this approach into medical student education may be an effective, organized means of addressing the gap left by an underdeveloped men's mental health perspective in medical student education.

To demonstrate some preliminary observations on this hypothesis, a vignette involving one medical student's experience in working with a male child through this framework is provided. The value of hands-on experience in this perspective and actual clinical work with a very disruptive male child is explored. The student's reactions are detailed and the comfort and assistance that the RFP-C approach's organizing principles provided are discussed.

This chapter will be followed by a discussion of the similar but distinct utility of this approach in general psychiatric resident education from an even more detailed exploration of a psychiatric resident's exposure in the program (Chapter 11). The objective of the current chapter is to retain a focus on medical student education through the path that we have outlined in this introduction and onto which we will now proceed.

The Problem of Medical Student Recruitment into Child and Adolescent Psychiatry

Attempting to teach child and adolescent psychiatry in the medical student curriculum has been previously described as a frustrating experience for educators [2]. The authors of this perspective, having reviewed many articles on the topic, feel that it is safe to generalize this frustration among many of the educators that teach medical students. The crux of the frustration is described as the sense of the need for additional child psychiatrists coupled with the fact that many if not most of these trainees will experience a decrease in interest in child and adolescent psychiatry through the course of the medical education and, ultimately, many who expressed initial interest will fail to pursue the field.

The dearth of child and adolescent psychiatrists relative to the need and the specifics of the real need for recruitment is worth summarizing. The authors cite the United States Bureau of Health Profession's 2000 projections that by 2020 there would be a need for about 12,000 child psychiatrists in the United States that would not be met by the supply of just over 8,000 [3]. They additionally cite the United States Surgeon General's 2000 report [4] which advocated for improve medical education infrastructure to address this crisis in need yet seemed to go largely unrealized in the ensuing decade and a half. The shortage of child and adolescent psychiatrists in rural communities is additionally directly cited [5] and the severe negative consequences of this shortage referenced.

Yet, medical students do seem to become less interested in child and adolescent psychiatry training over there educational years despite educators attempts to reach them. Attempts do seem to include addressing their possible concerns about the field as well as their potential anxieties, but data suggests that the current strategies are insufficient. Indeed, despite the great need, the number of residents in child and adolescent training as of a 2006 *Journal of the American Academy of Child and Adolescent Psychiatry* article had not changed significantly since 1995 [5].

The authors of the article which details what a frustrating experience this may be for educators go into detail hypothesizing the many potential reasons. They write,

> "What influences may mitigate migration into the field? Is it the perceived excessive workload, fear of deal- ing with uncertainty, concerns about the limitations of the evidence base of the treatments, or worries about endless calls, e-mails, pages, and reports? Or is it simply lack of experience working with children, adolescents, and families; lack of exposure to sufficiently talented and passionate role models; and difficulties in carrying cases over time to witness the joys of change and deep gratitude of parents, teachers, and the children themselves? Could it be that sheer lack of exposure to sound teaching in medical schools and residencies focusing specifically on child and adolescent psychiatry is responsible for lack of recruitment into the field?" (p.173)

Among these many plausible determinants, the medical student belief that child psychiatry is emotionally stressful and that there is a lack of family and social supports for children is especially important. These beliefs were the most common negative perceptions about child psychiatry in a survey of 223 medical students who were administered self-report questionnaires pre- and post- general psychiatry clerkship. The perception of lack of family and social support actually doubled among the 223 students during medical students rotation through their psychiatric clerkship [7]. Understood from a slightly different perspective, this view seems to convey that medical students feel ineffective or unable to individually meet the needs of many children: work is both difficult, and the social supports and community network upon which recovery may depend does not seem to be available. This uncomfortable feeling or painful affect experienced by the medical students may lead them to avoid or disengage in an accurate appraisal of the field.

Medical Student Perspectives and Attempted Interventions to Improve Recruitment

Interventions at the level of residency training to try to bolster such recruitment, such as the creation of integrated adult and child psychiatry programs [6], have met partial success. There have also been designed interventions at the level of medical student education. Medical students often have very divergent positively- and negatively-valenced views of child psychiatry [7]. Active interventions employed during general adult clerkship experiences can directly change their perspective of child and adolescent psychiatry [8]. However, recruitment still remains low.

As an example of one such intervention at the level of medical student education [8], about one half of all medical students rotating through their general psychiatry clerkship were assigned to recieve a lecture on the topic of child and adolescent psychiatry. The intervention was found to produce positive changes in factual perceptions about training in child and adolescent psychiatry but was also found to be ineffective in enhancing medical students' motivation or interest in pursuing training. This is in contrast to the finding in this same study that there is a trend towards statistical significance ($p<0.10$) regarding a negative association between medical student interest in child psychiatry and not rotating through a child psychiatry clinical experience as part of their general adult psychiatry clerkship.

These findings support the hypothesis interventions involving child and adolescent psychiatry patient care may be an optimal means of improving recruitment in child and adolescent psychiatry over didactics or factual, intellectual conversations. Educators should consider which type of intervention may be optimal in promoting recruitment. Though any clinical exposure would likely be helpful to medical students, the active directive behavioral interventions utilized by child psychiatric training programs including in parent management training may not be optimal in motivating its use by child psychiatry practitioners. These options address aggression and tendency to go on the offensive in children with disruptive behavior disorders in a way in which these behaviors are meant to be "controlled:" the interaction becomes confrontation and a battle of wills. This can be off-putting and disturbing to mental health professionals, especially to medical student trainees who may be interested in child psychiatry, as it requires a mobilization and understanding of aggression that may be disturbing to acknowledge and thus promote avoidance of the field.

Another option in providing clinical experiences to medical students may be the RFP-C approach. RFP-C addresses the survey data that identifies emotional difficulty and inadequate child family and social support as the two main negative perspectives among medical students. The frame of RFP-C reinforces that even though working with some children who have been exposed to adversity owing to suboptimal psychosocial supports may induce very difficult feelings in the provider, these children are capable of being helped through working in a one on one individual approach.

It is important to note that the study which found this finding may have unique relevance to men's mental health. Given that there are generally negative attitudes about working with male children with externalizing disorders among even trained child and adolescent psychiatrists, it may be particularly useful for the field of Men's Mental Health for medical students to gain exposure to working with these boys.

Engaging in a psychiatry clinical year clerkship intervention that emphasizes the ability of mental health professionals to help these children in direct individual psychotherapy may mitigate the negative impact upon medical students of their common perspectives. It is possible that current exposures to child and adolescent psychiatry are not offering alternative, empowering perspectives, in addition to the usual agent of trying to form a relationship, and that this failure leads medical students to feel pessimistic about further exploration of training in the field. Direct involvement with male children and youths may help to demonstrate that though supports are certainly in need for this population, providers and practitioners can effect change in this population through a clerkship experience that emphasizes the power of individual psychotherapy in Men's Mental Health.

These issues of recruitment may be particularly important in ensuring that boys, male adolescents, and young men have access to a full set of care options, in a field which is become more and more dominated by women. It appears that affect-oriented approaches to oppositional defiant disorder may be more targeted to the male-specific neural underpinnings of the disorder rather than management-oriented approaches, because management approaches can be perceived by the patient as a wish by the mental health provider to develop dominance and promote the patient's submission.

EXAMPLE OF RFP-C APPLICATION IN STUDENT TEACHING

The second author (F.O.) was selected to participate in a longitudinal ambulatory care program throughout her third year clerkship. She would join the first author (T.R) Friday mornings at an inner-city community clinic where she would have two hours of direct patient contact under supervision. Because of the first author's interest in working with externalizing behaviors in a young male age group, usually these children fit these characteristics; they would often be diagnosed with ODD. On many occasions, they were recently discharged from an inpatient unit where they had been hospitalized for violence.

As an example of the quality of a common patient interaction in this setting, one seven year old boy used computer paper to fold origami of Japanese throwing stars (Shuriken) and “Freddy Krueger nails,” with which he generously adorned his fingers and held in group ready to throw and scratch. Notably, this creative artistry served the role of enabling him to avoid structured and unstructured drawing, which had been the request made of him. Though initially quite calm and methodological with his origami, with each subsequent question asked of him the child would launch a throwing star a the examiner's face. He later made a Chinese fortune teller as a playful response to avoid answering questions substantially; when the examiner would ask his question, the boy would move his Chinese fortune teller, flip a flap, and show the examiner that the flap simply read “die” or “you suck.”

The RFP-C approached helped the clinician to not feel aggravated or attacked by these behaviors but rather to appreciate the child's creativity in creating these defenses against the painful affects which may emerge by further discussion with the evaluating physician.

In a following week, the second author had the opportunity to attempt to engage this child in a second of two intake interviews. As the material above suggests, this was a difficult child to interview. The second author expressed some concerns over how to play with such a child. She felt she “wouldn't do it right” or would “do something wrong.” These negative feelings, which she was very brave to express, were absolutely in line with the commonly-reported medical student perspectives. As we have attempted to demonstrate in this paper, these feelings may be enlarged when working with a male child with disruptive behavior disorders where the child works to attack rather than experience his own anxiety or negative affect. In this case, the aggression portrayed by the child helped the medical student to feel she would be inept to work with the child, or even somehow might "do something wrong" and possibly hurt the child. The RFP-C framework helped the student in processing the idea of projected aggression and processing her own feelings as meaningful and understandable reactions to working with such a disruptive male child.

The medical student engaged in this discussion with the first author and was assigned the relevant text from the psychotherapy manual for RFP-C concerning the second evaluation visit with the child. The structure of the manual was reported by the student to be of help in preparing for the visit. When the second visit came, the medical student engaged in appropriate play therapy with the child which was nondirective and expressive and allowed the same conflicts observed in the child in the first session to unroll without distortion. Additionally, the medical student was able to refrain from making any experience-distant interpretations. She could and did identify moments where the play would be interrupted, could internally recognize the meaning for the interruptions, and could simply point out to the child the interruption of the play and ask, "Oh, what happened?"

To derive more systematic and quantitative measures of the education experience, the student was asked to complete ratings of self-efficacy via The Counselor Self-Efficacy Scale (CSES), The Valuable Elements in Psychotherapy Questionnaire (VEP-Q), and a scale designed to quantify perspectives of prognosis coupled with a vignette of a male child with a disruptive behavior disorder. The second author was then asked to author a brief blurb regarding the experience of filling out these questionnaires in light of what she had learned through the two-week interaction with this child and her self-directed reading usual the manual for RFP-C. Her response was to write,

> "My sessions with Dr. Rice coupled with the reading materials moderately influenced my ratings. Particularly in completing the VEP-Q and Prognosis. The manual helped me conceptualize externalizing behaviors which oppositional defiant disorder falls under (which is the working differential diagnosis of the same case at hand)."

Psychoanalysis and Its Relevance in Medical Student Education

The psychoanalytic background of RFP-C should be addressed. Asch and Marcus, in a chronologically dated but nonetheless highly contemporarily relevant article [9] concerning the difficulties from the psychoanalytic community's perspective on medical student education, writes,

> "The failure of psychoanalysts to play a continuing role in the departments of general psychiatry is pertinent. Psychoanalysts do not always value the general psychiatrist's tasks. Psychoanalysis does not always value cases of psychosis, near psychosis, medically ill patients with psychological problems, short-term psychotherapy, reparative

> work of any nature—in other words, the tasks of general psychiatrists in hospital settings. Analysts may feel that interest in applied analysis serves only to dilute and threaten the main body of analytic work. But as psychoanalysts in the United States concentrate more on the better-functioning neurotic patient and move into increasingly private outpatient office practice, they are also isolating themselves from the hospital setting in which academic departments find their home." (p.1049)

It is notable that, in the creation of RFP-C, the emphasis upon children with a greater disturbed (externalizing) presentation rather than a neurotic (internalizing) presentation is very consistent with avoiding this bias that has been the traditional psychoanalytic resistance to involvement in medical student education. The emphasis on evidence based medicine through pre- and post-measure evaluations, a targeted behavior approach, its foundation in neurobiological correlates through hypothesized synchrony with the implicit emotion regulation framework [10], and its short-term makes clear that RFP-C overcomes many of the failures of traditional psychoanalysis in its integration in medical student teaching. There is great value to this. In fact, the two major issues which initially interest medical students in training in child and adolescent psychiatry are actually psychoanalysis and pediatrics [2]. This may be surprising as, over the course of engagement in contemporary academic psychiatry, many forget or fail to see altogether that psychoanalysis is valued by medical students, potentially until their interests are not developed or nourished. RFP-C, in addition to its role in supporting interest in child and adolescent psychiatry as well as in men's mental health, may be a valuable tool for retaining interest among trainees in psychoanalytic perspectives. Ultimately this may be one avenue towards a goal of reopening a dialogue between psychoanalysis and contemporary academic psychiatry.

Regarding the interest in psychoanalysis and pediatrics, Schlozman and Beresin write,

> "The relationship is obvious: both psychoanalytic studies and pediatrics are steeped in developmental inquiry. Is it possible that students and residents do not experience a strong focus in human development throughout their formative years? Have we neglected a real emphasis on the human condition that places growth and change in children, families, and communities on the back burner? One would expect that the basic and clinical research on brain plasticity, attachment theory, developmental psychopathology, family systems, evidence-based psychotherapies, psychopharmacology, and the environmental, community, and cultural influences on health and illness would provide a newfound excitement in child and adolescent psychiatry" (p. 173).

This perspective is valuable, as it emphasizes the need for psychoanalytic thinking to be flexible and integrated into the rest of the body of psychiatry and contemporary neuroscience. RFP-C achieves this. There is precedent that such interventions are effective for education, as this approach is very similar to a group [12] which aimed to integrate neuroscience with humanistic psychiatry.

Conclusion

Male toddlers, preschool- and school-aged boys, male adolescents, and young male adults with externalizing behaviors may be a particularly difficult population for medical students to reach. If exposed to these children in a safe environment where an emphasis on understanding the meanings of their behaviors is consistently and persistently pursued, medical students my learn to better tolerate the uncomfortable feelings which they cite as one of the two most significant negative perceptions of child psychiatry. Emphasizing an individual psychotherapeutic approach may also help medical students to not identify the lack of family and social support for children as severe as a significant shortcoming of the field. Reducing or even eliminating these negative believes may have major consequences for recruitment into child psychiatry.

There are certainly implications of this hypothesis for the field of Men's Mental Health. If this field is to grow in the area of child and adolescent psychiatry, it will have to work towards reducing recruitment problems as well. Simultaneously, these issues are highly specific to Men's Mental Health. Often, the raw aggression in these presentations can be off-putting or even disturbing to mental health professionals, especially those who have yet to build confidence through gained experience in working with male youth. This may be particularly true with prospective mental health professionals, including medical students. RFP-C may offer a means to help medical students to overcome their negative beliefs and enter the field at higher rates. Our vignette aims to demonstrate the value of this perspective, and further testing are underway.

References

[1] Beresin E. Innovation and inspiration in child and adolescent psychiatric education. *Acad. Psychiatry* (2008) 32: 429-432. doi:10.1176/appi.ap.32.5.346.

[2] Schlozman SC, Beresin E V. Frustration and opportunity: Teaching child and adolescent psychiatry throughout medical education. *Acad. Psychiatry* (2010) 34:172–174. doi:10.1176/appi.ap.34.3.172.

[3] United States Department of Health and Homeland Security Resources and Services Administration. Physician supply and demand: projections to 2020. Available at http://bhpr.hrsa.gov/healthworkforce/reports/physiciansupplydemand/physiciansupplyprojections.htm.

[4] United States Public Health Servicie. Report of the Surgeon General's Conference on Children's Mental Health: A National Agenda Item. Available at www.surgeongeneral.gov/topics/cmh/cmhreport.pdf. Washington, DC (2000).

[5] Thomas CR, Holzer CE. The continuing shortage of child and adolescent psychiatrists. *J. Am. Acad. Child. Adolesc. Psychiatry* (2006) 45:1023–1031. doi:10.1097/01.chi.0000225353.16831.5d.

[6] Stubbe D, Martin A, Bloch M, Belitsky R, Carter D, Ebert M, Friedman A, Giese A, Kirwin P, Ross RG, et al. Model curriculum for academic child and adolescent psychiatry training. *Acad. Psychiatry* (2008) 32:366–376. doi:10.1176/appi.ap.32.5.366.

[7] Martin VL, Bennett DS, Pitale M. Medical Students' Perceptions of Child Psychiatry: Pre- and Post-Psychiatry Clerkship. (2005).

[8] Martin VL, Bennett DS, Pitale M. Medical students' interest in child psychiatry: A clerkship intervention. *Acad. Psychiatry* (2007) 31:225–227. doi:10.1176/appi.ap.31.3.225.

[9] Asch SS, Marcus ER. The Current Status of Psychoanalysis in Medical Student Education in the United States: A Preliminary Overview. *J. Am. Psychoanal. Assoc.* (1988) 36:1033–1057. doi:10.1177/000306518803600409.

[10] Rice TR, Hoffman L. Defense mechanisms and implicit emotion regulation: a comparison of a psychodynamic construct with one from contemporary neuroscience. *J. Am. Psychoanal. Assoc.* (2014) 62:693–708. doi:10.1177/0003065114546746.

[11] Sudak DM, Goldberg D a. Trends in psychotherapy training: A national survey of psychiatry residency training. *Acad. Psychiatry* (2012) 36:369–373. doi:10.1176/appi.ap.11030057.

[12] Griffith JL. Neuroscience and Humanistic Psychiatry: a Residency Curriculum. *Acad. Psychiatry* (2014) 38:177–184. doi:10.1007/s40596-014-0063-5.

In: Neurobiology of Men's Mental Health
Editors: Leo Sher and Timothy R. Rice
ISBN: 978-1-63483-191-8
© 2015 Nova Science Publishers, Inc.

Chapter 11

REGULATION-FOCUSED PSYCHOTHERAPY FOR CHILDREN IN GENERAL PSYCHIATRIC EDUCATION: AN ACCOUNT FROM THE PERSPECTIVE OF MEN'S MENTAL HEALTH

Timothy R. Rice*[*,1], MD, Raj Loungani[2], MD, MA
Emma Gaines[3],MA, Tracy Prout[3], PhD
and Leon Hoffman[1], MD
[1]Department of Psychiatry, Icahn School of Medicine at Mount Sinai, New York, NY, US
[2]Department of Psychiatry, State University of New York Downstate, New York, NY, US
[3]Ferkauf Graduate School of Psychology, Yeshiva University, New York, NY, US

ABSTRACT

As detailed in the preceding chapter, male toddlers, preschool- and school-aged boys, male adolescents, and young male adults often present with externalizing behaviors. On many occasions, these children are treated with antipsychotics to manage their aggression and externalizing behaviors. This may not be an optimal means by which to help these children. Discomfort with this practice model may be a key reason for which recruitment to child and adolescent psychiatry training is historically problematic, creating a significant deficit in providers relative to need and ultimately leaving child and adolescents without sufficient access to care. Without trained providers practices are suboptimal and there is no base for which providers can be trained in practices specific to men's mental health in child and adolescent populations. In this chapter we demonstrate the capability of a structured educational intervention in child and adolescent psychiatry to promote factors in an adult psychiatry resident that are related to pursuing further

* Corresponding author: Timothy R. Rice, M.D. Icahn School of Medicine at Mount Sinai, Box 1230, One Gustave L. Levy Place, New York, NY 10029, USA, Email: timothy.rice@mssm.edu

training in child and adolescent psychiatry. These factors were drawn from the validated measures of the Counselor Self-Efficacy Scale (CSES), The Valuable Elements in Psychotherapy Questionnaire (VEP-Q), and a scale designed to quantify perspectives of prognosis coupled with a vignette of a male child with a disruptive behavior disorder. The intervention, self-directed reading in the practice manual *Regulation Focused Psychotherapy for Children*, is shown through pre- and post-measures to increase these factors. A subjective narrative provided by the adult resident offers a qualitative assessment of change after the intervention. Implications for adult psychiatry residency education and training in both child and adolescent psychiatry as well as psychodynamic psychiatry and psychoanalysis are discussed.

Keywords: men's mental health, residency education, child and adolescent psychiatry, regulation focused psychotherapy for children

INTRODUCTION

The provision of mental health care to male children and adolescents which recognizes and addresses mental health concepts specific to male development and psychobiology requires a sufficient population of providers. Yet, there is a significant shortage of child and adolescent psychiatrists. In the United States, there is only one child psychiatrist for every 2,238 children with a diagnosable mental disorder [1]. The geographic distribution of practitioners is additionally skewed, leading to very deprived areas [2]. To compensate for this, often non-specialists without training in child and adolescent psychiatry treat youth with psychiatric disorders, evoking therapeutic, ethical, and related non-pharmacologic considerations in this practice [3]. Providers who lack training may not understand socioemotional development and may not always follow best practices in the care of children and adolescents; in fact, in the serious matter of prescribing children antipsychotics, this is the confirmed case [4].

This issue of antipsychotic medication to address aggression is of particular relevance to men's mental health in child, pre-adolescent, and adolescent populations, as aggression and the disruptive behavior disorders are a particularly male phenomenon and reaction in childhood [5]. These agents are frequently used in this population [6] without a sound evidence base for this practice [7]. This trend towards increased antipsychotic prescriptions has increased significantly over the past decade [8], despite increasing concerns that children may be more vulnerable to the adverse effects of these medications [9, 10].

In the preceding chapter, we reviewed the means of addressing the shortage of child and adolescent psychiatrists through an emphasis on medical student education. We discussed that an educational approach which conveys male children's disruptive and aggressive behaviors as communications rather than provocations to elicit defensiveness may be a crucial factor in raising interest among physicians in psychiatry and, in specific, child psychiatry.

However, there is a second hurdle to increasing the population of child and adolescent psychiatrists. This hurdle is the transition from general adult psychiatry training to child and adolescent psychiatry training. This transition is an interesting one in that whereas a large proportion (nearly two-thirds) of early adult psychiatry residents express an interest in child and adolescent psychiatry training [11], only a very small proportion of these initial interests actually materialize into applications and matriculations. Interventions to try to bolster such recruitment, such as the creation of integrated adult and child psychiatry residency programs [12], have met only partial success. Exploring the determinants of this interest drain may be especially important to ensure that there are trained providers capable of providing mental health care to children and adolescents.

Our hypothesis, that interest in working in child and adolescent psychiatry requires a mobilization and understanding of aggression that is fundamental to male children with disruptive behavior disorders which may be disturbing to trainees to acknowledge and may thence promote avoidance of the field, may be equally relevant to adult psychiatry residents.

For example, nearly one in five (19%) of general psychiatry residents reported disliking their child and adolescent psychiatry experience in general residency [11]. In this study, this was the highest cited factor for not choosing child and adolescent psychiatry in this study behind the somewhat vague and avoidant response of "preferring to work with adults" [11]. When the reasons behind disliking their child and adolescent psychiatry rotation were explored further in a smaller discussion group, the study authors found that,

> "they were 'turned off' by being predominantly exposed to seriously-disturbed children with a high preponderance of externalizing symptoms emphasizing medication management only." (p.188)

One can infer that these "seriously-disturbed" children with externalizing behaviors are often male, and that the emphasis on medication management is a direct reference to feeling draw into a coercive battle of symptom control rather than into communication and understanding with these children. Parents, teachers, social service and health care workers, and others want to control these kids and they do not have the tools. On a systems level, the outcome of this unfortunate situation is the current practices of escalating antipsychotic use in these children without a clear evidence base and with serious medical and psychological consequences. But, as we will show, this may be a key issue in recruitment into child and adolescent psychiatry, a deficit which has real implications for the development of a men's mental health emphasis in child and adolescent populations.

Proposal

We propose that the best means to address this deficiency is to implement an educational experience for adult psychiatry residents which promotes a psychoanalytically-informed orientation to the disruptive behavior disorders of male children. This proposal is in alignment with the goals of men's mental health in that it incorporates an understanding of male psychobiology and development to promote an approach that is specific to the problems and needs of male children. In addition to providing trainees with an experience in the specialty field of men's mental health, it is proposed that such an intervention would address a core deficiency in the recruitment of adult psychiatry residents into the field of child and adolescent psychiatry. Addressing this deficiency thus additionally promotes the provision of men's mental health in child and adolescent populations by ensuring that there is a healthy population of providers available to train in and to learn a nuanced understanding of the psychobiological and developmental needs of male children.

Disclaimer

For these reasons, whereas in the former chapter we have tried to demonstrate that educational interventions in medical students are worthwhile, we will focus here on training in general psychiatry residents. These companion articles are not oriented to be in competition for a limited number of educational resources that would advance men's mental health and the field of child and adolescent psychiatry as a whole; instead, attention to both are needed and warrant their own separate discussions. As we will discuss in greater detail below, the pool of potential child and adolescent psychiatrists come from general psychiatry residents, therefore we focus in this chapter on this population instead of medical students.

Background

Child and adolescent psychiatry is a young field which has had consistent difficulties with recruitment [13]. The first academic appointment to a child psychiatry position was granted only in 1935, to Leo Kanner at the Johns Hopkins Medical School. The American Academy of Child Psychiatry, the parent organization of the current American Academy of Child and Adolescent Psychiatry (AACAP) was not established until 1953. The American Board of Psychiatry and Neurology (ABPN) did not recognize the field until 1959, after which the Accreditation Council for Graduate Medical Education (ACGME) established standards for training in the field.

Another major milestone was the creation of the "Triple Board" Residency Program in 1986, enabling medical students to directly commit from medical school to training in adult psychiatry, pediatrics, and child and adolescent psychiatry. However, to date, the number of triple board residency programs remains small at only about 10 in total, which recruit only about 20 residents each year into this pathway [14, 15].

The "traditional" path, by which child and adolescent psychiatrists achieve their training through a fellowship in child and adolescent psychiatry after completing a residency in adult psychiatry, remains the predominant pathway into the practice of child and adolescent psychiatry. Across the world it is recognized that training in adult psychiatry which allows adult residents adequate exposure to child and adolescent psychiatry may be necessary to strengthen recruitment [16].

ACGME requirements for training in adult psychiatry specify training in child and adolescent psychiatry as a core requirement. The amount of training is two month full time equivalents (FTE) in "organized clinical experience...supervised by child and adolescent psychiatrists who are certified by the ABPN or who are judged by the Review Committee to have equivalent qualifications" [17]. However, it remains that while nearly two-thirds of adult psychiatry residents initially express interest in specializing in child and adolescent psychiatry, only a small number actually do. It thus remains important that the exposure to child and adolescent psychiatry training in adult residency be a *good* exposure. It is highly problematic that nearly one in five residents negatively experience child and adolescent psychiatry in training; even neutral or mixed feelings about training is not enough to bolster recruitment. Recruitment is crucial for a healthy base capable of implementing the practices of men's mental health in child and adolescent populations.

Methods

In alignment with our proposal, we sought to evaluate the effectiveness in a structured educational experience in promoting factors in an adult psychiatry resident that were likely to encourage recruitment into child and adolescent psychiatry. The educational intervention was a self-directed immersion in the psychotherapy manual of Regulation Focused Psychotherapy for Children (RFPC) [18]. This manual details an approach to children with externalizing behaviors which offers understanding the meaning of the behaviors in lieu of attempting to control them. It is particularly adept in the provision of men's mental health in child and adolescent populations in that it directly addresses the aggression and externalizing behaviors which are so characteristic of male children and pre-adolescents. The psychotherapy program is a short-term structured approach which integrates pre- and post-measures and which has been designed to permit evidence-based testing in the near future. Its background is of the psychodynamic and psychoanalytic orientation. It was developed to integrate with transdiagnostic dimensional brain-based models of psychopathology by focusing on the emotion regulation system and thus integrates a psychotherapy approach with affective neuroscience. This is in accordance with the National Institute of Mental Health's Research Domain Criteria (RDoC) project [19, 20].

The participant was a third year adult psychiatry resident (R.L.) with preexisting interest in child and adolescent psychiatry. Baseline measurements of factors assessed by the authors to be related to the pursuit of training in adolescent psychiatry were measured in the resident through self-report questionnaires. The resident read the RFPC manual and then repeated the self-report questionnaires blinded from his prior scores. He reviewed the differences between the scores and wrote a narrative to detail his perspective on the differences between the pre- and post-intervention scores as they related of the experiences of this intervention.

The measures were the Counselor Self-Efficacy Scale (CSES), Valuable Elements of Psychotherapy Questionnaire (VEP-Q), and a scale designed to quantify perspectives of prognosis adapted from the longer Clinical Judgment Scale coupled with a vignette of a male child with a disruptive behavior disorder [21–24]. These measures were chosen through a literature search by the third author (E.G.,). These specific measures were chosen because of their strong psychometric properties, as well as their accessibility and ease of administration to participants. The CSES is a 20 item self-report questionnaire that measures skill and knowledge competency in the practice of therapy. It has demonstrated good reliability with a Chronbach's alpha of 0.91 and test-retest reliability of 0.85. The CSES also has high construct-related validity when compared with other established measures of self-efficacy and other assessments of self-efficacy based on theory and counselor development models. The VEP-Q is a self-report questionnaire with 17 items intended to assess views regarding how to best conduct therapy. This measure yields three major factors: psychodynamic therapy (PDT), cognitive behavior therapy (CBT), as well as a common factor (CF). The reliability of the measure is good to excellent, with test-retest reliability high for all scales, ranging from 0.82 to 0.96, as well as Cronbach's alpha ranging from 0.77 to 0.86. There is high construct validity in differentiating among the orientations. Seven items from the Clinical Judgment Scale were adapted to evaluate views on prognosis of a hypothetical client using a self-report format and Likert scale. Items included likelihood of client making substantial progress and clinician selection for caseload, as well as client motivation for change, capacity for insight, severity of impairment, and clinician expectations of course of treatment.

Items assessing the resident's self-efficacy in the practice of therapy and views on the prognosis of a hypothetical child patient with a disruptive behavior disorder were felt to be perhaps of greatest interest in evaluating the impact of the educational intervention on the adult resident. We were particularly interested in practitioner confidence, self-reported competency, as well as optimism in efficacy in working with male children with disruptive behavior disorders. We expected that the educational intervention would increase the adult resident's self-reported confidence and competency in therapeutic work as well as positivity regarding treatment for children with externalizing behaviors. The vignette was chosen to evaluate these factors with the hypothesis in mind that this intervention would encourage the provider to meet these children in a non-coercive stance and incite progress. The therapist orientation questionnaire was chosen to determine whether any shift in perspective would occur through the education. In the future, theoretical orientation may be viewed as a mediator or moderator in analyses of a larger cohort, with the hypothesis that a pre-existing orientation to psychodynamic psychiatry may facilitate greater impact of the educational intervention given its psychoanalytic nature and background in psychodynamic psychiatry.

RESULTS

The resident completed both pre- and post-intervention measures following his immersion in the manual. A summary of his pre- and post-intervention scores is provided in Figure 1. On the measures assessing the adult resident's self-reported confidence and competency as a clinician and his views regarding the hypothetical client's prognosis, there was an increase between the pre- and post-test conditions, representing change in the expected direction. While the adult's resident's views on the scale of theoretical orientation did not produce change in the expected direction, these results may be due to factors external to this educational intervention.

The resident produced the following narrative:

> "As a third-year adult psychiatry resident - a psychiatrist-in-training about one year away from being a full-fledged psychiatrist - who is interested in utilizing a psychotherapeutic (particularly a psychoanalytic) approach whenever possible over merely prescribing medications that also carry adverse side effects and impeding effects on brain development we don't even yet fully understand, and that often just mask the externalized symptoms rather than address their underlying causes, I much appreciate Drs. Leon Hoffman, Tim Rice, and Tracy Prout's manualized approach to treating children with externalizing behavioral disorders. Reading and implementing principles from RFP-C in my work with children (and adults for that matter) has made me much more optimistic about entering a career in child & adolescent psychiatry. In fact, examining and attempting to understand the meaning behind disruptive behaviors has made me more effective and has added much joy to my work with clients of all ages, since everyone to a certain extent displays defenses to be able to avoid, tolerate, or cope with internal conflicts and unpleasant emotions.
>
> As Shaw, Lewis, and Katyal described, a critical drawback of child and adolescent psychiatry rotations and practice is the emphasis in both inpatient and outpatient settings on treating the externalizing symptoms of "seriously-disturbed children" with primarily medication management – often causing undue side effects and some level of potential stunting of cognitive development -- rather than spending more effort addressing and

treating the *cause* of such behaviors. What makes matter's worse is weary parents often consent for these medications with the best of intentions for their children, whether it's in the best interest of their kids who may or may not wish or assent to take the meds. And if kids do not take these meds voluntarily? They may just get an intramuscular injection of the same pharmacological compound against their will – a traumatizing experience to be sure -- whether or not it's truly helpful and often with a questionable indication. We in psychiatry and pediatrics utilize these medications for far more than their FDA indications, with a less than stellar evidence base. Clearly, viewing externalizing behaviors through the lens of being improvable through non-pharmacological means, as with the RFP-C approach, is a far less invasive and more effective approach that may garner more buy-in from clients and parents.

Take the case of this 6-year-old boy, whose ongoing treatment has been directly influenced by the RFP-C approach. This child with a history of ADHD presented to outpatient treatment after numerous suspensions from school over the past few months due to threatening to kill himself, becoming aggressive by hitting, kicking, and stabbing his classmates and then himself with a pencil, throwing items from the class toy cabinet, and increasingly throwing tantrums and being disruptive at school and home while ignoring the guidance of his teacher, mother, and others. Now, during my second-year inpatient and outpatient child and adolescent psychiatry rotations - or even earlier in the third year for that matter - I was taught to and ultimately followed through with airing toward a focus on simply explaining the nature of ADHD and prescribing a stimulant and perhaps even a mood stabilizer or atypical antipsychotic to treat the disruptive behaviors of such a patient, with perhaps an adjunctive behavioral plan at home and school.

I wouldn't even have considered that this child and his family would be amenable to attending weekly psychotherapy, let alone a psychoanalytic approach, and certainly didn't expect a young child and his family could gain insight into why he engaged in such destructive behaviors and could work through his defenses to translate their heightened understanding into genuine therapeutic change. However, after reading the RFP-C manual and implementing its principles over the past couple months, that's exactly what has transpired through an empathic, non-judgmental analytic stance, conversing and interacting with the child while engaging in some form of play - the primary language of young children, whether it be Checkers, Legos, a science experiment kit, drawing, or so one - and an ongoing collaboration with the boy's family and school teacher and guidance counselor. Positive change has not only been elicited by interpreting the potential meanings behind the boy's actions, which as we learn in the manual are often a way of expressing his feelings that he was having much difficulty in verbalizing, but also by deconstructing the problematic and positive ways in which he relates to others (particularly his parents, teacher, and peers) and the world at large by examining his conveyance of feelings, attitudes, yearnings, or ideas he has displayed toward me as the therapist. By exploring and bringing to the surface the underlying meanings and often the primitive defenses against the boy's underlying anxieties leading to his disruptive externalizing behaviors, and doing so by working in the here-and-now transference, more meaningful and in turn more intense, lasting change for the better is possible.

Indeed, as illustrated by the results of the CSES, Clinical Judgment Scale, and Likert Scale, studying and practicing principles used in the RFP-C has strengthened my confidence, competence, support, and optimism of a positive result regarding treating externalizing behaviors with psychoanalytical principles.

The VEP-Q results remaining fairly stable before and after reading the manual is likely the result of me having a stronger psychodynamic perspective - though just as experienced in cognitive-behavioral therapy - in the first place. This, coupled with the great overlap of psychodynamic and CBT principle (e.g., insight roughly equating cognitive restructuring, unconscious material uncovered by an interpretation being analogous to core beliefs that represent the deeper meaning underlying automatic thoughts), could explain the lack of change in VEP-Q. All in all, RFP-C has and will continue to serve as an extremely valuable approach to improving externalizing behaviors more safely and effectively by first seeking to uncover the meanings of such behaviors, rather than by merely administering medications or attempting to control children's behaviors otherwise."

Scale	Pre-Test	Post-Test
Prognosis: Appropriateness for therapy (0 Least appropriate – 4 Most appropriate)	3	4
Prognosis: Selection for your caseload (0 Least likely to select – 4 most likely to select)	3	4
Prognosis: Severity of impairment (0 Most severely impaired – 4 least severely impaired)	2	3
Prognosis: Motivation for change (0 Least motivated – 4 most motivated)	1	2
Prognosis: Capacity for insight (0 Least capacity – 4 most capacity)	1	3
Prognosis: Likelihood of making substantial progress (0 Least likelihood – 4 Most likelihood)	3	4
Prognosis: Number of therapy sessions required for this patient to make substantial progress	11-15	6-10
Prognosis: Number of sessions you expect this patient will attend therapy	6-10	21+
CSES	84	94
VEP-Q: Psychodynamic Therapy Factor	29	26
VEP-Q: Cognitive Behavior Therapy Factor	25	26
VEP-Q: Common Factor	29	27

Figure 1. Summary of pre- and post-intervention results.

Discussion

This preliminary data demonstrates that a structured intervention at the level of adult psychiatry residency training may influence factors related to recruitment into child and adolescent psychiatry. The personal narrative provided by the adult resident is further evidence of this truth. Given that the provision of men's mental health in a child and adolescent population requires a population of practitioners who are trained to address the specific presentations and needs of children and adolescents, this may be a very important intervention. Because it has been designed to address the specific factor which may impede adult residents from pursuing training, namely, the discomfort associated with entering into "battles of will" and coercive cycles to control and manage disruptive behaviors through the use of psychotherapy and/or punishment, there may be significant benefit to this approach.

It is a unique training experience in men's mental health in that it targets a need of these male children that is specific to their psychobiology and underlying developmental needs. Indeed, affect-oriented approaches to externalizing disorders such as oppositional defiant disorder (ODD) may be more in fitting with the male-specific neural underpinnings of the disorder rather than management-oriented approaches [25].

Psychoanalytic Considerations

In addition to promoting recruitment in to child and adolescent psychiatry, it is believed that such an orientation may also promote recruitment into advanced studies in psychodynamic psychiatry and/or psychoanalysis. In the setting of an environment which emphasizes evidence based treatments and short-term care, fewer and fewer adult psychiatry residency graduates choose further training in psychodynamics and psychoanalysis, yet graduates consistently convey sustained support for further training in psychodynamics within residency education [26]. Compared with medical students, psychiatry residents may indeed be more available for emersion in an analytically-informed approach: in one study when directly compared with medical students, psychiatry residents were more likely to be tolerant of ambiguity, interpersonally and socially related, empathic, and interested in a developmental perspective [11]. All these qualities may be important for receptiveness to fundamental principles of psychoanalytically-informed ideas and further psychoanalytic education [27].

Psychoanalytically-informed education is rare in medical schools and, has historically faced challenges, even nearly three decades ago when economic and cultural/political burdens were less significant [28]. It is difficult to encourage trainees to be open to such training and often problematic to find psychoanalysts willing to leave the comfort of practice and enter the academic environment to teach [28]. The emphasis on men's mental health in the approach presented in this chapter may as a tertiary product address both these barriers by incorporating psychoanalytically-informed ideas within the broader structure of men's mental health. This structuring may enable both students and teachers to rise above the historical conflicts between academia and psychoanalysis and permit engagement where there is need as evidenced by trainees and graduates' persistent support for further training in psychodynamics during residency [26].

The emphasis on evidence-based medicine in RFP-C, by virtue of its development with an emphasis on permitting reliable replication, short-term duration, and incorporation of pre- and post-measures to enable trials, may additionally promote resident openness to psychoanalytic ideas. A major concern about psychoanalysis among residents is that the evidence base is "skimpy" [29]. A psychodynamic intervention which is evidence-based would help address this concern. The neurobiological emphasis in RFPC is also valuable. There is a precedent for folding psychoanalytic and developmental ideas into mainstream residency education by integrating neuroscience with "threatened" or vulnerable topics [30] within residency curricula. The concordance of RFP-C with affective neuroscience permits a potential for a similar marriage in this intervention within the topic of men's mental health in male children and adolescents.

Conclusion and Summary

We have proposed that the best means to address this deficiency is to implement an educational experience for adult psychiatry residents which promotes a psychoanalytically-informed orientation to the disruptive behavior disorders of male children. This proposal is in alignment with the goals of men's mental health in that it incorporates an understanding of male psychobiology and development to promote an approach that is specific to the problems and needs of male children. In addition to providing trainees with an experience in the specialty field of men's mental health, we have proposed that such an intervention would address a core deficiency in the recruitment of adult psychiatry residents into the field of child and adolescent psychiatry. The benefits to recruitment into psychodynamic psychiatry and psychoanalysis have also been presented, an opportunity which surveys suggest are highly and consistently valued by psychiatrists [20]. Addressing this deficiency thus additionally promotes the provision of men's mental health in child and adolescent populations by ensuring that there is a population of providers available to train in and to learn a nuanced understanding of the psychobiological and developmental needs of male children. Interventions such as that described in this chapter may play a crucial role in the reduction of suboptimal adherence to best practice guidelines and promote the best interests of male children and adolescents.

References

[1] Thomas CR, Holzer CE. The continuing shortage of child and adolescent psychiatrists. *J. Am. Acad. Child. Adolesc. Psychiatry*. 2006;45:1023–31.

[2] Thomas CR, Holzer CE. National distribution of child and adolescent psychiatrists. *J. Am. Acad. Child. Adolesc. Psychiatry*. 1999;38:9–15; discussion 15–6.

[3] Dell ML. Child and Adolescent Depression: Psychotherapeutic, Ethical, and Related Nonpharmacologic Considerations for General Psychiatrists and Others Who Prescribe. *Psychiatric Clinics of North America*. 2012. p. 181–201.

[4] Rettew DC, Greenblatt J, Kamon J, Neal D, Harder V, Wasserman R, et al. Antipsychotic Medication *Prescribing in Children Enrolled in Medicaid. Pediatrics*. 2015 Mar 2;

[5] Rice TR, Rihmer Z, Golier J, Sher L. Future Directions Concerning the Impact of Childhood and Adolescent Adversities in the Field of Menâ€TMs Mental Health: The New York Declaration. *Front Public Heal*. 2015 Jan 19;3(January):1–5.

[6] Findling RL. Atypical antipsychotic treatment of disruptive behavior disorders in children and adolescents. *J. Clin. Psychiatry*. 2008 Jan;69 Suppl 4:1–2.

[7] Pringsheim T, Gorman D. Second-generation antipsychotics for the treatment of disruptive behaviour disorders in children: a systematic review. *Can. J. Psychiatry*. 2012 Dec;57(12):722–7.

[8] Olfson M, Blanco C, Liu S-M, Wang S, Correll CU. National Trends in the Office-Based Treatment of Children, Adolescents, and Adults With Antipsychotics. *Arch. Gen. Psychiatry*. 2012 Aug 6;71(1):81–90.

[9] Correll CU. Metabolic side effects of second-generation antipsychotics in children and adolescents: a different story? *J. Clin. Psychiatry*. 2005 Oct;66(10):1–2.

[10] Maayan L, Correll CU. Weight gain and metabolic risks associated with antipsychotic medications in children and adolescents. *J. Child. Adolesc. Psychopharmacol.* 2011 Dec;21(6):517–35.

[11] Shaw J a., Lewis JE, Katyal S. Factors affecting recruitment into child and adolescent psychiatry training. *Acad. Psychiatry*. 2010;34(June):183–9.

[12] Stubbe D, Martin A, Bloch M, Belitsky R, Carter D, Ebert M, et al. Model curriculum for academic child and adolescent psychiatry training. *Acad. Psychiatry*. 2008;32(October):366–76.

[13] Kim WJ, Bechtold D, Brooks BA, Enzer N, Joshi P, King C, et al. Child and Adolescent Psychiatry Workforce: A Critical Shortage and National Challenge. *Acad. Psychiatry*. 2003;27:277–82.

[14] Gleason MM, Fritz GK. Innovative training in pediatrics, general psychiatry, and child psychiatry: Background, outcomes, and experiences. *Academic Psychiatry*. 2009. p. 99–104.

[15] Gray DD, Bilder DA, Leonard HL, Anders TF. Triple Board Training and New "Portals" into Child Psychiatry Training. *Child and Adolescent Psychiatric Clinics of North America*. 2007. p. 55–66.

[16] Tateno M, Kato T, Nakano W, Teo AR, Nakagawa A, Miyajima K, et al. Attitudes of early-career psychiatrists in Japan toward child and adolescent psychiatry and their career decision. *Psychiatry Clin. Neurosci.* 2010;64:199–201.

[17] Accreditation Council for Graduate Medical Education. *Psychiatry Program Requirem*ents, 2014 update. 2014;

[18] Hoffman L, Rice T. Manual for Regulation-Focused Psychotherapy for Children with Externalizing Behaviors (RFP-C): A Psychodynamic Approach. New York, NY: Routledge; 2015.

[19] Insel T, Cuthbert B, Garvey M, Heinssen R, Pine DS, Quinn K, et al. Research domain criteria (RDoC): toward a new classification framework for research on mental disorders. *Am. J. Psychiatry*. 2010;167:748–51.

[20] Insel TR. The NIMH Research Domain Criteria (RDoC) Project: Precision Medicine for Psychiatry. *Am. J. Psychiatry*. 2014 Apr 1;171(4):395–7.

[21] Melchert T, Hays V, Wiljanen L, Kolocek A. The Counselor Self-Efficacy Scale Items. *J. Couns. Dev*. 1996;74:640–4.

[22] Larsson BPM, Kaldo V, Broberg AG. Theoretical orientation and therapists' attitudes to important components of therapy: a study based on the valuable elements in psychotherapy questionnaire. *Cogn. Behav. Ther.* 2010;39(3):161–72.

[23] Houts AC, Graham K. Can religion make you crazy? Impact of client and therapist religious values on clinical judgments. *J. Consult. Clin. Psychol.* 1986;54(2):267–71.

[24] Spitzler RL, Gibbon M, Skodol AE, Williams JBW, First MB. DSM-IV-TR casebook: A learning companion to the diagnostic and statistical manual of mental disorders (4th ed., text rev.). Washington, DC: American Psychiatric Publishing, Inc.; 2002.

[25] Cavanagh M, Quinn D, Duncan D, Graham T, Balbuena L. Oppositional Defiant Disorder Is Better Conceptualized as a Disorder of Emotional Regulation. *J. Atten. Disord.* 2014 Jan 30;1–9.

[26] Petersen T, Fava M, Alpert JE, Vorono S, Sanders KM, Mischoulon D. Does psychiatry residency training reflect the "real world" of psychiatry practice? A survey of residency graduates. *Acad. Psychiatry*. 2007;31(August):281–9.

[27] Purcell SD. Becoming related: the education of a psychoanalyst. *Psychoanal Q*. 2014 Oct;83(4):783–802.

[28] Asch SS, Marcus ER. The Current Status of Psychoanalysis in Medical Student Education in the United States: A Preliminary Overview. *J. Am. Psychoanal Assoc.* 1988;36:1033–57.

[29] Yager J, Mellman L, Rubin E, Tasman A. The RRC mandate for residency programs to demonstrate psychodynamic psychotherapy competency among residents: A debate. *Acad. Psychiatry*. 2005;29(3):339–49.

[30] Griffith JL. Neuroscience and Humanistic Psychiatry: a Residency Curriculum. *Acad. Psychiatry*. 2014;38:177–84.

Section Two: Suicide and Homicide in Men

In: Neurobiology of Men's Mental Health
Editors: Leo Sher and Timothy R. Rice
ISBN: 978-1-63483-191-8
© 2015 Nova Science Publishers, Inc.

Chapter 12

SEX DIFFERENCES IN SUICIDE RATES FROM THE PERSPECTIVE OF PSYCHIATRIC CARE TAILORED TO MEN

Masahito Fushimi, MD, PhD*
Akita Prefectural Mental Health and Welfare Center, Japan

ABSTRACT

In most countries, though the suicide rate is higher among males as a reflection of more successful attempts among men, there are more failed attempts among females. Awareness, or consciousness, of gender is understood through a complex biological and psychosocial process, based on gender identity and gender roles. It is necessary to understand this complex process to determine why there are gender differences in suicide rates.

Hypotheses for the gender difference in suicide rates include the following: "Men have a sociocultural code of conduct, according to which they believe they should resolve problems by themselves without seeking help from those around them," "Women have more failed attempts because they have a tendency to exhibit suicidal behavior as a means of communication; in other words, they tend to use suicidal behaviors as a way of making others change their behaviors towards the women, or as a way of seeking help," "Because women prefer means that involve less pain and bodily disfigurement, they have more failed attempts," and "There is a sociocultural norm/belief that it is masculine to succeed and feminine to fail." In considering gender differences in suicide rates, the opportunities for precautions to be put in place to prevent completion of suicide attempts may be related to gender inequality and differences in values; however, there is also a need to investigate these differences from a neurobiological perspective, and that is the aim of this paper. One possible explanation is that, compared to women, men's neurobiological ability to control impulsivity is weaker, making it easier for them to attempt to resolve problems through aggressive and high-risk means, such as suicide. Furthermore, men tend to choose more lethal means than women do, which leads to increased success rates. The relationship between sex hormones, such as testosterone, and mood and cognitive function has also been discussed in other theories.

* Corresponding Author address: Email: fushimi@sings.jp; fushimi-jscn@umin.ac.jp

The idea that men's mental health should be considered from a neurobiological perspective is a countermeasure to the assumption that men's vulnerability stems from issues such as reduced functioning of the serotonergic system, which is related to impulsivity. Measures targeting psychological states that lead to suicide attempts related to problems stemming from serotonergic system functioning abnormalities—such as inherently low impulse control, alcohol and drug abuse, and organic brain damage—need to be improved. Therefore, we should focus on improving impulsivity control and using psychiatric management, such as psychiatric medication, to suppress or reduce suicidal behavior and suicide attempts. Utilizing post-suicide attempt crisis intervention and follow-up is also essential.

Keywords: gender differences, males, serotonergic system, sex hormones, suicidal behavior, testosterone

INTRODUCTION

More than 90% of suicide victims have a psychiatric disorder, and it is uncommon for suicide to occur in an individual who is not psychiatrically ill [1–3]. Research shows that between 50–70% of suicides are not committed in a rational state but rather in a condition of clinical and deep depression, in which cognitive distortions are present and rational judgment is absent [2–4]. Most of these suicides occur within the context of a major depressive episode, with major depression forming the background of upwards of half of all suicides [5]. Schizophrenia, personality disorders, alcoholism, and substance abuse are other psychiatric disorders frequently associated with suicide [2, 3]. As the findings of previous studies indicate, mental illness, especially affective disorders with comorbidity of personality disorders in 40–50% and other psychiatric disorders in even more cases, is the predominant factor found in suicides of both genders [6–8]. Substance misuse disorders are, however, generally more common in male suicides [9], as is the case for individuals with schizophrenia [10]. Eating disorders, especially anorexia nervosa, also carry a high risk of suicide and most sufferers are female [11].

SEX DIFFERENCES IN SUICIDALITY

Previous epidemiological studies of risk factors for suicide have revealed that there are important gender differences in suicidal behavior. It is thought that these reflect not only differences in etiology but also other important variations by gender in relation to risk, the nature of suicidal behavior, and its prevention and treatment [6]. In some European countries, there are increasing patterns of morbidity and suicide mortality attributable to depression. Other countries have shown decreases in depression morbidity and suicide mortality. Still other countries, like Sweden, have shown increasing depression morbidity but a steady or decreasing suicide mortality. Namely, despite the increasing prevalence of depression and the continuous existence of psychosocial and existential depressogenic factors, some countries show stable or decreasing suicide rates [4]. These differences can perhaps be attributed to psychosocial background factors relevant to the multifactorial etiology of depression and suicide, but most importantly may be attributed to factors of accessibility to adequate diagnosis and treatment of depression (i.e., increasing and sometimes ready access to the detection, treatment, and monitoring of depression) [4, 12]. The above considerations are helping us to develop ways to improve the well-established antisuicidal preventive effect of the treatment of depression, by using education programs as a way of relieving the burden of depression and suicide [2, 4].

In many countries, although depression has consistently been found to be two to three times higher in women, suicide rates are up to four to five times higher in men, which may be explained by poor recognition, diagnosis, and treatment of depression in males [13, 14]. China is one exception, with very high completed suicide rates among females [13, 14]. In times of stress, social transition, and loss of identity, males seem to be more vulnerable than females are. This, combined with male alexithymia and inability to recognize their own depression and need for help, may be a reason for one of the striking paradoxes in general health care today; that is, the fact that male mortality is so high, especially in times of transition and stress, despite low male morbidity rates. Female morbidity generally tends to be much higher; however, despite this, females live longer than males do and are not as sensitive to stress and transitional phenomena [4].

Men frequently do not seek help for problems and those who do are less likely than women are to be given a diagnosis of a mental health problem [4]. Men value independence and decisiveness, and avoid acknowledging a need for help, viewing this as a weakness. In contrast, women value interdependence, and they more readily consult friends and accept help. Women consider decisions in a relationship context, taking many factors into consideration, and feel freer to change their minds [9, 14, 15]. It is possible that men may manifest more "atypical" symptoms of depression, leading to misdiagnosis, and the differences in the incidence of mental health problems may be attributable to factors such as differing exposure to life events, coping styles, or hormones. Men and women are equally likely to suffer from mental health problems; however, the types of mental health problems men suffer from differ in frequency from those experienced by women. Whereas women are more likely to be diagnosed with depression and anxiety, men have a higher incidence of substance misuse disorder and antisocial behavior diagnoses. Despite the numerous existing theories, thus far the reasons for these differences remain unclear.

In recent years, several Western European countries have experienced an increase in suicide rates in males, particularly in the younger age groups [14, 16]. In contrast, suicide rates of females have declined, especially in older women, or remained fairly stable, particularly in the young [14]. This pattern is especially marked in the United Kingdom (UK), with an overall rise in male rates and a decrease in female rates, suggesting that causal factors and, possibly, protective factors have different roles in the two genders [14, 17, 18]. Social factors, especially those linked to changes in gender roles, seem the most likely explanation [14, 19]. Further possible explanations proposed for the recent increase in suicides among young males include social and economic changes that have made it more feasible for females to leave unsatisfactory relationships. These changes to the relationship dynamic may have exposed a greater number of vulnerable males to the risk of suicidal acts [14, 19]. Today, in societies undergoing sociological and economic transition, especially in the former Soviet Union and other Eastern European countries, a dramatic decrease in life expectancy, especially in males, can be seen. In these countries, suicide mortality and death due to external causes, such as homicide, traffic accidents, and forms of risk-taking behavior, as well as mortality from cardiovascular diseases, increased dramatically during the 1990s, whereas the corresponding figures were stable in Western European countries [4].

In the UK, because the suicide rate for men is three to four times greater than that for women, men have been specifically targeted. Suicide rates amongst older men, who were previously in the highest risk group for suicide, and amongst women have been falling; however, the suicide rate in young men has been rising. Important risk factors for suicide are male gender, living alone, unemployment, alcohol and drug misuse, previous deliberate self-harm, and mental illness.

In contrast to suicide, rates of deliberate self-harm (DSH) are usually higher in females than males, with previous studies of suicidal behavior demonstrating this pattern in many countries [3, 14, 20]. There has, however, been an increase in rates of DSH in males in some countries; for example, in the UK, this trend has been particularly marked in young males [14, 21]. The higher rate of DSH in females, along with the stronger association between DSH and suicide in males, suggests that acts of DSH by females are more often based on non-suicidal motivation [14, 22, 23]. In females, the appeal function, whereby DSH is used to communicate distress or to modify the behavior and reactions of other people, seems more common, while in males, DSH is more often associated with greater suicidal intent. Moreover, it is reported that in community samples, females have greater suicidal ideation than males do [14, 24]. It is well recognized that males tend to use violent means of both suicide and DSH more often than do females, and greater suicidal intent, aggression, knowledge regarding violent means, and less concern about bodily disfigurement, are all likely explanations for the increased likelihood of violent means of suicide in males, compared to females [3, 14]. Males commit suicide between two and four times as frequently as females, while for attempted suicide the sex ratio is almost exactly reversed [3, 5, 15]. Various theories have been proposed to explain the higher rate of completed suicide in males [15] but these do not account for the higher incidence of attempted suicide in females [3, 15]. One explanation for the gender difference in both types of suicidal behavior is that males use more lethal methods and are, therefore, more successful. However, these differences do not provide a complete answer because when the method is held constant, gender differences remain [15]. Furthermore, gender differences in suicide and attempted suicide apply cross-culturally, even though methods of suicide are not necessarily more lethal for males than for females in all countries [15].

Research into the factors that protect against suicide is scarce, and the protective role in females of having a young child has been identified, confirming previous research findings regarding childbearing, which indicate that pregnancy is also a time of reduced risk [6, 14, 25]. Furthermore, being a parent of a young child, rather than married status per se, appeared to explain the apparent protective effect of marriage for women in the Danish study, whereas in men marriage appeared to be a protective factor in its own right (or single status a risk factor) [6, 14, 25]. The risk factors of suicide differed according to gender, with unemployment playing a more important role in predicting suicide for men than it does for women. The hypothesis that suicide is considered more masculine than attempted suicide is supported in some social evaluations of suicidal behavior, and the hypothesis that suicide is considered more potent than attempted suicide is partially supported [15]. Gender differences in suicidal behavior can be viewed in part as a product of specific cultural expectations. Once a person has reached a crisis point whereby he or she sees suicidal behavior as the main possible response to a problem, social acceptability may influence the chosen response. If suicide is socially evaluated as more masculine than attempted suicide, and/or if it is less socially acceptable for males than for females to attempt suicide, then it is plausible to suppose that these cultural values may constrict the male's choice options to either suicide or another, more culturally acceptable response. In other words, due to social pressures against attempted suicide, males would tend to bypass the less drastic solution of attempting suicide, and go directly to suicide [15]. There is an assumption that social expectations of suicidal behavior vary as a function of the gender role of the suicidal person, and that most males—or at least most males who commit suicide—want to be masculine rather than feminine [15].

NEUROTRANSMITTER AND SUICIDALITY

Considerable scientific evidence exists about the linkage between serotonin metabolism and aggression, acting out, and outbursts of anger, as associated features of major depression, especially in males [1, 4]. There is also evidence with regard to the significance of serotonin levels in acute suicidality and stress-related conditions [1, 4]. Furthermore, there is increasing clinical support for the usefulness of serotonin-active drugs in the treatment of abusive and violent behaviors, acting out, sociopathic behavior, and impulsiveness, as found in these patients [1, 4]. It corresponds with a form of temporary personality disorder and seems to confirm that the breakdown of fight, flight, and coping abilities, which to a great extent seems related to serotonergic decompensation and which we call depression, may have a different phenomenology in males and females [1, 4]. Even data from research on antidepressive drugs with animals in provoked learned helplessness situations show a distinction between male and female behavior [4]. It has been observed that male depression syndrome is characterized by temporary decreases in stress tolerance and impulse control, and increases in acting out, aggression, indecisiveness, irritability, substance abuse, antisocial features, morning discomfort, depressive thought content, and hereditary factors regarding suicide, depression, and alcoholism [4, 26]. Further research is needed to clarify the relationship between male depression and suicidal syndrome, which is likely caused by a breakdown in the serotonin-mediated ability to cope, fight, and flee. Transmitter turnover and receptors, and their relationship with the type of suicidal behavior or the underlying psychiatric disorder, should be the focus of further research.

In previous literature describing postmortem studies of suicide victims, most of the work has concentrated on the serotonergic and noradrenergic systems [1]. Other neurotransmitters (e.g., γ-aminobutyric-acid, acetylcholine, dopamine, and peptide modulator and transmitters) have been studied to a far lesser degree; therefore, this review has been confined to the serotonergic and noradrenergic systems. Most studies have focused on the cerebral cortex, in particular the prefrontal cortex, because of its hypothesized role in behavioral restraint, and the general consensus is that the prefrontal context is responsible for receptor binding changes in suicide victims [1].

Moreover, reduced serotonergic and noradrenergic functioning in the brains of suicide victims has been noted [1]. In relation to this, Mann et al. (1996) described that the 5-hydroxytryptamine (5-HT) and postsynaptic serotonin 5-HT_2 and 5-HT_{1A} receptors are increased, and the 5-HT transporter binding (or a related binding site) is decreased [1]. 5-HT_{1C} sites are present only in very low concentrations in prefrontal cortex and are more abundant, but appear unaltered, in the hippocampus of suicide victims. Because 5-HT_2 and 5-HT_{1A} receptors are located predominantly on local neurons in the cortex and the 5-HT transporter is located on the presynaptic axon terminal of serotonergic neurons that innervate the cortex, the most parsimonious conclusion is that there is reduced serotonergic innervation and consequent upregulation of postsynaptic receptor sites in suicide victims. However, lesion studies in rodents have largely failed to demonstrate such upregulation, so the precise mechanism in suicide victims is unclear. Further, in the brain stem of suicide victims, 5-HT and 5-hydroxyindoleacetic acid (5-HIAA) levels appear to be lower. It remains to be determined whether this reduction is because of fewer serotonin-synthesizing neurons in the dorsal and median raphe that project to the cerebral cortex (and elsewhere), or whether there is reduced capacity for serotonin transmitter synthesis [1].

The role of norepinephrine (NE) is less clear than that of serotonin. NE levels are reported to be increased in the cortex, but there are insufficient data to determine whether 3-methoxy-4-hydroxyphenylglycol (MHPG) is similarly changed, and as such it remains unclear whether suicide victims have altered NE turnover or activity [1]. In the brain stem, NE and MHPG concentrations seem to be reduced, and there are fewer pigmented noradrenergic neurons. One study has reported the seemingly counterintuitive finding of increased activity of the NE biosynthetic enzyme tyrosine hydroxylase [1]. These findings represent a snapshot of brain function at the moment of suicide and, therefore, include the following factors: effects of genetics, development, and early life processes; the associated psychiatric disease and any treatment; environmental stresses; and the artifacts of postmortem delay [1]. Age, gender, method of death, and drug use can be controlled for, while use of a psychological autopsy method and psychiatric control subjects are needed to control for psychiatric illness and long-term treatment effects [1, 4].

Sex Hormone and Suicidality

Testosterone, an androgen hormone, regulates male sexual development, and influences muscle size, levels of red blood cells, bone density, a sense of well-being, mood, cognition, and sexual and reproductive functioning. Multiple lines of evidence suggest that testosterone may affect mood and behavior in both males and females, and that there is an association between testosterone and suicidal behavior [27, 28]. Variations between the results of studies of the relationship between testosterone and suicidal behavior could be related to many factors, including differences in sample selection criteria and differences in time intervals between suicide attempts and blood sampling [29]. The best-known neurobehavioral effects of testosterone are on sexual function and aggression. Furthermore, testosterone and other androgens may be involved in the pathophysiology of mood disorders and suicidal behavior. However, the role of testosterone in the pathophysiology of suicidal behavior remains unclear. Some observations suggest that suicidal behavior may be related to blood levels of testosterone; for example, testosterone levels were found to correlate with the number of suicide attempts in patients with bipolar disorder [30, 31].

Sher (2012) suggested that the link between testosterone and the neurobiology of suicidal behavior may be related to the following factors: a direct influence of testosterone on suicidality via certain brain mechanisms; testosterone influencing aggression and, consequently, suicidality; testosterone influencing mood and, consequently, suicidality; and testosterone influencing cognition and, consequently, suicidality [30]. Further, it has been directly suggested that higher testosterone levels are associated with suicide in young people [32]. In young men, higher testosterone levels were associated with provoked verbal and physical aggression, which suggests that impulsive aggression correlates with higher testosterone levels [30, 31, 32]. In addition, multiple lines of evidence indicate that strong aggressiveness is associated with suicidal behavior [31]; as such, high aggression levels can predict suicidal acts [30]. Therefore, aggression may mediate the effect of high testosterone levels on suicidal behavior in adolescents and young adults.

Testosterone deficiency is a widely recognized hormonal alteration associated with male aging. Further, in older people, depressed mood and impaired cognition are associated with decreased testosterone levels, whereas higher blood levels of testosterone are associated with better mood and cognitive functioning [30, 31, 33, 34].

It has been observed that older men with a testosterone deficiency frequently have depressive symptoms, and depressed men often have lower blood testosterone levels compared to those of non-depressed individuals [30, 31, 33]. Further, hypogonadal men frequently display lower cognitive abilities compared to those of men with normal testosterone levels [30, 31].

It has been reported that higher blood levels of testosterone in older men are associated with a lower risk of future Alzheimer's disease development; while low blood testosterone levels are associated with Alzheimer's disease [34]. In addition, both depression and cognitive impairment have been found to be associated with suicidal behavior, and may mediate the effect of decreased testosterone levels on suicidality in older men [30, 31]. Sher (2013) suggested that the treatment of hypogonadism in older men may improve mood and cognition, and consequently, reduce suicidal behavior, and, further, that it is reasonable to propose that suicidal behavior in young men is associated with high testosterone levels, whereas suicidality in older men is associated with decreased testosterone secretion [27, 28]. This indicates that the effects of testosterone on suicidality in men should be studied separately in young and older individuals.

Additionally, one study has been conducted to examine whether there is a relationship between testosterone levels and clinical parameters in bipolar suicide attempters [29]. Sher et al., (2014) showed that, at baseline, testosterone levels were positively correlated with the number of previous major depressive episodes and suicide attempts, and that higher baseline testosterone levels predicted suicide attempts during the follow-up period. Further, number of major depressive episodes, maximum lethality of suicide attempts, and testosterone levels were higher for men compared to women [29]. However, current suicidal ideation scores were higher in women than for men. Controlling for sex, we found that testosterone levels positively correlated with the number of both manic episodes and suicide attempts. Sher et al., (2014) concluded that their study results suggest that testosterone levels may be related to the course of bipolar disorder and suicidal behavior [29]. There is considerable evidence suggesting that testosterone may play a role in the pathophysiology of mood disorders in females, and demonstrating that blood testosterone levels may predict suicidal behavior in women with bipolar disorder [29].

Prevention of Male Suicide

Suicide prevention strategies understandably include ensuring that clinicians and others likely to encounter people at risk possess adequate risk-assessment skills [4]. The increasing rate of suicide in men is of great concern, especially since the suicide rate is decreasing amongst other populations. The reason for the increased suicide rate in men remains unclear. It may be that depression in men is not recognized, though it is also true that men are poor users of health services and, therefore, may not present in order to receive appropriate management [4, 14]. Men also tend to use more effective ways of committing suicide, which may explain the increased rate of success compared to that for women [3, 15]. Another reason is that men may present with different symptoms of depression, which are missed during the early stages of a depressive illness, in particular. In men, a depressive illness may manifest as increased levels of stress, aggression, and alcohol use [4, 14, 15]. Although the majority of those with depression are seen by general practitioners (GPs), few are diagnosed correctly, and in some cases depression is insufficiently treated and leads to suicide [2, 4, 14, 15]. Another problem is related to reaching depressed and suicidal patients who may be seen by the medical system with somatic complaints, may not be seen by the medical system at all, or are actively avoiding the medical system. After making contact with a GP, decreases were reported in the number of female suicide victims who were depressed and used drastic methods; however, male suicides were not affected by the GP's improved ability to diagnose and treat depression [4, 14, 15]. The difficult-to-reach groups were males, adolescents, aggressive and sociopathic acting-out individuals, and those with alexithymia, who are incapable of insight into their own depression and helplessness and are also in need of support and treatment [4]. A scale was constructed with a focus on the male depression syndrome, which often is missed when using conventional depression diagnostic criteria [4]. This diagnostic system is presently being subjected to scientific validation [4].

Today, in the society of some developed and developing countries, increasing psychosocial pressure can be found. The multifactorial pathogenesis of depression and suicidality means that better education, increasing accessibility, and increasing quality of medical care, leading to better diagnosis and treatment of depression, sometimes have a limited effect. An educational intervention focused on faster and more accurate diagnosis and improved treatment of depression, thereby counteracting depression morbidity and depression-related suicide mortality, is clearly a medical intervention, directed to components of the medical system. As such, it should always be accompanied by other social or network approaches focused on the multifactorial background factors behind the process of becoming depressed and suicidal, such as unemployment, transition, and financial problems [2, 4]. In the UK, to try to reduce the number of deaths by suicide in young men, suicide prevention strategies recommend that clinical staff should have improved risk management skills, that measures should be introduced to reduce alcohol and drug use, and that support should be readily available during a crisis [4, 14, 15]. The aim is to promote the mental health of students, support the development of college counseling services, and promote the recognition of suicide risk in primary care [4]. Unfortunately, men, especially young men, use primary care and other statutory and non-statutory services differently than women do. Men are much less likely to consult their GP than women are, and are also less likely to use counseling services [4, 14, 15]. More females than males seek help from GPs for mental health problems, which may explain why the apparent benefits of an educational program in detection and treatment of depression for general practitioners were found to be confined to females, more of whom were treated for depression and fewer of whom committed suicide [4, 14, 15, 26].

While improved detection and management of psychiatric disorders is undoubtedly a key factor in the prevention of suicidal behavior in males as well as females, there is increasing evidence that alterations in socioeconomic conditions are also very relevant to suicide prevention in males. Unemployment, retirement, being single, and sickness absence are significant risk factors for suicide. While debate continues over the relative contribution of unemployment to suicidal behavior, some of the findings add to the weight of evidence that occupational factors are particularly important in influencing suicide by males [2, 35]. Increased occupational instability has been proposed as one factor behind the recent increase in suicides in young males [19]. Workplace initiatives may be important to help identify and support men suffering from depression. Initiatives in places where men pursue recreational activities may also have a part to play, as will raising awareness and decreasing the stigma of depression. There is the possibility that such a multifaceted approach will help reverse this worrying trend. Although the predictive power of schedules to assist risk assessment is unlikely to ever be substantial, the question has been raised as to whether different, if overlapping, risk-assessment schedules are required for the two genders [12, 14].

Little research attention has been paid to possible gender differences in response to treatment of people at risk of suicidal acts. Clinical evidence suggests that male compliance with depression treatment plans is poorer than female compliance [2, 14]. There is also some indication that fewer male than female DSH patients benefit from the treatments they are offered [14, 36]. While this may reflect differences in overall attitudes to seeking and taking help, it could also result from the style of therapy that is available. Gender differences in verbal communication abilities and the reluctance of many males to share emotional problems may make some of the usual talking therapies less attractive to some males, at least initially. Treatment programs that have more of a practical emphasis, perhaps focused on problem solving, could prove more successful in engaging males at risk of suicide.

CONCLUSION

Considerable gender differences mean that the risk of suicidal behavior, and its nature, causes, prevention, and treatment should be studied from the perspective of gender. In addition to the epidemiological studies to disentangle more of the social and economic associations with suicidal behavior in each gender, further investigations of neurobiological factors related to risk of suicidal behavior should be conducted from a gender perspective. Furthermore, gender differences in treatments for suicidal behavior should be investigated to develop gender-specific approaches. A history of depression is as strong a suicide predictor as earlier suicide attempts. However, postvention for male depression and suicidal attempts by means of follow-up is poor both in psychiatry and primary care. Thus, an important strategy for preventing depression-related morbidity and suicide mortality is enabling clinicians and others likely to encounter people at risk, to diagnose and treat depression. Identification of the brain systems most affected, or predictive, of suicide may then provide the opportunity for effective intervention. Much remains to be learned through further systematic neurobiological studies with regard to preventing suicide in males. Gender differences in suicidal behavior clearly merit more research attention to generate information that can guide clinical practice and prevention strategies in ways that will prove most effective for preventing suicidal behavior in both genders.

REFERENCES

[1] Mann JJ, Underwood MD, Arango V: Postmortem studies of suicide victims, in Biology of Schizophrenia and Affective disease. Edited by Watson SJ, Washington, DC, American Psychiatric Press, 1996, pp 197-221.

[2] Fushimi M, Sugawara J, Shimizu T: Suicide patterns and characteristics in Akita, Japan. *Psychiatry Clin. Neurosci.* 2005; 59:296-302.

[3] Fushimi M, Sugawara J, Saito S: Comparison of completed and attempted suicide in Akita, Japan. *Psychiatry Clin. Neurosci* 2006; 60:289-295.

[4] Rutz W: The role of primary physicians in preventing suicide: possibilities, short-comings, and the challenge in reaching male suicides, in Suicide Prevention, Resources for the Millennium. Edited by Lester D, Philadelphia, Brunner-Routledge, 2000, pp173-187.

[5] Murphy GE: Why women are less likely than men to commit suicide. *Compr. Psychiatry* 1998; 39:165-175.

[6] Qin P, Agerbo E, Westergård-Nielsen N, Eriksson T, Mortensen PB: Gender differences in risk factors for suicide in Denmark. *Br. J. Psychiatry* 2000; 177:546-550.

[7] Henriksson MM, Aro HM, Marttunen MJ, Heikkinen ME, Isometsä ET, Kuoppasalmi KI, Lönnqvist JK: Mental disorders and comorbidity in suicide. *Am. J. Psychiatry* 1993; 150:935-940.

[8] Foster T, Gillespie K, McClelland R: Mental disorders and suicide in Northern Ireland. *Br. J. Psychiatry* 1997; 170:447-452.

[9] Murphy GE: Psychiatric aspects of suicidal behaviour: substance abuse, in The International Handbook of Suicide and Attempted Suicide. Edited by Hawton K, van Heeringen K. Chichester, John Wiley & Sons, 2000, pp 135-146.

[10] De Hert M, Peuskens J: Psychiatric aspects of suicidal behaviour: schizophrenia, in The International Handbook of Suicide and Attempted Suicide. Edited by Hawton K, van Heeringen K. Chichester, john Wiley & Sons, 2000, pp 121-134.

[11] Harris EC, Barraclough B: Suicide as an outcome for mental disorders. A meta-analysis. *Br. J. Psychiatry* 1997; 170:205-228.

[12] Goldney RD: Prediction of suicide and attempted suicide, in The International Handbook of Suicide and Attempted Suicide. Edited by Hawton K, van Heeringen K. Chichester, John Wiley & Sons, 2000, pp 585-596.

[13] Cheng ATA, Lee C-S: Suicide in Asia and the Far East, in The International Handbook of Suicide and Attempted Suicide. Edited by Hawton K, van Heeringen K. Chichester, John Wiley & Sons, 2000, pp 29-48.

[14] Hawton K: Sex and suicide. Gender differences in suicidal behaviour. *Br. J. Psychiatry* 2000; 177: 484-485.

[15] Linehan MM: Suicide and attempted suicide: study of perceived sex differences. *Percept. Mot. Skills* 1973; 37:31-34.
[16] Cantor CH: Suicide in the Western World, in The International Handbook of Suicide and Attempted Suicide. Edited by Hawton K, van Heeringen K. Chichester, John Wiley & Sons, 2000, pp 9-28.
[17] Hawton K: By their own young hand. BMJ 1992; 304:1000.
[18] Kelly S, Bunting J: Trends in suicide in England and Wales, 1982-96. Popul Trends 1998; 92:29-41.
[19] Hawton K: Why has suicide increased in young males? *Crisis* 1998; 19:119-124.
[20] Schmidtke A, Bille-Brahe U, DeLeo D, Kerkhof A, Bjerke T, Crepet P, Haring C, Hawton K, Lönnqvist J, Michel K, Pommereau X, Querejeta I, Phillipe I, Salander-Renberg E, Temesváry B, Wasserman D, Fricke S, Weinacker B, Sampaio-Faria JG: Attempted suicide in Europe: rates, trends and sociodemographic characteristics of suicide attempters during the period 1989-1992. Results of the WHO/EURO Multicentre Study on Parasuicide. *Acta Psychiatr. Scand.* 1996; 93:327-338.
[21] Hawton K, Fagg J, Simkin S, Bale E, Bond A: Trends in deliberate self-harm in Oxford, 1985-1995. Implications for clinical services and the prevention of suicide. *Br. J. Psychiatry* 1997; 171:556-560.
[22] Hawton K, Fagg J: Suicide, and other causes of death, following attempted suicide. *Br. J. Psychiatry* 1988; 152:359-366.
[23] Hawton K, Arensman E, Wasserman D, Hulten A, Bille-Brahe U, Bjerke T, Crepet P, Deisenhammer E, Kerkhof A, De Leo D, Michel K, Ostamo A, Philippe A, Querejeta I, Salander-Renberg E, Schmidtke A, Temesvary B: Relationship between attempted suicide and suicide rates among young people in Europe. *J. Epidemiol. Community Health* 1998; 52:191–194.
[24] Paykel ES, Myers JK, Lindenthal JJ, Tanner J: Suicidal feelings in the general population: a prevalence study. *Br. J. Psychiatry* 1974; 124:460-469.
[25] Appleby L: Suicidal behaviour in childbearing women. *International Review of Psychiatry* 1996; 8:107- 115.
[26] Rutz W: Improvement of care for people suffering from depression: the need for comprehensive education. *Int. Clin. Psychopharmacol* 1999; 14 (suppl 3):S27-33.
[27] Sher L: High and low testosterone levels may be associated with suicidal behavior in young and older men, respectively. *Aust. N Z J Psychiatry* 2013; 47: 492-493.
[28] Sher L: Low testosterone levels may be associated with suicidal behavior in older men while high testosterone levels may be related to suicidal behavior in adolescents and young adults: a hypothesis. *Int. J. Adolesc. Med. Health* 2013; 25:263-268.
[29] Sher L, Grunebaum MF, Sullivan GM, Burke AK, Cooper TB, Mann JJ, Oquendo MA: Association of testosterone levels and future suicide attempts in females with bipolar disorder. *J. Affect Disord* 2014; 166:98-102
[30] Sher L: Testosterone and suicidal behavior. *Expert Rev. Neurother* 2012; 12: 257-259.
[31] Sher L, Grunebaum MF, Sullivan GM, Burke AK, Cooper TB, Mann JJ, Oquendo MA: Testosterone levels in suicide attempters with bipolar disorder. *J. Psychiatr. Res.* 2012; 46:1267-1271.
[32] Roland BC, Morris JL Sr, Zelhart PF: Proposed relation of testosterone levels to male suicides and sudden deaths. *Psychol. Rep.* 1986; 59:100-102.

[33] Amore M, Innamorati M, Costi S, Sher L, Girardi P, Pompili M: Partial androgen deficiency, depression and testosterone supplementation in aging men. *Int. J. Endocrinol.* 2012; 280724.
[34] Chu LW, Tam S, Wong RL, Yik PY, Song Y, Cheung BM, Morley JE, Lam KS: Bioavailable testosterone predicts a lower risk of Alzheimer's disease in older men. *J. Alzheimers Dis.* 2010; 21:1335-1345.
[35] Platt S, Hawton K: Suicidal behaviour and the labour market, in *The International Handbook of Suicide and Attempted Suicide.* Edited by Hawton K, Heeringen KV. Chichester, John Wiley & Sons, 2000, pp 309-384.
[36] Hawton K: Attempted suicide, in Cognitive Behaviour Therapy: Science and Practice. Edited by Clark DM, Fairburn C. Oxford, Oxford University Press, 1997, pp 285-312.

In: Neurobiology of Men's Mental Health
Editors: Leo Sher and Timothy R. Rice
ISBN: 978-1-63483-191-8
© 2015 Nova Science Publishers, Inc.

Chapter 13

MALE SUICIDAL BEHAVIOR

Diego de la Vega[1], MD Samuel Aidan Kelly[2] , PA and Lucas Giner[3*], MD

[1]Unidad de Salud Mental Comunitaria de Huelva.
Servicio Andaluz de Salud
Huelva, Spain
[2]New York, NY, US
[3]Departament of Psychiatry
University of Sevilla
Seville, Spain

ABSTRACT

Suicide poses a major mental health challenge. Numerous studies have reported gender differences in suicidal behavior, though the specific influence of gender remains unclear. Time and again the same "gender paradox" has been seen: while women make more suicide attempts (two to three times as often), men have higher rates of completed suicide.

In the following sections we will review the risk factors and other features more frequently associated with male suicide. Those risk factors are classified into three broad groups: mental disorders, psychological dimensions, and socioeconomic stressors. Considering these factors together can help explain the gender differences in suicidal behavior. We suggest that the higher rates of completed suicide among males may be due to a confluence of psychological characteristics which together lead to suicidality, reluctance to seek help, greater intent to die, and more violent suicide methods.

We also demonstrate a distinct need for studies directly comparing the specific suicide risk factors affecting men and women. By better understanding these differences, preventative programs could be specifically targeted according to the most relevant risk factors.

* Corresponding Author address: Lucas Giner, Departamento de Psiquiatria, Universidad de Sevilla, Email: lginer@us.es

Keywords: male, suicide, gender paradox, treatment seeking

INTRODUCTION

Suicide poses a major mental health challenge. It is estimated that each year more than one million die by suicide worldwide, and suicide attempt rates appear 10 to 20 times higher [1]. Increasing awareness and knowledge of the mortality and morbidity associated with suicidal behavior ranks high on the World Health Organization's agenda.

Literature suggests that suicide is associated with various risk factors including heritable (genetic) traits [2], mental disorders [3] and physical disorders [4, 5] in addition to psychological [6, 7], sociodemographic [8-10] and socioeconomic factors [11, 12].

Studies usually do not analyze the influence of gender in the risk factors of suicidal behavior.

It therefore remains unclear if risk factors in males can be generalized for females and vice versa.

Preventative programs need to account for this discrepancy. With an understanding of the role gender plays in suicidal behavior, such programs could be tailored to meet the specific needs of their target populations. Traditional masculinity is connected to stoicism, autonomy, and success.

This promotes maladaptive strategies such as emotional inexpressiveness, reluctance to seek help, and alcohol abuse [13]. These factors and others contribute to the higher male proportion of death by suicide.

To better understand this issue, this chapter will examine the role that gender plays in suicidal behavior.

The Gender Paradox

Several classifications have attempted to categorize the continuum from suicidal ideation to completed suicide. The Columbia Suicidality Classification Scheme, for instance, classifies suicidal behaviors as suicidal (completed suicide, suicide attempt, suicidal ideation, preparatory actions), indeterminate (self-injurious behavior with unknown intent or outcome), or non-suicidal (self-injurious behavior without suicidal intent, other) [14].

Such distinctions are important for both clinical care and research of suicidal behavior. In addition to very different final results between the completed suicide and the suicide attempt, they may also involve different influences. Studying the characteristics that discriminate suicide completers from suicide attempters may help to reach a better understanding of the relationship between these different suicidal behaviors. While several risk factors have been associated with suicidal behavior, the relative weight of each factor could differ in suicide attempt compared to completed suicide [15-16].

Numerous studies have reported gender differences in suicidal behavior [17, 18]. While women make more suicide attempts (two to three times as often) [19], men have higher rates of completed suicide (with certain exceptions such as Chinese [20] and Indian [21] populations).

Table I. Ratios of male/female suicidal acts by country

Country	Suicide attempts	Completed suicide
U.S.A. (22)	0.73	3.83
Canada (23)	0.66	3.37
Australia (24)	0.61	2.99
Japan (25)	0.66	2.08

This gender gap can vary between countries, although as we can see it is fairly constant in a given region over time.

Table II. Suicidal behavior in the United States (rates among 100,000 people) (22)

	Nonfatal self-harm episodes		Completed suicide	
Year	Male	Female	Male	Female
2008	100.56	146.17	19.03	4.78
2009	104.36	139.05	19.23	4.87
2010	131.24	169.29	19.78	4.99
2011	131.69	180.56	20.00	5.22

These tables reflect the well-known gender paradox: men have higher completed suicide rates than women, although women make more suicide attempts.

Risk Factors

Many risk and protective factors for suicide have been identified. Some risk factors are clinical (mental disorders, physical health problems), some are sociodemographic (age, sex, religion, social support, economic situation, profession, adverse events) and some are personal (genetics, number of previous attempts, impulsivity, level of suicidal intent, hopeless). While all of these factors can be important, the relative weight of each may differ between men and women.

Unfortunately, most research examining risk factors for suicidality does not consider gender differences [26]. This may be seen as a limitation since it assumes that risk factors identified in these studies act similarly among males and females. In fact, gender distinctions have been reported in various mental disorders. Depression, a well-known risk factor for suicide, is more prevalent among women than men (risk ratios roughly 2:1) [27]. Depression is also much more commonly seen in female suicide than in male suicide (OR 12.95 vs. 6.56) [28]. However, other studies of suicidal behavior could not find gender differences with respect to Axis I or Axis II disorders [26]. Nevertheless, there is a clear difference in male vs. female rates of completed and attempted suicide. This difference alone underscores the need to investigate the possible gender dependence of suicide risk factors [29].

In the following sections we will review the risk factors and other features more frequently associated with male suicide. Those risk factors can be classified into three broad groups: mental disorders, psychological dimensions and socioeconomic stressors.

Mental Disorders

Suicide is very strongly associated with the presence of mental disorders. In an extensive meta-analysis of 3,275 suicides from different geographical origins, Arsenault et al., found that 87.3% had been diagnosed with a mental disorder prior to death [3]. Actual rates may be still higher as psychological autopsy studies have been suggested to underestimate the overall rate of psychopathology [3]. While the frequent presence of mental disorders is seen in both men and women, suicide studies have found gender differences regarding the form of psychopathology (3). Among males, there is a higher prevalence of personality disorders (OR 2.01,p<0.0001), childhood disorders (OR 4.95,p<0.0001) and substance abuse disorders (OR 3.58,p<0.0001), particularly alcoholism (OR 2.19,p<0.0001). Suicide among females is more commonly associated with diagnosis of affective disorders, mainly depression (OR 0.53,p<0.0001).

It has been suggested that population studies and diagnostic systems are over-diagnosing major depression in women and/or under diagnosing it in men [29, 30]. Moreover, men are less likely to look for help than women. Women more frequently contact primary care providers within the year before suicide than men (100% women, 78% men) [31]. Taken together, these studies suggest that mental disorders are frequently under diagnosed in males, which could diminish rates of treatment and recovery [29]. A recent study analyzed suicide-risk discussions among 48 primary-care physicians and 1,776 depressed adult patients. While rates of suicidal ideation were found to be similar between males and females, suicide risk discussions occurred in only one encounter with a male patient as compared to in twelve encounters with female patients [32].

In addition to the possible under diagnosis and undert reatment of mental disorders, it has been suggested that even men who receive treatment could obtain less benefit from pharmacological therapies. Although most studies do not find a difference between genders regarding the clinical response to antidepressants [29], several have observed a better response by women to SSRIs and SNRIs [33-35].

Alcoholism

According to the World Health Organization, alcohol is more closely associated with aggressive behavior than any other psychotropic substance [36]. Research conducted in Europe has found that alcohol abuse is more prevalent in men than in women (5.2% vs. 1.7% respectively) [37].

Several studies show a strong link between alcohol and depression, with alcohol abuse acting as either a precursor or a response to the onset of depression. In relation to suicide, alcoholism can be considered both a direct and indirect risk factor. Alcohol is as an emotional disinhibitor and promotes impulsive behavior [38]. Furthermore it facilitates other risk factors including depression, family conflicts, and work problems. In a study of more than four thousand suicide attempts, Haw [39] found a relation between relatively low intent to die and alcohol abuse in males (35.2% men vs. 20.2% women, χ^2=14.6, p<0.0001). The authors concluded that alcohol may exert a disinhibiting effect and could thereby facilitate suicidal acts in males with relatively low suicidal intent.

It is estimated that individuals with an alcohol use disorder have 60–120 times greater suicide risk than those without any psychiatric disorder [38]. A study comparing 636 individuals who attempted or completed suicide found that completers were five times more likely to be alcohol abusers (OR=5.5; 95% CI=2.3–14.2. This impact of alcohol was relatively subtle in women (7.5% attempters, 11.1% completers p= ns), and only reached statistical significance among men (17.0% attempters, 30.7% completers p=0.007) [40].

Depression

While depression is a major risk factor for suicide in both genders, it has been found to occur more frequently in female suicide [3]. However the real impact of depressive disorders in male suicide could be underestimated. It has been reported that major depressive disorder imposes a greater suicide risk on men than on women, with lifetime risks of 7% and 1%, respectively [30]. This gender dependence is even greater among young males, where the male to female ratio was 10:1 for youths under 25 and 5.6:1 for adults. While it is true that depressive disorders are more frequently observed in women [30], this may be due to an under diagnosis of male depression. Numerous studies have found that depression often differs between men and women in prevalence, symptomatology, course, comorbidity and treatment response [33, 41, 41-43]. Rutz [44] has even proposed the existence of a specific male depression syndrome. The symptoms that describe this syndrome include several suicide risk factors: higher aggressiveness, low impulse control, antisocial behavior, impulsivity, mood instability during depression, and family history of suicide or alcoholism [45, 46].

Personality Disorders

Numerous studies have examined the relation between personality disorders and suicide. Repeat suicide attempters in particular show higher prevalence of personality disorders, especially borderline personality disorder (OR 4.37, IC 95%: 1.33-14.29) [47]. Among suicide attempters, comorbid Axis I and personality disorders pose an estimated six-times greater suicide risk compared to Axis I disorders alone [48].

As Arsenault et al., reported [3], personality disorders are more commonly seen in male suicide. When compared with suicide attempters, suicide completers have been found to more frequently meet two characteristics: being male (OR=9.6; 95% CI=4.42-20.9) and presenting narcissistic personality disorder (OR: 21.4; 95% CI: 6.8–67.7) [40]. Narcissistic personality disorder seems to be a unique risk factor compared to other cluster B personality disorders. Although suicide attempters diagnosed with narcissistic personality disorder do not differ from attempters without a cluster B personality disorder on impulsivity measurements (using the Barratt Impulsiveness Scale; $t = -0.33$, $df = 442$, $p = 0.738$), they do make suicide attempts characterized by higher lethality compared to those with other cluster B personality disorders (using the Beck Suicide Intent Scale $t = -4.24$, $df = 439$, $p<0.001$) [49].

Additionally, antisocial personality disorder is more frequently diagnosed in men and among substance abusers [50-53]. In a study comparing 158 alcohol-related suicides with non-alcohol-related suicides, alcohol-related suicides presented higher prevalence of antisocial personality disorder (OR=4.68) [54].

PSYCHOLOGICAL DIMENSIONS

While much has been written about factors that could explain the gender gap, few studies have focused on psychological characteristics which could mediate gender differences in suicide risk.

Aggression and Impulsivity

Aggression and impulsivity are linked to suicidal vulnerability across diagnoses [55, 56]. Mann et al., proposed a stress–diathesis model in which suicide results from the combination of constitutional predisposition (such as a tendency towards aggression or impulsivity), environmental stressors (such as an interpersonal loss) and a psychiatric illness (such as a mood disorder) [57]. Remarkably, studies have reported suicide to be more closely linked to traits like aggression and impulsivity than to severity of depression or psychosis [57].

Differences between men and women regarding aggressive behavior are well recognized and have been documented in several reviews [58-60]. The gender difference in physical aggression appears in children as young as 1 to 2 years of age and remains consistent into adulthood [61]. Several studies reporting higher levels of aggression among males [56, 62] have also seen a strong relationship between the level of aggression and the lethality of suicide attempts. Baca et al., reported that the lethality of attempts committed by depressed subjects was correlated to their level of lifetime aggressive behavior (measured by the Brown-Goodwin Scale) [63].

Male overrepresentation in socially problematic and criminal behaviors has been linked to impulsivity. Cross et al., [64] published a meta-analysis examining gender differences in impulsivity and distinguished between reward hypersensitivity, punishment hyposensitivity and inadequate effortful control. They found that men showed greater sensation seeking on questionnaire measures and behavioral risk-taking tasks than women. However, there was not a significant difference on measures of deficits in effortful control between the sexes. The results showed a more important gender difference in motivational rather than effortful or executive forms of behavior control.

Impulsivity itself is closely associated with suicide, including both attempt and completion [65-68]. Specifically, it has been suggested that impulsivity could facilitate the transition from suicidal thought to suicidal acts [69].

It is possible that these psychological risk factors are mediated by neurochemical abnormalities. Several serotonergic abnormalities have been associated with suicidal behavior: lower levels of serotonin metabolite 5-hydroxyindoleacetic acid (5-HIAA) in cerebrospinal fluid, decreased serotonin and/or 5-HIAA in brainstem serotonin nuclei, and altered serotonin receptor and transporter binding [70, 71]. These alterations can increase suicide risk in three ways: promoting impulsive or disinhibited behavior, impairing decision-making, and causing mood dysregulation or depression [72]. Moreover, their effects seem to be independent of psychiatric diagnosis; lower cerebrospinal fluid (5-HIAA) has been found in suicide attempters with major depression, schizophrenia, and personality disorders, but not in non-suicidal subjects with the same psychiatric diagnoses [38, 73]. Several studies have reported gender differences in the serotonin system of suicidal and non-suicidal subjects [74,

75], although the results are still inconsistent. When Anisman et al., investigated 5-HT receptor activity, they found that 5-HT_{2A} and 5-HT_{1A} mRNA expression was elevated among suicidal male subjects relative to control subjects, but not among female subjects [76]. It seems that impulsive–aggressive traits could be part of a developmental cascade that increases suicide risk among a certain subset (mainly suicides among young people and suicide characterized by particularly impulsive–aggressive features) [77]. This subset of suicides could be the result of a dysfunctional developmental cascade initiated early in life (with environmental stressors, especially childhood abuse or neglect) and mediated by excessive impulsive–aggressive behaviors.

Socioeconomic Features and Stressful Events

Psychosocial stressors such as abuse, divorce, separation from a partner, financial difficulties or gender-identity conflicts are associated with an increase in suicide risk [29]. These risk factors affect suicidality differently in males and females, and could explain part of the gender paradox.

Sexual Abuse

A history of childhood sexual abuse is one of the most important suicide risk factors. Childhood abuse (sexual, emotional, or physical) together with post-traumatic stress disorder is associated with increased suicidality: individuals with these histories present with more frequent and serious attempts along with higher levels of suicidal intent than those who have not suffered abuse [78]. The association between childhood sexual abuse and suicide is stronger in men than in women [79]. Martin et al., [80] reported that sexually abused boys more frequently presented with a history of suicide attempts than did abused girls (55% Vs 29%). The authors suggest this could be due to a more severe male response to sexual abuse. In addition, the impact of sexual abuse history could be even higher than recognized since it has been reported that males may be more reluctant to report these events [81].

Work-Related Factors

Occupational activities can be considered a source of personal satisfaction, but they can also be associated with psychological distress. Certain work-related features are constantly associated with suicide risk, and their influence might be different among men and women. Several studies suggest that unemployment is associated with an increase in suicide rates [82, 83]. Stuckler et al., reported that every 1% increase in unemployment was associated with a 0.79% rise in suicide rates at younger ages (both in males and females). This influence is different among gender and age groups [11, 83]. Young (working age) men are especially sensitive, while old-age suicide mortality (above 65years old) responds much less to unemployment [83]. This could be explained by male identity being highly connected to profession, which leads to great suffering in the context of unemployment. It is also well known that some professions carry a greater risk of suicide [84]. Military suicide has become a frequent issue in scientific publications. It is estimated that active duty suicides increased from a rate lower than among civilians to one exceeding that of civilians in 2008, and male veterans are believed to present twice the risk of dying from suicide as their civilian counterparts [85]. In a study designed to analyze the risk factors predicting suicide in the 12 months after US Army inpatient treatment for a psychiatric disorder (53,769 hospitalizations of active duty soldiers), male sex was the single strongest predictor (OR: 7.9; 95% CI: 1.9-32.6) [86].

Widowhood

The lower mortality of married people (versus unmarried) is a consistent empirical finding [87]. Therefore, it is not surprising that after the end of a marriage due to the death of one spouse, mortality of the other could increase, even by suicide. The suicide risk of widowed persons increases after bereavement, especially during the first year [88]. This effect is dramatically more important in males as seen in the relative suicide risks during the first year after loss of a partner: men 15.5 (CI 95%: 10.2–23.6), women 7.9 (CI 95%: 3.5–17.8) [89-92]. Men seem to need more time than women to recover [88], while women are more likely to preserve social and familial relations that they can turn to for support after losing their marriage partner. In fact, some studies did not observe any increased risk for suicide after widowhood among women [93].

Divorce or Separation

Studies on the effect of divorce and separation on suicidal behavior indicate that relationship breakdown constitutes a major risk factor for both suicidal ideation and completion [94]. Individual and ecological studies have reported a higher suicide risk in divorced individuals and in countries with higher divorce rates [95, 96]. Once again, separated males are at a particularly increased risk of developing suicidal behaviors during the separation process [94, 97]. Several psychosocial risk factors for suicide in separated males have been described including lower education and separation-related shame and stress from legal negotiations, particularly regarding property and finances [98].

Old Age

In some regions suicide rates reflect the Hungarian pattern, characterized by an increasing number of suicides among the elderly, especially males [99; 100]. This is crucial, since the decreasing physical health in older people can lead to death more probably when committing a suicide attempt [101].

It is also worth noting the difficulty men can face upon retirement, typically at age 65. Since men remain the primary economic providers in most households, the moment of transition from working life to retirement can be particularly difficult due to the loss of interpersonal contact and self-esteem [29].

Table III. Most frequent suicide method (%) according to the WHO mortality database (103)

Men							Women						
Otherpoisoning	Pesticides	Hanging	Drowning	Firearms	Falls	Other	Otherpoisoning	Pesticides	Hanging	Drowning	Firearms	Falls	Other
6.6	3.6	68.7	0.0	12.6	0.2	8.2	22.7	12.6	41.2	0.8	9.2	0.8	12.6
0.7	1.7	49.1	1.5	37.6	2.4	7.0	3.4	4.1	38.0	4.2	25.9	10.3	14.1
2.0	8.3	52.4	0.9	22.1	1.8	12.6	6.5	16.0	37.6	2.3	13.4	3.9	20.4
10.2	0.4	44.4	2.3	21.6	4.7	16.3	34.3	05	36.8	4.0	3.8	6.5	14.1
7.1	0.3	20.4	0.9	60.6	1.9	8.8	31.0	0.5	16.9	2.1	35.7	3.4	10.5
2.5	1.9	42.0	0.7	25.4	10.3	17.2	8.9	2.9	31.1	2.1	9.1	21.9	24.0
1.3	2.5	68.7	2.6	0.2	8.1	16.5	2.9	4.3	59.9	7.8	0.0	12.5	12.7
0.4	37.5	39.2	3.2	0.4	9.5	9.8	0.8	42.8	26.0	3.8	0.1	18.5	8.1
8.0	1.1	45.4	1.3	11.5	3.6	29.1	26.5	0.7	36.4	3.9	2.6	4.6	25.3
6.4	1.0	48.4	1.9	11.2	2.5	28.6	19.7	0.4	42.5	4.4	2.2	6.4	24.3
8.6	1.0	48.9	3.9	22.1	4.9	10.6	26.3	2.0	29.2	12.4	4.1	12.4	13.5
8.0	1.3	55.5	2.1	10.3	7.4	15.5	22.0	2.0	38.9	7.2	1.4	14.1	14.4
7.0	4.6	70.3	1.4	4.0	4.9	7.8	28.1	7.0	43.4	4.5	0.6	9.9	6.5
11.7	1.4	47.9	6.6	4.4	7.8	20.2	24.0	1.8	33.6	11.0	0.6	10.7	18.3
3.5	2.6	52.7	3.9	7.1	18.4	11.8	8.3	5.4	29.4	7.6	0.9	36.9	11.5

FEATURES OF MALE SUICIDE

Method of Suicide

The method employed to commit suicide depends on variables such as availability, age, geographical location, historical moment and cultural acceptance of certain methods. There is reason to believe it may also depend on gender [102].

There is clear variation by country in the methods employed for completed suicide. Table III summarizes the most common suicide methods (% of total suicides by country) used by men and women in different areas.

Proportions of the different methods vary not only from one country to another, but also within the same country between males and females.

Although worldwide men are more likely to commit suicide than women, differences between countries remain. While in European countries the ratio is 3-4:1, in Asian countries it is less than 2:1 [104].

These differences could be influenced by the methods of suicide employed. A recent study compared the suicide methods and gender ratios of Eastern (Korea and Japan) and Western countries (USA and Canada). It found that the higher gender ratio in Western countries (United States 3.9 or Australia 3.3 vs Korea 1.8 or Japan 2.7) may correlate with a greater influence of gender on the choice of suicide method [105].

In fact, these differences in suicide method offer a potential explanation for the large gap in suicide completion rates between males and females [102]. It is widely accepted that while women typically use less aggressive such as poisoning, men tend to use much more immediately lethal methods. High risk methods are more violent and irreversible, reducing the chance of either surviving or aborting the attempt [106].

Homicide-suicide (also called murder-suicide or extended suicide) is the murder of one or more individuals followed by the suicide of the perpetrator [107]. Although the rates of homicide-suicide are relatively low, it bears mentioning that the perpetrator is almost always male, while most victims are female (especially intimate partner or family members). Furthermore they frequently present risk factors related to completed suicide: males, history of mental disorder, previous suicide attempts, influence of alcohol, recent relation breakdown, history of violence, unemployment, and firearms accessibility.

Previous Suicide Attempts

Previous suicide attempts are frequently associated with completed suicide [10].

However 56% of suicide victims die from their first suicidal act. As would be expected with more aggressive methods, this is much more frequent among males than females (62% and 38% respectively) [108]. Some studies report gender differences in the proportion of previous attempters among suicide completers, specifically that men were less likely to have made previous attempts (20% of men and 43.5% of women) [15]. Some authors also suggest that the presence of previous suicide attempts can be considered a stronger predictor of completed suicide among women than among men [109]. While these gender differences could be explained predominantly by method choice, other factors such as level of aggression and suicidal intent must be considered.

Level of Suicidal Intent

It is important to understand the patient's wish to die in order to determinate the severity of the suicide attempt. This is termed suicidal intent. It is widely accepted that higher suicidal intent is a strong indicator of determination to perform a completed suicide. One instrument to measure this factor is the Beck Suicide Intent Scale (SIS), which assesses the degree of intent to die and provides an indication of subjects who might be at increased risk of suicide [39]. Our group has found that completed suicide cases obtained higher scores on the SIS compared to attempted cases (13.97 vs. 21.76, $p<0.001$) [110]. Sun and Jia observed that completed suicide with higher SIS scores tend to be more violent [111]. Furthermore, it has been found that higher levels of suicidal intent during an attempt increase the likelihood of early mortality, both by suicide and other causes [112].

Considering these data together, the evidence suggests that the level of suicide intentionality identifies the severity of the suicidal act. Several studies have investigated the level of suicidal intent in attempters, finding males obtain higher total SIS scores than females [39, 113, 114]. This implies that the gender gap may be partially explained by a greater intention to die in male suicidal behavior than female.

TREATMENT SEEKING AND STIGMA

Stigma is defined asa distinguishing mark of social disapproval. Hence, social stigma would be the rejection of individuals due to features that distinguish them from others members of society. This rejection can lead to discrimination and prejudice toward the person. When the rejection is of oneself, due to acceptance of the prejudiced perceptions held by the others, we consider this self-stigma.

Both social stigma and self-stigma are closely associated with mental illness. They cause significant suffering (due to low self-esteem, embarrassment, depression) and can limit help-seeking. Diala et al., observed that those with negative beliefs toward mental health treatment were at least five times less likely to seek help [115].

This has been described as another source of self-stigma. When a person sees himself as in need of help, this can cause a different stigma, the self-stigma of seeking help [116]. Men are pushed by cultural expectations to be more stoic, self-reliant and unwilling to seek help [117]. A heightened sense of masculinity is associated with the rejection of help-seeking, partly mediated by self-stigma [117]. This phenomenon has also been observed in suicide. Younes found that male suicide completers had less recent contact with their PCPs [15], which highlights male rejection of help-seeking and the difficulty of reaching suicidal men.

Several studies have shown that men are less likely to seek help in a crisis situation, dubbed "help negation": a negative association between suicidal ideation and help-seeking [118]. This circumstance has been observed in completed suicides, attempted suicide [119], and even prior to engaging in suicidal behaviors when only suicidal ideation is present [120].

In fact, several suicide prevention studies have reported that attempts at intervention decreased suicide rates only for women [121]. Examples of this are 1) the Gotland study in Sweden that was based on systematic education and training of primary care providers in the detection and treatment of depression [122]; 2) the Tele Help-Tele Check Service in Italy, which provided supportive services to at-risk elders [123]; and 3) the program implemented in the city of Matsudai, Japan, that was based on depression screening and health education through primary care and public health nursing [124; 125]. None of these intervention attempts achieved significant improvement in suicide rates among males. In the Japanese study, for instance, the female risk of completing suicide was reduced by 70% (age-adjusted IRR: 0.30; 95% CI: 0.14-0.67), but there was no change in the risk for males [125].

DISCUSSION

Men commit suicide between three to four times as often as women. However for suicide attempts an opposite ratio is seen, with women making many more attempts. This gender difference for completed and attempted suicides is stable for centuries across the majority of the countries. Many studies have been done to explain why gender is such a significant predictor for suicide, but an explanation of the gender ratio remains unresolved. The current overview aimed to summarize the factors that could contribute to overrepresentation of males in completed suicide and refine the profile of male suicide completers.

How can the influence of gender explain this phenomenon? Gender definition is complex and must not be reduced to the biological differences between both sexes. The World Health Organization defines it as the "socially constructed roles, behaviors, activities, and attributes that a given society considers appropriate for men and women." Gender roles specify what is expected of men and women. Although these roles are constantly changing, they retain a significant influence on the gender differences observed in risk factors, prevalence of mental disorders and help-seeking behavior.

Adherence to traditional masculinity increases suicide risk by negatively influencing mental state, social support, help-seeking behaviors, and the acceptability of suicide [126]. Furthermore this negative influence extends beyond suicide or mental disorders. It has been stated that men exhibit far poorer health practices than women, including almost every health risk behavior: substance use, lack of medical screening and awareness, failure to take appropriate safety precautions, etc. These behaviors seem to derive largely from traditional expectations of masculinity, and together they substantially increase male morbidity and mortality [127]. This seems to indicate that a more substantial disparity in perceived gender roles would correlate with an increased gender gap in suicidal behaviors [17].

Male rates for completed suicide at first attempt are much higher than those of females. This gender difference may relate to a greater intent to die on the part of male suicide attempters. One theory suggests that women are more likely to employ suicidal behaviors as a means of expressing discomfort and the need for help [128]. Certainly it is frequently observed that suicidal men score higher on suicide intent scales compared to suicidal women.

The higher male completion rate on first attempt can also be explained by gender differences in the choice of suicide method. Several arguments have been formulated to explain the source of this difference in method choice. Some authors suggest that the answer lies in different accessibility to more lethal methods [129], such as guns. Undoubtedly the availability of a suicide method is a predictor of suicide attempts [130] and women are less likely to own or to be familiar with firearms. However that explanation on its own seems inadequate since it cannot be applied in countries with restricted access to guns where the gender difference remains. Nevertheless, this could offer an interesting explanation for the use of drugs by women to commit suicide: depression is more frequently diagnosed in women, making them more likely to be treated with medication, and several studies have found that being on psychiatric treatment increases the risk of committing suicide with these medications [131]. Another explanation suggests that women prefer to use methods that cause less physical disfigurement [132]. It has been observed than men more often opt to shoot themselves in the head, which may explain the higher suicide completion rate; this is a higher lethality form of the same suicide method. Similar reasoning has been used to explain the higher lethality of suicide attempts by rural Chinese women compared to Western women. Both groups frequently use poisoning, but while Chinese women use pesticides, women from the US and Europe employ the much less lethal method of drug overdose.

All told, it can be reasonably concluded that while higher suicide completion rates in males are influenced by the choice of more lethal methods, this choice is not the only cause of the gender gap and other factors have to be considered [101]. Psychological and psychiatric differences between men and women could be important to understand the gender paradox. We have seen that males are frequently associated with traits that might predispose to suicide, mainly impulsivity and aggressiveness. Differences in diagnosis have also been reported, indicating that substance-related disorders, personality disorders and childhood disorders are significantly more common in male suicides. This suggests that the psychopathology underlying suicide may be different in males and females. All mental illnesses are linked to both social and self-stigma, and men might be particularly unable to seek assistance since they often regard help-seeking as unsuitable to males. More than 90% of suicides occur in the context of a psychiatric disorder (1), and the lack of treatment could amplify the intent to die [133]. This is particularly true for depression, a well-known risk factor for suicide. The importance of this problem is further evident when we consider that psychotherapy and medication can be highly effective in the treatment of mental disorders such as depression [134]. Whatever the reason, the fact remains that depressed or suicidal males are more reluctant than females to seek professional help. This reluctance to seek assistance can have tragic consequences. It is not uncommon for the first evidence of male depression to be the completed suicide itself. Equivalently suicidal women may be spared this fate either by directly seeking treatment or by making an uncompleted suicide attempt; this makes their symptoms known, allowing them to be connected with appropriate mental health care while their male counterparts go untreated. Fortunately, emerging evidence shows that improved information about depression is associated with a reduction of the stigma [117]. Therefore we have it within our power to fight the stigma and assist the process of treatment-seeking.

CONCLUSION

The higher rates of completed suicide among males could be due to the confluence of psychological characteristics which together lead to suicidality, reluctance to seek help, greater intent to die and more violent suicide methods. In any case, there is a clear need for studies directly comparing the specific suicide risk factors affecting men and women. By better understanding these differences, preventative programs could be specifically targeted according to the most relevant risk factors.

References

[1] Hawton K, van HK. Suicide. Lancet 2009 Apr 18;373(9672):1372-81.

[2] Sokolowski M, Wasserman J, Wasserman D. Genome-wide association studies of suicidal behaviors: A review. *Eur. Neuropsychopharmacol.* 2014 Oct;24(10):1567-77.

[3] Arsenault-Lapierre G, Kim CD, Turecki G. Psychiatric diagnoses in 3275 suicide: a meta-analysis. *BMC Psychiatry* 2004 Nov 4;4(1):37.

[4] MacLean J, Kinley DJ, Jacobi F, Bolton JM, Sareen J. The relationship between physical conditions and suicidal behavior among those with mood disorders. *J. Affect. Disord.* 2010 Nov 13.

[5] Scott KM, Hwang I, Chiu WT, Kessler RC, Sampson NA, Angermeyer M, et al. Chronic physical conditions and their association with first onset of suicidal behavior in the world mental health surveys. *Psychosom. Med.* 2010 Sep;72(7):712-9.

[6] Lau R, Morse CA, Macfarlane S. Psychological factors among elderly women with suicidal intentions or attempts to suicide: a controlled comparison. *J. Women Aging* 2010 Jan;22(1):3-14.

[7] Roberts RE, Roberts CR, Xing Y. One-year incidence of suicide attempts and associated risk and protective factors among adolescents. *Arch. Suicide Res.* 2010;14(1):66-78.

[8] Shelef A, Hiss J, Cherkashin G, Berger U, Aizenberg D, Baruch Y, et al. Psychosocial and medical aspects of older suicide completers in Israel: a 10-year survey. *Int. J. Geriatr. Psychiatry* 2014 Aug;29(8):846-51.

[9] Ministerio de Sanidad, Política Social e Igualdad. Guía de Práctica Clínica de Prevención y Tratamiento de la Conducta Suicida. 2012.

[10] Wasserman D, Rihmer Z, Rujescu D, Sarchiapone M, Sokolowski M, Titelman D, et al. [The European Psychiatric Association (EPA) guidance on suicide treatment and prevention]. *Neuropsychopharmacol* Hung 2012 Jun;14(2):113-36.

[11] Cordoba-Dona JA, San SM, Escolar-Pujolar A, Martinez-Faure JE, Gustafsson PE. Economic crisis and suicidal behaviour: the role of unemployment, sex and age in Andalusia, southern Spain. *Int. J. Equity* Health 2014;13:55.

[12] Barth A, Sogner L, Gnambs T, Kundi M, Reiner A, Winker R. Socioeconomic factors and suicide: an analysis of 18 industrialized countries for the years 1983 through 2007. *J. Occup. Environ. Med.* 2011 Mar;53(3):313-7.

[13] Moller-Leimkuhler AM. The gender gap in suicide and premature death or: why are men so vulnerable? *Eur. Arch. Psychiatry Clin. Neurosci.* 2003 Feb;253(1):1-8.

[14] Posner K, Oquendo MA, Gould M, Stanley B, Davies M. Columbia Classification Algorithm of Suicide Assessment (C-CASA): classification of suicidal events in the FDA's pediatric suicidal risk analysis of antidepressants. *Am. J. Psychiatry* 2007 Jul;164(7):1035-43.

[15] Younes N, Melchior M, Turbelin C, Blanchon T, Hanslik T, Chan CC. Attempted and completed suicide in primary care: Not what we expected? *J. Affect. Disord.* 2014 Sep 1;170C:150-4.

[16] Parra U, I, Blasco-Fontecilla H, Garcia-Pares G, Giro BM, Llorens CM, Cebria MA, et al. Attempted and completed suicide: not what we expected? *J. Affect. Disord.* 2013 Sep 25;150(3):840-6.

[17] Canetto SS, Sakinofsky I. The gender paradox in suicide. *Suicide Life Threat Behav* 1998;28(1):1-23.

[18] Li Y, Dai Q, Torres ME, Zhang J. Gender-specific association between iron status and the history of attempted suicide: implications for gender paradox of suicide behaviors. *Prog. Neuropsychopharmacol. Biol. Psychiatry* 2007 Oct 1;31(7):1429-35.

[19] Lemieux AM, Saman DM, Lutfiyya MN. Men and suicide in primary care. *Dis. Mon.* 2014 Apr;60(4):155-61.

[20] Cheng AT, Lee SC. Suicide in Asia and the far East. In: Hawton K vHK, editor. Suicide and attempted suicide.Baffins Lane, Chichester (England): John Wiley & Sons Ltd; 2000. p. 29-48.

[21] Aaron R, Joseph A, Abraham S, Muliyil J, George K, Prasad J, et al. Suicides in young people in rural southern India. *Lancet* 2004 Apr 3;363(9415):1117-8.

[22] Centre for disease control and prevention. Nonfatal Injury Reports, 2001 - 2013. [cited 2014 Dec 10]. Available from: http://webappa.cdc.gov/sasweb/ncipc/nfirates2001.html.

[23] Tanya Navaneelan. Suicide rates: An overview.[cited 2014 Dec 10].Available from:http://www.statcan.gc.ca/pub/82-624-x/2012001/article/11696-eng.htm.

[24] National Media Initiative. Facts and stats about suicide in Australia. [cited 2014 Dec 10]. Available from: http://www.mindframe-media.info/for-media/reporting-suicide/facts-and-stats.

[25] Yamamura T, Kinoshita H, Nishiguchi M, Hishida S. A perspective in epidemiology of suicide in Japan. *Vojnosanit Pregl.* 2006 Jun;63(6):575-83.

[26] McGirr A, Seguin M, Renaud J, Benkelfat C, Alda M, Turecki G. Gender and risk factors for suicide: evidence for heterogeneity in predisposing mechanisms in a psychological autopsy study. *J. Clin. Psychiatry* 2006 Oct;67(10):1612-7.

[27] Kessler RC. Epidemiology of women and depression. J. Affect. Disord. 2003 Mar;74(1):5-13.

[28] Yoshimasu K, Kiyohara C, Miyashita K. Suicidal risk factors and completed suicide: meta-analyses based on psychological autopsy studies. *Environ. Health Prev. Med.* 2008 Sep;13(5):243-56.

[29] Schrijvers DL, Bollen J, Sabbe BG. The gender paradox in suicidal behavior and its impact on the suicidal process. *J. Affect. Disord.* 2012 Apr;138(1-2):19-26.

[30] Blair-West GW, Cantor CH, Mellsop GW, Eyeson-Annan ML. Lifetime suicide risk in major depression: sex and age determinants. *J. Affect. Disord.* 1999 Oct;55(2-3):171-8.

[31] Luoma JB, Martin CE, Pearson JL. Contact with mental health and primary care providers before suicide: a review of the evidence. *Am. J. Psychiatry 2002* Jun;159(6):909-16.

[32] Vannoy SD, Robins LS. Suicide-related discussions with depressed primary care patients in the USA: gender and quality gaps. A mixed methods analysis. *BMJ Open* 2011 Jan 1;1(2):e000198.

[33] Khan A, Brodhead AE, Schwartz KA, Kolts RL, Brown WA. Sex differences in antidepressant response in recent antidepressant clinical trials. *J. Clin. Psychopharmacol* 2005 Aug;25(4):318-24.

[34] Khan A, Schwartz KA, Kolts RL, Brown WA. BMI, sex, and antidepressant response. *J. Affect. Disord.* 2007 Apr;99(1-3):101-6.

[35] Kornstein SG, Schatzberg AF, Thase ME, Yonkers KA, McCullough JP, Keitner GI, et al. Gender differences in treatment response to sertraline versus imipramine in chronic depression. *Am. J. Psychiatry* 2000 Sep;157(9):1445-52.

[36] WHO Expert Committee on Problems Related to Alcohol Consumption. Second report. *World Health Organ. Tech. Rep. Ser.* 2007;(944):1-7, back.

[37] Rehm J, Anderson P, Barry J, Dimitrov P, Elekes Z, Feijao F, et al. Prevalence of and potential influencing factors for alcohol dependence in europe. *Eur. Addict. Res.* 2015;21(1):6-18.

[38] Sher L. Alcohol and suicide: neurobiological and clinical aspects. *ScientificWorld Journal* 2006;6:700-6.

[39] Haw C, Hawton K. Life problems and deliberate self-harm: Associations with gender, age, suicidal intent and psychiatric and personality disorder. *J. Affect Disord.* 2008 Jul;109(1-2):139-48.

[40] Giner L, Blasco-Fontecilla H, Mercedes Perez-Rodriguez M, Garcia-Nieto R, Giner J, Guija JA, et al. Personality disorders and health problems distinguish suicide attempters from completers in a direct comparison. *J. Affect. Disord.* 2013 Nov;151(2):474-83.

[41] Williams JB, Spitzer RL, Linzer M, Kroenke K, Hahn SR, deGruy FV, et al. Gender differences in depression in primary care. *Am. J. Obstet. Gynecol.* 1995 Aug;173(2):654-9.

[42] Kessler RC, McGonagle KA, Swartz M, Blazer DG, Nelson CB. Sex and depression in the National Comorbidity Survey. I: Lifetime prevalence, chronicity and recurrence. *J. Affect. Disord.* 1993 Oct;29(2-3):85-96.

[43] Kornstein SG, Schatzberg AF, Thase ME, Yonkers KA, McCullough JP, Keitner GI, et al. Gender differences in chronic major and double depression. *J. Affect. Disord.* 2000 Oct;60(1):1-11.

[44] Rutz W. Improvement of care for people suffering from depression: the need for comprehensive education. *Int. Clin. Psychopharmacol.* 1999 Jun;14 Suppl 3:S27-S33.

[45] Rutz W. Preventing suicide and premature death by education and treatment. *J. Affect. Disord.* 2001 Jan;62(1-2):123-9.

[46] Azorin JM, Belzeaux R, Fakra E, Kaladjian A, Hantouche E, Lancrenon S, et al. Gender differences in a cohort of major depressive patients: further evidence for the male depression syndrome hypothesis. *J. Affect. Disord.* 2014 Oct;167:85-92.

[47] Dirks BL. Repetition of parasuicide--ICD-10 personality disorders and adversity. *Acta. Psychiatr. Scand.* 1998 Sep;98(3):208-13.

[48] Foster T, Gillespie K, McClelland R. Mental disorders and suicide in Northern Ireland. *Br. J. Psychiatry* 1997 May;170:447-52.

[49] Blasco-Fontecilla H, Baca-Garcia E, Dervic K, Perez-Rodriguez MM, Lopez-Castroman J, Saiz-Ruiz J, et al. Specific features of suicidal behavior in patients with narcissistic personality disorder. *J. Clin. Psychiatry* 2009 Nov;70(11):1583-7.

[50] Khan S, Okuda M, Hasin DS, Secades-Villa R, Keyes K, Lin KH, et al. Gender differences in lifetime alcohol dependence: results from the national epidemiologic

survey on alcohol and related conditions. *Alcohol. Clin. Exp. Res.* 2013 Oct;37(10):1696-705.

[51] Yang M, Mamy J, Zhou L, Liao YH, Wang Q, Seewoobudul V, et al. Gender differences in prevalence and correlates of antisocial personality disorder among heroin dependent users in compulsory isolation treatment in China. *Addict. Behav.* 2014 Mar;39(3):573-9.

[52] Helzer JE, Pryzbeck TR. The co-occurrence of alcoholism with other psychiatric disorders in the general population and its impact on treatment. *J. Stud. Alcohol.* 1988 May;49(3):219-24.

[53] Regier DA, Farmer ME, Rae DS, Locke BZ, Keith SJ, Judd LL, et al. Comorbidity of mental disorders with alcohol and other drug abuse. Results from the Epidemiologic Catchment Area (ECA) Study. *JAMA* 1990 Nov 21;264(19):2511-8.

[54] Chachamovich E, Ding Y, Turecki G. Levels of aggressiveness are higher among alcohol-related suicides: results from a psychological autopsy study. *Alcohol* 2012 Sep;46(6):529-36.

[55] Soloff P, White R, Diwadkar VA. Impulsivity, aggression and brain structure in high and low lethality suicide attempters with borderline personality disorder. *Psychiatry Res.* 2014 Jun 30;222(3):131-9.

[56] Carballo JJ, Garcia-Nieto R, Harkavy-Friedman J, de Leon-Martinez V, Baca-Garcia E. Aggressiveness across development and suicidal behavior in depressed patients. Arch. Suicide Res. 2014;18(1):39-49.

[57] Mann JJ, Waternaux C, Haas GL, Malone KM. Towards a clinical model of suicidal behavior in psychiatric patients. *Am. J. Psychiatry* 1999;156:181-9.

[58] Knight GP, Fabes RA, Higgins DA. Concerns about drawing causal inferences from meta-analyses: an example in the study of gender differences in aggression. *Psychol. Bull* 1996 May;119(3):410-21.

[59] Bettencourt BA, Miller N. Gender differences in aggression as a function of provocation: a meta-analysis. *Psychol. Bull* 1996 May;119(3):422-47.

[60] Archer J. Sex differences in aggression in real-world settings: a meta-analytic review. *Review of General Psychology* 2004;8(4):291-322.

[61] Potegal M, Archer J. Sex differences in childhood anger and aggression. *Child. Adolesc. Psychiatr. Clin. N. Am.* 2004 Jul;13(3):513-vii.

[62] Dalca IM, McGirr A, Renaud J, Turecki G. Gender-specific suicide risk factors: a case-control study of individuals with major depressive disorder. *J. Clin. Psychiatry* 2013 Dec;74(12):1209-16.

[63] Baca-Garcia E, Oquendo MA, Saiz-Ruiz J, Mann JJ, de LJ. A pilot study on differences in aggression in New York City and Madrid, Spain, and their possible impact on suicidal behavior. *J. Clin. Psychiatry* 2006 Mar;67(3):375-80.

[64] Cross CP, Copping LT, Campbell A. Sex differences in impulsivity: a meta-analysis. *Psychol. Bull* 2011 Jan;137(1):97-130.

[65] Soloff PH, Lynch KG, Kelly TM, Malone KM, Mann JJ. Characteristics of suicide attempts of patients with major depressive episode and borderline personality disorder: a comparative study. *Am. J. Psychiatry* 2000 Apr;157(4):601-8.

[66] Fazaa N, Page S. Personality style and impulsivity as determinants of suicidal subgroups. *Arch. Suicide Res.* 2009;13(1):31-45.

[67] Brent DA, Oquendo M, Birmaher B, Greenhill L, Kolko D, Stanley B, et al. Familial transmission of mood disorders: convergence and divergence with transmission of suicidal behavior. *J. Am. Acad. Child. Adolesc. Psychiatry* 2004 Oct;43(10):1259-66.
[68] Brent DA, Bridge J, Johnson BA, Connolly J. Suicidal behavior runs in families. A controlled family study of adolescent suicide victims. *Arch. Gen. Psychiatry* 1996 Dec;53(12):1145-52.
[69] Klonsky ED, May A. Rethinking impulsivity in suicide. Suicide Life Threat Behav. 2010 Dec;40(6):612-9.
[70] Mann JJ. Neurobiology of suicidal behaviour. *Nat. Rev. Neurosci* 2003 Oct;4(10):819-28.
[71] Oquendo MA, Mann JJ. The biology of impulsivity and suicidality. Psychiatr. *Clin. North Am.* 2000 Mar;23(1):11-25.
[72] Mann JJ, Currier DM. Stress, genetics and epigenetic effects on the neurobiology of suicidal behavior and depression. *Eur. Psychiatry* 2010 May 5.
[73] Placidi GP, Oquendo MA, Malone KM, Huang YY, Ellis SP, Mann JJ. Aggressivity, suicide attempts, and depression: relationship to cerebrospinal fluid monoamine metabolite levels. *Biol. Psychiatry* 2001 Nov 15;50(10):783-91.
[74] Roggenbach J, Muller-Oerlinghausen B, Franke L, Uebelhack R, Blank S, Ahrens B. Peripheral serotonergic markers in acutely suicidal patients. 1. Comparison of serotonergic platelet measures between suicidal individuals, nonsuicidal patients with major depression and healthy subjects. *J. Neural. Transm* 2007;114(4):479-87.
[75] Pivac N, Muck-Seler D, Barisic I, Jakovljevic M, Puretic Z. Platelet serotonin concentration in dialysis patients with somatic symptoms of depression. *Life Sci.* 2001 Apr 13;68(21):2423-33.
[76] Anisman H, Du L, Palkovits M, Faludi G, Kovacs GG, Szontagh-Kishazi P, et al. Serotonin receptor subtype and p11 mRNA expression in stress-relevant brain regions of suicide and control subjects. *J. Psychiatry Neurosci.* 2008 Mar;33(2):131-41.
[77] Turecki G. Dissecting the suicide phenotype: the role of impulsive-aggressive behaviours. *J. Psychiatry Neurosci.* 2005 Nov;30(6):398-408.
[78] Lopez-Castroman J, Jaussent I, Beziat S, Guillaume S, Baca-Garcia E, Olie E, et al. Posttraumatic Stress Disorder following childhood abuse increases the severity of suicide attempts. *J. Affect. Disord.* 2014 Aug 15;170C:7-14.
[79] Rhodes AE, Boyle MH, Tonmyr L, Wekerle C, Goodman D, Leslie B, et al. Sex differences in childhood sexual abuse and suicide-related behaviors. *Suicide Life Threat Behav.* 2011 Jun;41(3):235-54.
[80] Martin G, Bergen HA, Richardson AS, Roeger L, Allison S. Sexual abuse and suicidality: gender differences in a large community sample of adolescents. *Child. Abuse Negl.* 2004 May;28(5):491-503.
[81] Bebbington PE, Cooper C, Minot S, Brugha TS, Jenkins R, Meltzer H, et al. Suicide attempts, gender, and sexual abuse: data from the 2000 British Psychiatric Morbidity Survey. *Am. J. Psychiatry* 2009 Oct;166(10):1135-40.
[82] Stuckler D, Basu S, Suhrcke M, Coutts A, McKee M. The public health effect of economic crises and alternative policy responses in Europe: an empirical analysis. *Lancet* 2009 Jul 25;374(9686):315-23.
[83] Breuer C. Unemployment and suicide mortality: evidence from regional panel data in Europe. *Health Econ.* 2014 Jun 17.

[84] Centers for Disease Control and Prevention. National occupational mortality surveillance (NOMS). [cited 2014 Dec 7]. Available from: http://www.cdc.gov/niosh/topics/noms/.

[85] Lazar SG. The mental health needs of military service members and veterans. *Psychodyn Psychiatry* 2014 Sep;42(3):459-78.

[86] Kessler RC, Warner CH, Ivany C, Petukhova MV, Rose S, Bromet EJ, et al. Predicting Suicides After Psychiatric Hospitalization in US Army Soldiers: The Army Study to Assess Risk and Resilience in Servicemembers (Army STARRS). *JAMA Psychiatry* 2014 Nov 12.

[87] Rendall MS, Weden MM, Favreault MM, Waldron H. The protective effect of marriage for survival: a review and update. *Demography* 2011 May;48(2):481-506.

[88] Erlangsen A, Jeune B, Bille-Brahe U, Vaupel JW. Loss of partner and suicide risks among oldest old: a population-based register study. *Age Ageing* 2004 Jul;33(4):378-83.

[89] Roskar S, Podlesek A, Kuzmanic M, Demsar LO, Zaletel M, Marusic A. Suicide risk and its relationship to change in marital status. *Crisis* 2011;32(1):24-30.

[90] Corcoran P. The impact of widowhood on Irish mortality due to suicide and accidents. *Eur. J. Public Health* 2009 Dec;19(6):583-5.

[91] jdacic-Gross V, Ring M, Gadola E, Lauber C, Bopp M, Gutzwiller F, et al. Suicide after bereavement: an overlooked problem. *Psychol. Med.* 2008 May;38(5):673-6.

[92] Stroebe M, Stroebe W, Abakoumkin G. The broken heart: suicidal ideation in bereavement. *Am. J. Psychiatry* 2005 Nov;162(11):2178-80.

[93] Li G. The interaction effect of bereavement and sex on the risk of suicide in the elderly: an historical cohort study. *Soc. Sci. Med.* 1995 Mar;40(6):825-8.

[94] Scourfield J, Evans R. Why Might Men Be More at Risk of Suicide After a Relationship Breakdown? Sociological Insights. *Am. J. Mens* Health 2014 Aug 26.

[95] Cutright P, Fernquist RM. Three explanations of marital status differences in suicide rates: social integration, marital status integration, and the culture of suicide. *Omega* (Westport) 2007;56(2):175-90.

[96] Fernquist RM. Does the level of divorce or religiosity make a difference? Cross-national suicide rates in 21 developed countries. *Archives of Suicide Research* 2014;(7):265-77.

[97] Corcoran P, Nagar A. Suicide and marital status in Northern Ireland. *Soc. Psychiatry Psychiatr. Epidemiol.* 2010 Aug;45(8):795-800.

[98] Kolves K, Ide N, De Leo D. Suicidal ideation and behaviour in the aftermath of marital separation: gender differences. *J. Affect Disord.* 2010 Jan;120(1-3):48-53.

[99] Fuse T. Modernization and suicide in Japan: a comparative approach. *Cult. Dev.* 1980;12(1):45-80.

[100] Schmidtke A, Sell R, Lohr C. [Epidemiology of suicide in older persons.]. *Z Gerontol. Geriatr.* 2008 Feb;41(1):3-13.

[101] Cibis A, Mergl R, Bramesfeld A, Althaus D, Niklewski G, Schmidtke A, et al. Preference of lethal methods is not the only cause for higher suicide rates in males. *J. Affect. Disord.* 2012 Jan;136(1-2):9-16.

[102] Callanan VJ, Davis MS. Gender differences in suicide methods. Soc *Psychiatry Psychiatr. Epidemiol.* 2012 Jun;47(6):857-69.

[103] Ajdacic-Gross V, Weiss MG, Ring M, Hepp U, Bopp M, Gutzwiller F, et al. Methods of suicide: international suicide patterns derived from the WHO mortality database. *Bull World Health Organ*. 2008 Sep;86(9):726-32.

[104] Yip PS. Age, sex, marital status and suicide: an empirical study of east and west. *Psychol. Rep*. 1998 Feb;82(1):311-22.

[105] Hee AM, Park S, Ha K, Choi SH, Hong JP. Gender ratio comparisons of the suicide rates and methods in Korea, Japan, Australia, and the United States. *J. Affect. Disord.* 2012 Dec 15;142(1-3):161-5.

[106] Denning DG, Conwell Y, King D, Cox C. Method choice, intent, and gender in completed suicide. *Suicide Life Threat Behav*. 2000;30(3):282-8.

[107] Panczak R, Geissbuhler M, Zwahlen M, Killias M, Tal K, Egger M. Homicide-suicides compared to homicides and suicides: systematic review and meta-analysis. *Forensic. Sci. Int*. 2013 Dec 10;233(1-3):28-36.

[108] Isometsa ET, Lonnqvist JK. Suicide attempts preceding completed suicide. Br. J. Psychiatry 1998 Dec;173:531-5.

[109] Tsirigotis K, Gruszczynski W, Tsirigotis M. Gender differentiation in methods of suicide attempts. *Med. Sci. Monit*. 2011 Aug;17(8):H65-H70.

[110] Giner L. Diferencias en la conducta suicida. Estudio comparativo entre los intentos de suicidio y los suicidios consumados. Tesis Doctoral.: Universidad Autónoma de Madrid; 2010.

[111] Sun SH, Jia CX. Completed suicide with violent and non-violent methods in rural Shandong, China: a psychological autopsy study. *PLoS ONE* 2014;9(8):e104333.

[112] Suominen K, Isometsa E, Ostamo A, Lonnqvist J. Level of suicidal intent predicts overall mortality and suicide after attempted suicide: a 12-year follow-up study. *BMC Psychiatry* 2004 Apr 20;4(1):11.

[113] Diaz FJ, Baca-Garcia E, Diaz-Sastre C, Garcia RE, Blasco H, Braquehais CD, et al. Dimensions of suicidal behavior according to patient reports. *Eur. Arch. Psychiatry Clin. Neurosci*. 2003 Aug;253(4):197-202.

[114] Kumar CT, Mohan R, Ranjith G, Chandrasekaran R. Characteristics of high intent suicide attempters admitted to a general hospital. *J. Affect. Disord.* 2006 Mar;91(1):77-81.

[115] Diala C, Muntaner C, Walrath C, Nickerson KJ, LaVeist TA, Leaf PJ. Racial differences in attitudes toward professional mental health care and in the use of services. *Am. J. Orthopsychiatry* 2000 Oct;70(4):455-64.

[116] Tucker JR, Hammer JH, Vogel DL, Bitman RL, Wade NG, Maier EJ. Disentangling self-stigma: are mental illness and help-seeking self-stigmas different? *J. Couns Psychol*. 2013 Oct;60(4):520-31.

[117] Latalova K, Kamaradova D, Prasko J. Perspectives on perceived stigma and self-stigma in adult male patients with depression. Neuropsychiatr. *Dis. Treat* 2014;10:1399-405.

[118] Clark DC. "Rational" suicide and people with terminal conditions or disabilities. *Issues Law Med* 1992;8(2):147-66.

[119] Suominen KH, Isometsa ET, Ostamo AI, Lonnqvist JK. Health care contacts before and after attempted suicide. *Soc. Psychiatry Psychiatr. Epidemiol.* 2002 Feb;37(2):89-94.

[120] Wilson CJ, Deane FP, Marshall KL, Dalley A. Adolescents' suicidal thinking and reluctance to consult general medical practitioners. *J. Youth Adolesc*. 2010 Apr;39(4):343-56.

[121] Conwell Y. Suicide prevention in later life: a glass half full, or half empty? *Am. J. Psychiatry* 2009 Aug;166(8):845-8.

[122] Rutz W, von KL, Pihlgren H, Rihmer Z, Walinder J. Prevention of male suicides: lessons from Gotland study. *Lancet* 1995 Feb 25;345(8948):524.

[123] De Leo D, Carollo G, Dello BM. Lower suicide rates associated with a Tele-Help/Tele-Check service for the elderly at home. *Am. J. Psychiatry* 1995 Apr;152(4):632-4.

[124] Oyama H, Sakashita T, Ono Y, Goto M, Fujita M, Koida J. Effect of Community-based Intervention Using Depression Screening on Elderly Suicide Risk: A Meta-analysis of the Evidence from Japan. *Community Ment. Health J.* 2008 Mar 25.

[125] Oyama H, Goto M, Fujita M, Shibuya H, Sakashita T. Preventing elderly suicide through primary care by community-based screening for depression in rural Japan. *Crisis* 2006;27(2):58-65.

[126] Houle J, Mishara BL, Chagnon F. An empirical test of a mediation model of the impact of the traditional male gender role on suicidal behavior in men. *J. Affect. Disord.* 2008 Apr;107(1-3):37-43.

[127] Courtenay. Behavioral factors associated with disease, injury, and death among men: Evidence and implications for prevention. *The Journal of Men's Studies* 2000;(9):81-142.

[128] Hawton K. Sex and suicide. Gender differences in suicidal behaviour. *Br. J. Psychiatry* 2000 Dec;177:484-5.

[129] McIntosh JL, Jewell BL. Sex difference trends in completed suicide. Suicide Life Threat Behav. 1986;16(1):16-27.

[130] Stack S. Suicide: a 15-year review of the sociological literature. Part I: cultural and economic factors. *Suicide Life Threat Behav*. 2000;30(2):145-62.

[131] Jick H, Kaye JA, Jick SS. Antidepressants and the risk of suicidal behaviors. *JAMA* 2004 Jul 21;292(3):338-43.

[132] Wintemute GJ, Parham CA,Beaumont JJ,Wright M,Drake C. Mortality among recent purchasers of handguns. *N. Engl. J. Med.* 1999.

[133] Murphy GE. Why women are less likely than men to commit suicide. *Compr. Psychiatry* 1998 Jul;39(4):165-75.

[134] Reynolds CF, III, Dew MA, Pollock BG, Mulsant BH, Frank E, Miller MD, et al. Maintenance treatment of major depression in old age. *N. Engl. J. Med.* 2006 Mar 16;354(11):1130-8.

In: Neurobiology of Men's Mental Health
Editors: Leo Sher and Timothy R. Rice

ISBN: 978-1-63483-191-8
© 2015 Nova Science Publishers, Inc.

Chapter 14

PLASMA TESTOSTERONE DIFFERENCES BETWEEN SUICIDE ATTEMPTERS AND COMMUNITY CONTROLS IN BOTH MEN AND WOMEN OF CHINA

Jie Zhang[1,*], PhD and Cun-Xian Jia[2], PhD

[1] Shandong University School of Public Health Center for Suicide Prevention Research, Jinan, China
State University of New York Buffalo State, Buffalo, New York, US
[2] Shandong University School of Public Health Center for Suicide Prevention Research, Jinan, China

ABSTRACT

To understand how physiological markers of men and women characterize the differences in suicide attempt, we hypothesize that attempted suicides have higher level of testosterone than their non-suicidal counterparts, which should be true of both men and women. In Shandong Province of China, 245 rural suicide attempters aged 16-50 years were consecutively recruited from October 1, 2009 to March 31, 2011. They were compared with 245 age and gender matched community controls. Blood sample was extracted at the time of interview for both patients and controls, and the plasma testosterone level was tested for comparisons. The testosterone level was significantly higher for male suicide attempters than for their community counterparts, with the OR of 1.211 (95%CI: 1.055-1.391), even when some other factors were adjusted, such as education years, family SES, marital status, coping skills, impulsivity, and psychiatric disorders. There was also a tendency that high testosterone level was related to female suicide attempt, with the OR being 1.209 (95%CI: 0.719-2.031). Testosterone, as a form of physiological marking, is somewhat another predictor of attempted suicide. Besides other factors, it might be the testosterone, not gender, which explains the suicide risks and the gender ratio of the rates. The lack of sample size limited this current study to

* Corresponding Author: Dr. Jie Zhang, Professor of Sociology, Department of Sociology, State University of New York College at Buffalo, 1300 Elmwood Avenue, Buffalo, New York 14222, USA. Telephone: 1-716-878-6425; FAX: 1-716-878-4009; Email: zhangj@buffalostate.edu.

establish a significant and positive relationship between testosterone and suicide risk in women as found in men.

Keywords: testosterone, suicide, suicide attempt, gender, china

INTRODUCTION

Testosterone (T) is a steroid hormone from the androgen group that circulates in the human body and affects cells, the nervous system, and the brain. It is present to a lesser extent in females, who may be more sensitive than males to its effects (Torjesen & Sandnes, 2004). Animal studies have indicated a link between incidents of aggression and the individual level of circulating testosterone, but the relationship is much less clear in humans. For example, Dabbs and Hargrove (1997) found that people with high testosterone were more restless and energetic, but not more aggressive. In non-humans, testosterone is associated mainly with dominance (Montoya, Terburg, Bos, & Honk, 2012). In most species, males are more aggressive than females. In humans, males engage in crime and especially violent crime more than females (Van Goozen, 2005). Also, the involvement in crime usually rises in the early teens to mid-teens which happen at the same time as testosterone levels rise (Wright, Ellis, & Beaver, 2009). However, men and women differ in ways other than testosterone, for example socialization. One of the reasons for the relatively high rate of suicide in China might be that Chinese women have a different social role from women in the West, and indeed these roles might vary by context (urban vs. rural) (Zhang, 2010). Similarly, the environment a teenager is in might have more influence on their tendency to violence than their T level. This study tries to disentangle the effects of socialization and human T levels on gendered behaviors.

Suicide as an aggressive deviance happens more in men than in women worldwide, but the gender ratio tends to be inverse in the Chinese suicide rates (Phillips, Li, & Zhang, 2002), and close to 1 in the suicide rates among Asian populations (Hvistendahl, 2012). Likewise, the gender differences in sports records are relatively larger in the West than in the East (Zhang, 2000). Suicidologists have been pondering on the relationship between testosterone and suicidal behavior since decades ago, suggesting that the gender difference in the rates of fatal suicidal actions is attributed to gender differences in testosterone levels (Lester, 1993). Thus, beyond the social and psychiatric factors that contribute to the gender difference, we want to know how physiological characteristics of men and women mark the differences in suicidal behaviors, as the pattern is across cultural and regardless of mental health status. If higher level of testosterone is related to more aggression by men, it may also explain the higher suicide rates among men than women. If testosterone plays such an important role in violent and deviant behaviors, we hypothesize that suicidal people have higher level of testosterone than their non-suicidal counterparts, which should be true of both men and women. As the gender ratio for the Chinese male and female suicide rates is so close in comparison with the gender ratios found in the West (Phillips et al., 2002), and the sports records (another aggressive presentation that requires high level of androgen) are much closer between Chinese men and women than between Western men and women (Zhang, 2000), we suspect that the gender ratio of male to female testosterones among the Chinese is also smaller than that in the West.

In a recent study, Sher et al., (2012) tested and found the association between testosterone level and bipolar disorders as well as suicidal behavior. Plasma testosterone was assayed using a double antibody radioimmunoassay procedure. The number of major depressive episodes, the maximum lethality of suicide attempts, and the testosterone levels were higher in men compared to women. Controlling for gender, they found that testosterone levels positively correlated with the number of manic episodes and the number of suicide attempts. Their findings were consistent with previous observations of the association between testosterone levels and parameters of mood and behavior. This study suggested that testosterone levels may be related to the course of bipolar disorder and suicidal behavior (Sher et al., 2012). However, the positive relationship between testosterone level and suicide may not be true for all ages of people. For example, high T levels may be associated with suicidal behavior in young men, and low T levels may be associated with suicidal behavior in older men (Sher, 2013).

However, in another recent study by Gustavsson et al., (2003), testosterone in cerebrospinal fluid (CSF) was analyzed in men with a recent suicide attempt, diagnostically subdivided into groups according to DSM-III-R axis I and II diagnosis and mode of suicidal behavior, but the researchers found the patients had lower CSF testosterone levels than aggressive violent patients in other studies. Patients with depression or dysthymia showed higher CSF testosterone levels than the rest. The researchers suggested that suicide attempts may be mediated by different biological variables than aggression (Gustavsson et al., 2003).

Further, there is observation that low testosterone is associated with male suicide (Tripodianakis, Markianos, Rouvali, & Istikoglou, 2007). In order to find further evidence, Perez-Rodriguez et al., (2011) studied a sample of 112 male suicide attempters with 37 male blood donors as non-suicidal controls. They found that the mean testosterone levels were 5.1 ± 2.9 ng/ml in the attempters and 4.6 ± 1.6 ng/ml in the controls, but the group differences in testosterone levels were not significant when they studied the interaction with time of extraction of the blood ($F = 0.37$; d.f. = 2; $p = 0.70$) or when matched by age and time of extraction ($t = -0.74$; d.f. = 26; $p = 0.47$). They suggested further research to include serial determinations with a minimal time gap between the attempt and the blood extraction and controls within the same time periods, as the individual testosterone levels may vary between the normal time and the time of tension release, such as acting suicide (Perez-Rodriguez et al., 2011).

With inconsistencies in the findings, we planned to re-test the relationship between the levels of testosterone and suicide attempt with a Chinese sample. We aimed to investigate the differences in the testosterone levels between suicide attempters and their community controls, for men and women, with the hypothesis that, both for men and women, the testosterone level is higher for the suicide attempters than for the community controls.

METHODS

Study Population and Procedure

Six rural counties (Ju'nan, Lijin, Ningyang, Penglai, Tengzhou, and Zoucheng) in Shandong Province were randomly selected for the data collection. Shandong is a typical agricultural and industrial province of China located in the East Coast of China with about half of the 80 million populations living in rural areas. Villagers aged 15-70 years who had attempted suicide from October 1, 2009 to March 31, 2011 and had been admitted to emergency room for treatment were consecutively recruited for study.

A systematic random sampling was practiced to selected community controls in the same areas with the age and gender matched to the suicide attempters. The local Centers for Disease Control and Prevention (CDCs) assisted in the recruitment of cases and controls: connecting the research team with the local medical institutions and ensuring the informed consent by each subject before the interview and procedure. A face to face interview with each of the cases and controls was followed by blood sample drawn after the signature on the informed consent was obtained. To ensure the accuracy of the information and protection of both subjects and the interviewers, audio tape recording was suggested unless the subject disapproved. Each interview and procedure was completed in subject's home or village doctor's clinic office, without the presence of anyone else. In this study, a total of 245 (73 male and 172 female) suicide attempters aged 16-50 years and their paired controls (n=245: 73 males and 172 females) were obtained as the study sample.

Testosterone Measurement

Blood sample was taken by medical staff on the same day of the interview. Ethylenediaminetetraacetic acid disodium salt was used for anticoagulation. Iodine [125I]

Testosterone Radioimmunoassay Kit was employed for measuring the plasma Testosterone. The kits were produced by Beijing North Biological Technology Institute (Approval number: S10940093). In an tube, 50 ul plasma of each subject or standard sample or quality sample was added in the measurement tube, and 100 ul Iodine [125I] Testosterone, 100 ul rabbit anti-Testosterone antibody were added into the same measurement tube. The added materials were well shaken up, washed bath for one hour in 37 °C. Then 500 ul donkey anti rabbit immunization separating medium was added in the same tube, well shaken up, then centrifuged after 15 minutes for laying up in room temperature.

The upper clear liquid was absorbed and discarded, the left was used to obtain the radioactivity number by γ calculator. The radiation-immunization technology instrument (SN-684) was produced by Shanghai Nuclear Institute Rihuan Guangdian Instrument Limited Corporation. The γ numbers were auto-transformed into concentration (ng/ml) by computer. The whole procedure was carried out in the same laboratory by the same technicians with the same kits, and the technicians were blind of the status of each blood sample in terms of gender and the case-control.

Social Psychological and Demographic Measures

Depression was measured by CES-D, a self-report depression scale for research in the general population (Radloff, 1977), which has 20 items with four-point Likert scale from 0=rarely or no day per week to 3=most or all of the days per week.

The total score can range from 0 to 60. Anxiety was measured by STAI, which is a 40-item measure that indicates the intensity of feelings of anxiety and distinguishes between state anxiety and trait anxiety (Spielberger, 1983). We used in this study only the items on the trait anxiety that consists of 20 statements with each item having 4 responses from 1 (never) to 4 (almost always). The total score of anxiety ranges from 20 to 80. Impulsivity was measured by the 23-item Dickman Impulsivity Instrument (DII) (Dickman, 1990) with response options to each item being true (1) or false (0). The DII is divided into two subscales: Functional Impulsivity (FI; 11 items) and Dysfunctional Impulsivity (DI; 12 items). Hopelessness of the subjects was assessed by Beck Hopelessness Scale (BHS) that includes 20 items with 5-point Likert responses with a total score ranging from 20 to 100 (A. T. Beck, Steer, Beck, & Newman, 1993; A.T. Beck, Weissman, Lester, & Trexler, 1974). Suicidal ideation was evaluated by the Beck Scale for Suicide Ideation (SSI) which has 19 items with each item response from 0 to 2, and the total score can range from 0 to 38 (Aaron T. Beck, 1991). Coping skills were measured by the Simplified Coping Style Questionnaire (SCSQ) (Xie, 1999) originated from the Coping Response Inventory (CRI) (Moos, Brennan, Fondacaro, & Moos, 1990) from which we included in this study two sub-scales: Approach Coping Scales (12 items) and Avoidance Coping Scales (8 items). The response options to each item were from 0= never to 3=often, with the total score ranging from 0 to 36 for approach coping, and from 0 to 24 for avoidance coping. Mental disorder was evaluated by the SCID -- Structured Clinical Interview for DSM-IV Axis I Disorders (Spitzer, Williams, Gibbon, & First, 1988) and the diagnoses for each suicide attempter and control was assessed by experienced psychiatrists who were on the research team.

All the above instruments were originally developed in the West. Translation and back translations were performed on each of them, and all scales were validated with various samples in Chinese societies before their applications in this current study (Gao, Zhang, & Jia, 2011; Kong, Zhang, Jia, & Zhou, 2007; Li & Zhang, 2012; Zhang et al., 2003; Zhang & Gao, 2012; Zhang, Norvilitis, & Jin, 2001; Zhang, Sun, Kong, & Wang, 2012).

The list of demographic variables includes gender, age, marital status, education years, and family social economic status (SES). To limit the effect of age on the variance of the testosterone level, we sampled the cases and controls approximately between 15 and 50 years of age. The group of ever married subjects includes those currently married, remarried, temporarily separate due to work location, and those who live together without marriage. Family SES is a measure of family annual income in Chinese Renminbi (RMB), which was exchanged at the time of the data collection with rate of US$ to 6.5 RMB.

Statistical Procedures

The variables were first carried out normality analysis. Median (minimum-maximum) was used for variables with non-normality distribution. n (%)was used for category variables. Z value for Mann-Whitney U test or Chi-square test was used for different characteristics comparison between attempted suicides and controls by gender. Different conditional logistic regression models were used for obtaining the OR values and their 95% confidential interval (CI) for the risk of testosterone level in attempted suicide. SPSS 16.0 was used for all the data analyses. The p values were set as two- tailed.

Ethical Issues

The research protocol was examined and approved by the human subject protection committee at Shandong University School of Public Health. Informed consent was obtained from each of the subjects before each administration of the data collection.

RESULTS

The reference value of plasma testosterone of the adults provided by Beijing North Biological Technology Institute was that for male adults (15-69 years of age), the mean and normal range were 5.75 (3.2-14.6) ng/ml.

For female adults (14-74 years of age), the mean and normal range were 0.48 (0.05-1.2) ng/ml. Comparisons of the testosterone levels between the suicide attempters and the community controls by gender are seen in Table 1.

Testosterone measurement literature in the West indicated that on an average in adult human males, the plasma concentration of testosterone was about 7-8 times greater than that of adult human females (Torjesen & Sandnes, 2004), but as the metabolic consumption of testosterone in males was greater, the daily production was about 20 times greater in men than in women (Southren et al., 1967; Southren, Tochimoto, Carmody, & Isurugi, 1965). Kratz and colleagues (2004) further reported that the range of the T values for American males was 9.36-37.10 ng/ml, and 0.21-2.98 ng/ml for American females.

For the Chinese community control sample, the T values ranged from 0.22 to 23.29 ng/ml for males, and from 0.05 to 5.15 ng/ml for females. It was also found that the average T value (ng/ml) was higher for the suicide attempters than for the community controls. Among them, the male attempters' T level was significantly higher than that of the male controls (5.42 vs. 4.06; $p=0.001$), and the female attempters' T level was also significantly higher than that of the female controls (0.26 vs. 0.22; $p=0.017$). Table 1 also shows the differences between attempters and controls on the demographic and psychological variables with the men and women of in the sample.

The bi-variate analyses shown in Table 1 indicate that education years, family SES, functional impulsivity, and approach coping were all protective factors against suicide attempt, while depression, anxiety, dysfunctional impulsivity, hopelessness, suicidal ideation and mental disorder were risk factors. Being married did not protect a Chinese rural man or woman from suicide risk, while in the West, marriage is a buffer of the self-injury act (Stack, 2000). To further illustrate the relationship between the testosterone levels and the demographic and clinical features of the suicide attempters and community controls, the bivariate analyses are presented in Table 2, separately by gender.

Table 3 shows the different regression models assessing the effect of the T level on the suicide act.

For men and women combined together, the T level was significantly and positively associated with suicide attempt in almost models (except Model 8), which indicates that the T level was an important predictor of suicide attempt after adjusting demographic and some psychological and psychiatric factors.

Table 1. Demographic and Clinical Features of the Suicide Attempters and Community Controls by Gender (N=490)

Variable	Suicide Attempters Median (min-max)/n (%)			Community Controls Median (min-max)/n (%)			Z*/Chi-Square for males (*p*)	Z*/Chi-Square for females (*p*)	Z*/Chi-Square for total (*p*)
	Male n=73	Female n=172	Total n=245	Male n=73	Female n=172	Total n=245			
Testosterone (ng/ml)	5.42 (0.84-22.81)	0.26 (0.00-0.39)	0.40 (0.00-22.81)	4.06 (0.22-23.99)	0.22 (0.05-5.15)	0.34 (0.05-23.99)	-3.258 (0.001)	-2.395 (0.017)	-2.029 (0.042)
Age	40 (17-50)	37 (16-50)	38 (16-50)	40 (15-53)	37 (14-52)	37 (14-53)	-0.116 (0.908)	-0.312 (0.755)	-0.321 (0.748)
Ever married (yes)	61 (83.6)	160 (93.0)	221 (90.2)	62 (84.9)	161 (93.6)	223 (91.0)	0.052 (0.820)	0.047 (0.829)	0.096 (0.757)
Education years	7 (0-14)	5 (0-14)	5.0 (0-14)	9 (5-17)	7.5 (0-17)	8 (0-17)	-4.606 (<0.001)	-3.780 (<0.001)	-5.476 (<0.001)
Family SES (*RMB*)	20,000 (0-200K)	20,149 (4K-400K)	20,149 (0-400K)	30,000 (5K-680K)	30,000 (1K-300K)	30,000 (1K-680K)	-3.397 (0.001)	-2.094 (0.036)	-3.376 (<0.001)
Depression (CES-D)	15 (0-51)	17 (0-54)	17 (0-54)	0 (0-37)	0 (0-28)	0 (0-37)	-7.566 (<0.001)	-12.044(<0.001)	-14.272 (<0.001)
Anxiety (TAI)	33 (23-67)	36 (20-62)	35 (20-67)	29 (20-59)	30 (20-50)	30 (20-59)	-5.066 (<0.001)	-6.385 (<0.001)	-8.097 (<0.001)
Functional impulsivity	7 (0-10)	6 (1-11)	6 (0-11)	7 (2-11)	7 (1-11)	7 (1-11)	-1.657 (0.097)	-2.168 (0.030)	-2.664 (0.008)
Dysfunctional impulsivity	3 (0-11)	2 (0-11)	2 (0-11)	1 (0-10)	2 (0-9)	2 (0-10)	-3.342 (0.001)	-2.514 (0.012)	-3.954 (<0.001)
Hopelessness (Beck HS)	47 (20-94)	49 (20-99)	48 (20-99)	28 (20-64)	30 (20-55)	29 (20-64)	-7.140 (<0.001)	-10.333(<0.001)	-12.544 (<0.001)
Suicide Ideation (Beck SSI)	0 (0-28)	0 (0-29)	0 (0-29)	0 (0-0)	0 (0-1)	0 (0-1)	-1.744 (0.081)	-2.374 (0.018)	-2.905 (0.004)
N of Days since attempt	336 (42-753)	296 (26-698)	305 (26-753)						
Approach coping	24 (8-35)	23 (8-36)	23 (8-36)	30 (8-36)	27 (12-36)	28 (8-36)	-5.454 (<0.001)	-6.533 (<0.001)	-8.412 (<0.001)
Avoidance coping	8 (0-15)	8 (0-18)	8 (0-18)	6 (1-17)	8 (0-19)	7 (0-19)	-2.345 (0.019)	-0.116 (0.907)	-1.305 (0.192)
Mental disorder (SCID)	25 (34.2)	41 (23.8)	66 (26.9)	4 (5.5)	4 (2.3)	8 (3.3)	18.976 (<0.001)	35.001 (<0.001)	53.546 (<0.001)

Notes:

1. * Z value for Mann-Whitney U test.

2. The bold figures are for correlations significant at 0.05 or lower probability level.

Table 2. Bivariate Correlations of Testosterone Levels with Demographic and Clinical Features of the Suicide Attempters and Community Controls by Gender

Variable	Suicide Attempters (r_p/p)			Community Controls (r_p/p)		
	Male (n=73)	Female (n=172)	Total (n=245)	Male (n=73)	Female (n=172)	Total (n=245)
Age	-0.193 (0.102)	-0.282 (0.000)	-0.078 (0.221)	-0.054(0.649)	-0.348 (<0.001)	-0.093 (0.149)
Gender (Male)	-	-	0.789 (<0.001)	-	-	0.767 (<0.001)
Ever married	-0.068 (0.570)	-0.123 (0.109)	-0.173 (0.007)	-0.114 (0.335)	-0.127(0.096)	-0.176 (0.006)
Education years	-0.127 (0.296)	0.092 (0.229)	0.218 (0.001)	-0.076 (0.521)	0.180 (0.018)	0.313 (<0.001)
Family SES	-0.041 (0.732)	-0.040 (0.603)	-0.081 (0.208)	-0.054 (0.652)	-0.065 (0.398)	0.072 (0.263)
CES-D	0.054 (0.654)	0.049 (0.523)	-0.003 (0.965)	0.212 (0.072)	0.261 (0.001)	0.143(0.026)
Anxiety	0.035 (0.767)	0.089 (0.247)	-0.009 (0.887)	0.115 (0.331)	0.210 (0.006)	-0.001 (0.984)
Functional impulsivity	0.015 (0.897)	-0.056 (0.469)	0.031 (0.635)	0.067 (0.574)	-0.174 (0.022)	-0.007 (0.912)
Dysfunction impulsivity	-0.070 (0.555)	-0.016 (0.833)	0.059 (0.358)	0.033 (0.781)	-0.178(0.019)	0.059 (0.354)
Hopelessness	-0.160 (0.188)	-0.122 (0.110)	-0.112 (0.082)	0.155 (0.195)	0.087 (0.258)	-0.025 (0.701)
Suicide ideation	0.005 (0.966)	0.066 (0.388)	0.009 (0.891)	-	0.123 (0.107)	0.028 (0.667)
Days between attempted suicide and obtaining blood sample (26-753days)	0.083 (0.484)	-0.069 (0.369)	-0.032 (0.613)	-	-	-
Approach coping	-0.037 (0.757)	0.022 (0.776)	0.031 (0.626)	-0.026 (0.828)	-0.007 (0.927)	0.138 (0.031)
Avoidance coping	0.117 (0.325)	-0.001 (0.991)	0.077 (0.228)	-0.112 (0.347)	-0.091 (0.237)	-0.116 (0.071)
SCID-mental disorder	-0.060 (0.612)	-0.030 (0.696)	0.065 (0.313)	-0.129 (0.278)	0.080(0.294)	0.083 (0.196)

Note: The bold figures are for correlations significant at 0.05 or lower probability levels.

Table 3. Different Models for Association between Testosterone Levels and Attempted Suicide by Gender with Conditional Logistic Regression

Variable	Male suicide attempters vs. male controls (n=146)			Female suicide attempters vs. female controls (n=344)			Total suicide attempters vs. total controls (n=490)		
	β	p	OR(95%CI)	β	p	OR(95%CI)	β	p	OR(95%CI)
Model 1	0.192	0.006	1.211 (1.055-1.391)	0.189	0.474	1.209 (0.719-2.031)	0.192	0.005	1.211 (1.060-1.384)
Model 2	0.105	0.167	1.111 (0.957-1.289)				0.146	0.035	1.157 (1.010-1.325)
Model 3	0.200	0.048	1.221 (1.002-1.489)				0.207	0.016	1.230 (1.039-1.455)
Model 4	0.326	0.049	1.385 (1.001-1.916)				0.216	0.026	1.241 (1.027-1.501)
Model 5	0.320	0.055	1.377 (0.994-1.909)				0.208	0.031	1.231 (1.020-1.487)
Model 6	3.799	0.612	44.655 (0.000-1E+008)				0.192	0.051	1.211 (0.999-1.449)
Model 7	4.234	0.233	68.960 (0.065-72726.858)				0.196	0.048	1.217 (1.002-1.479)
Model 8	0.133	0.942	1.142 (0.033-39.409)				0.154	0.132	1.167 (0.954-1.427)

Model 1: only testosterone
Model 2: further adjusting education years, family SES, and marital status
Model 3: further more adjusting mental disorders
Model 4: further more adjusting approaching coping, avoidance coping, functional impulsivity and dysfunctional impulsivity
Model 5: further more adjusting suicide ideation
Model 6: furthermore adjusting hopelessness
Model 7: furthermore adjusting anxiety
Model 8: furthermore adjusting depression

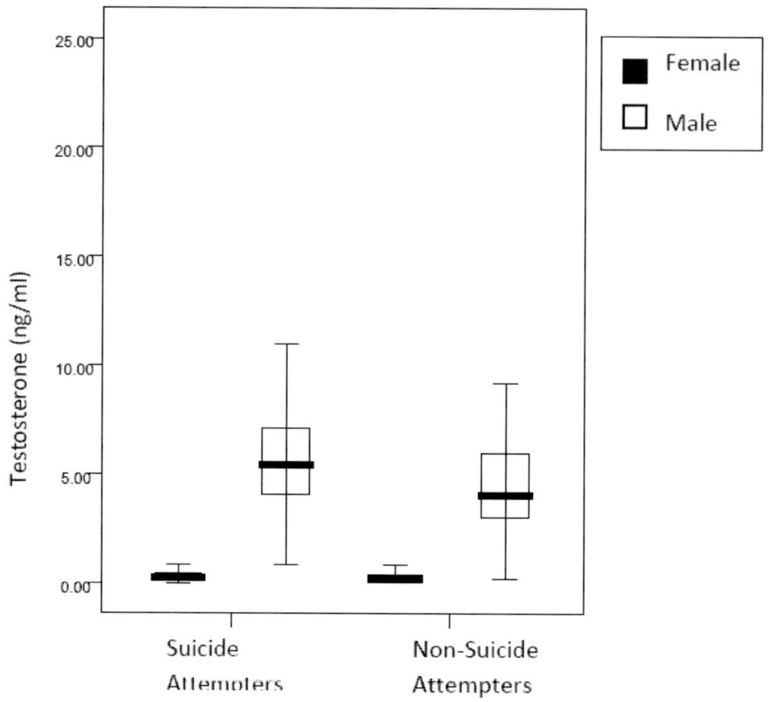

Figure 1. The Clustered Box-and-Shisker diagram of Comparing the testosterone levels (ng/ml) between the suicide attempt cases and the community controls by gender.

In Model 8, after further adjusting depression, the T level was not significantly associated with suicide attempt, and the reason might be that the T level is weaker than that of depression in the predicting suicide attempt, or there is some interaction between the T level and depression as effect on suicide attempt. While the pattern held true for men in some models (Model 1, Model 3, Model 4), which indicates that the T level might be an important predictor of male suicide attempt. However, its effects on suicide attempt might be weaker than that of psychological factors, or our male sample size is not enough to reach the significant level. For the women sample only, an increase of the T level was related to an increase the suicide attempt risk, but the association was not significant enough to make a strong claim.

The relationship between the T level and the suicide attempt and the difference between men and women in the sample are clearly illustrated in Figure 1, which shows that T levels in men were obviously higher than those in women for both suicide attempters and community controls. Further, it shows that the T level in male suicide attempters was obviously higher than that in the male controls. However, there was no significant difference in the T level between female suicide attempters and female community controls.

Discussion

In this study we used a case control method well designed and administered to collect and compare the data. We found that the testosterone levels of the suicide attempters were higher than those of the community controls, and this was separately true for men, women, and the total sample where men and women combined, although it was not significant at the .05 probability level for the female sample. Multiple regression models also confirmed the trend in the male and total sample after adjusting many other factors, such as education years, family SES, marital status, coping skills, impulsivity, and psychiatric disorders. The positive association between T levels and suicidal behavior in Chinese men supports the earlier findings in American samples (Sher, 2012).

The reason for the null finding in women between suicide attempters and community controls might be in the instrument we used for the test. RIA was used, but this is not suitable for measurement of the low levels of testosterone found in women (Rosner, Auchus, Azziz, Sluss, & Raff, 2007). People may argue that the T levels in women can be affected by the menstrual cycle in women, as T levels could increase 20-30% at mid-cycle (Van Anders & Watson, 2006).

Some of the early studies failed to establish the positive correlation between the testosterone level and suicide risk, partially because of the various timing for the blood sample withdrawal. As observed by Perez-Rodriguez et al., (2011), individual testosterone levels may vary between the normal time and the time of tension release, such as acting suicide. Also, as found in the studies of Mazur and Michalek (1998) and Gray et al., (2004), falling in love decreases men's testosterone levels while increasing women's testosterone levels, and sexual incidents lower men's testosterone levels. It has been evidenced that competitive sexual tension release reduces men's testosterone level, and afterward, the testosterone level tends to be normal. If suicidal acts are considered to be tension release, the normal testosterone level of a suicide attempter has to be assessed either before or months after the suicidal acts. In some of the previous studies, the researchers may not have waited long enough before they conducted the interview and blood withdrawal from the suicide attempters. The measure of the testosterone levels of the subjects should not be affected by the suicidal incident and should be as close to their normal status as possible. Therefore, a future study may compare the T levels of those who have suicide ideation or intent in comparison with those without suicide thought, to avoid the tension release effect and obtain the normal T level in the individuals. Also, funding permits, future study with the Chinese data should examine the degree of testosterone decrease as a function of tension release after a suicide attempt act.

Testosterone as a male hormone found in blood plasma can be a predictor of such aggressive behaviors as suicide attempt. This partially accounts for the reason why men are usually at higher risk of suicide than women worldwide. It is not only the positive association between the level of testosterone and suicide existing among men, there is also such a trend among women we have studied. In studying the suicide risk factors and searching for crisis intervention measures, we may need to take into consideration of this physiological feature of human life, beyond the social, psychological, and psychiatric limits.

The findings of this study and following research in this line may conclude that the person is at a high risk of suicide not because the person is a man but because the person has a high level of testosterone. Injection of testosterone plasma strengthens the performance of sports-people, and this mechanism works for both men and women.

One limitation of the study might be the lack of controlling for the extraneous variables in measuring the testosterone levels. The menstrual cycle in women as discussed above is one example. Another direction of study in this line concerns whether post-mortem testosterone levels can affect individuals who actually died in their attempts. Medications, BMI, chronic disease including diabetes mellitus, and the timing of the day are also know to affect androgen levels in men and women (Osuna, Gómez-Pérez, Arata-Bellabarba, & Villaroel, 2006). Another concern is the timing in a day of the blood sample extraction for both cases and controls. Blood sampling in the morning has often shown an age-related decrease in testosterone levels, while afternoon samples have failed to show such a decrease. Further, with almost a year since the failed attempt for some of the cases, the measured testosterone level could have reflected such factors as failure, return to a life of helplessness, return to a better life, etc. Many other factors other than prior suicidal tendencies might account for the testosterone levels. Therefore, future studies should shorten the period of time between the suicide attempt and the biological measurement. These findings suggested the possibility that the circadian rhythm in serum testosterone levels might be altered with normal aging in men (Bremner, Vitiello, & Prinz, 1983). Future studies may take these variables into consideration, although they may not significantly affect the findings in this current study.

With the preliminary findings from this current study, further research by the team will again look into the testosterone differences of men and women in general populations between East and West to find a clue to explain the different gender ratios both in suicide rates and in the sports performance. It was found that the gender differences in sports records are smaller for the Chinese athletes than for the American athletes (Zhang, 2000). Are the testosterone differences between men and women also smaller for the Chinese general populations than for the Americans?

References

Beck, A. T. (1991). *Beck Scale for Suicide Ideation.* San Antonio, TX: Psychological Corporation.

Beck, A. T., Steer, R. A., Beck, J. S., & Newman, C. F. (1993). Hopelessness, depression, suicidal ideation, and clinical diagnosis of depression. *Suicide and Life-Threatening Behavior, 23*(2), 139-145.

Beck, A. T., Weissman, A., Lester, D., & Trexler, L. (1974). The measurement of pessimism: the hopelessness scale. *Journal of Consulting and Clinical Psychology, 42*(6), 861-865.

Bremner, W. J., Vitiello, M. V., & Prinz, P. N. (1983). Loss of circadian rhythmicity in blood testosterone levels with aging in normal men. *Journal of Clinical Endocrinology and Metabolism, 56*(6), 1278-1281.

Dabbs, J. M., & Hargrove, M. F. (1997). Age, testosterone, and behavior among female prison inmates. *Psychosomatic Medicine, 59*(5), 477-480.

Dickman, S. J. (1990). Functional and dysfunctional impulsivity. *Journal of Personality and Social Psychology, 58*, 95-102.

Gao, Q., Zhang, J., & Jia, C. (2011). Psychometric properties of the Dickman Impulsivity Instrument in suicide victims and living controls of rural China. *Journal of Affective Disorders, 132*(2011), 368-374.

Gray, P. B., Campbell, B. C., Marlowe, F. W., Lipson, S. F., & Ellison, P. T. (2004). Social variables predict between-subject but not day-to-day variation in the testosterone of US men. *Psychoneuroendocrinology, 29*(9), 1153-1162. doi: http://dx.doi.org/10.1016/j.psyneuen.2004.01.008

Gustavsson, G., Träskman-Bendz, L., Higley, J. D., & Westrin, Å. (2003). CSF testosterone in 43 male suicide attempters. *European Neuropsychopharmacology, 13*(2), 105-109. doi: http://dx.doi.org/10.1016/S0924-977X(02)00131-1

Hvistendahl, M. (2012). Making Sense of a Senseless Act. *Science, 338*(6110), 1025-1027. doi: 10.1126/science.338.6110.1025

Kong, Y. Y., Zhang, J., Jia, S. H., & Zhou, L. (2007). Reliability and Validity of the Beck Hopelessness Scale for Adolescents. *Chinese Mental Health Journal, 21*(10), 686-689. (in Chinese).

Kratz, A., Ferraro, M., Sluss, P. M., & Lewandrowski, K. B. (2004). Case records of the Massachusetts General Hospital. Weekly clinicopathological exercises. Laboratory reference values. *New England Journal of Medicine, 351*(15), 1548-1563. doi: 10.1056/NEJMcpc049016

Lester, D. (1993). Testosterone and suicide: an explanation of sex differences in fatal suicidal behavior. *Personality and Individual Differences, 15*(3), 347-348. doi: 10.1016/0191-8869(93)90227-t

Li, Z., & Zhang, J. (2012). Coping Skills, Mental Disorders, and Suicide Among Rural Youths in China. *The Journal of Nervous and Mental Disease, 200*(10), 885-890.

Mazur, A., & Michalek, J. (1998). Marriage, Divorce, and Male Testosterone. *Social Forces, 77*(1), 315-330. doi: 10.1093/sf/77.1.315

Montoya, E., Terburg, D., Bos, P., & Honk, J. (2012). Testosterone, cortisol, and serotonin as key regulators of social aggression: A review and theoretical perspective. *Motivation and Emotion, 36*(1), 65-73. doi: 10.1007/s11031-011-9264-3

Moos, R. H., Brennan, P. L., Fondacaro, M. R., & Moos, B. S. (1990). Approach and avoidance coping responses among older problem and nonproblem drinkers. *Psychology and Aging, 5*(1), 31-40.

Osuna, J., Gómez-Pérez, R., Arata-Bellabarba, G., & Villaroel, V. (2006). Relationship between BMI, total testosterone, sex hormone-binding-globulin, leptin, insulin and insulin resistance in obese men. *Archives of Andrology, 52*(5), 355-361.

Perez-Rodriguez, M. M., Lopez-Castroman, J., Martinez-Vigo, M., Diaz-Sastre, C., Ceverino, A., Núñez-Beltrán, A., . . . Baca-Garcia, E. (2011). Lack of association between testosterone and suicide attempts. *Neuropsychobiology, 63*(2), 125-130.

Phillips, M. R., Li, X., & Zhang, Y. (2002). Suicide rates in China, 1995-99. *The Lancet, 359*(9359), 835-840. doi: 10.1016/S0140-6736(02)07954-0

Radloff, L. S. (1977). The CES-D Scale: A Self-Report Depression Scale for Research in the General Population. *Applied Psychological Measurement, 1*(3), 385-401. doi: 10.1177/014662167700100306

Rosner, W., Auchus, R. J., Azziz, R., Sluss, P. M., & Raff, H. (2007). Utility, Limitations, and Pitfalls in Measuring Testosterone: An Endocrine Society Position Statement.

Journal of Clinical Endocrinology & Metabolism, 92(2), 405-413. doi: 10.1210/jc.2006-1864

Sher, L. (2012). Testosterone and suicidal behavior. *Expert Review of Neurotherapeutics, 12*, 257-259.

Sher, L. (2013). High and low testosterone levels may be associated with suicidal behavior in young and older men, respectively. *Australian and New Zealand Journal of Psychiatry, 47*(5), 492-493.

Sher, L., Grunebaum, M. F., Sullivan, G. M., Burke, A. K., Cooper, T. B., Mann, J. J., & Oquendo, M. A. (2012). Testosterone levels in suicide attempters with bipolar disorder. *Journal of Psychiatric Research, 46*(10), 1267-1271.

Southren, A. L., Gordon, G. G., Tochimoto, S., Pinzon, G., Lane, D. R., & Stypulkowski, W. (1967). Mean Plasma Concentration, Metabolic Clearance and Basal Plasma Production Rates of Testosterone in Normal Young Men and Women Using a Constant Infusion Procedure: Effect of Time of Day and Plasma Concentration on the Metabolic Clearance Rate of Testosterone. *Journal of Clinical Endocrinology & Metabolism, 27*(5), 686-694. doi: 10.1210/jcem-27-5-686

Southren, A. L., Tochimoto, S., Carmody, N. C., & Isurugi, K. (1965). Plasma Production Rates of Testosterone in Normal Adult Men and Women and in Patients with the Syndrome of Feminizing Testes. *Journal of Clinical Endocrinology & Metabolism, 25*(11), 1441-1450. doi: 10.1210/jcem-25-11-1441

Spielberger, C. D. (1983). *Manual for the State-Trait Anxiety Inventory (Form Y)*. Palo Alto, CA: Consulting Psychologists Press, Inc.

Spitzer, R. L., Williams, J. B. W., Gibbon, M., & First, A. B. (1988). *Instruction Manual for the Structured Clinical Interview for DSM-III-R (SCID, 6/1/88 Revision)*. New York: Biometrics Research Department, New York State Psychiatric Institute.

Stack, S. (2000). Suicide: A 15-Year Review of the Sociological Literature, Part II: Modernization and Social Integration. *Suicide and Life-Threatening Behavior, 30*(2), 163-176. doi: 10.1111/j.1943-278X.2000.tb01074.x

Torjesen, P. A., & Sandnes, L. (2004). Serum Testosterone in Women as Measured by an Automated Immunoassay and a RIA. *Clinical Chemistry, 50*(3), 678-679. doi: 10.1373/clinchem.2003.027565

Tripodianakis, J., Markianos, M., Rouvali, O., & Istikoglou, C. (2007). Gonadal axis hormones in psychiatric male patients after a suicide attempt. *European Archives of Psychiatry and Clinical Neuroscience, 257*(3), 135-139.

Van Anders, S., & Watson, N. (2006). Menstrual cycle irregularities are associated with testosterone levels in healthy premenopausal women. *American Journal of Human Biology, 18*(6), 841-844.

Van Goozen, S. (2005). Chapter 14: Hormones and the Developmental Origins of Aggression. In R. E. Tremblay, W. W. Hartup & J. Archer (Eds.), *Developmental Origins of Aggression*: The Guilford Press.

Wright, J., Ellis, L., & Beaver, K. (2009). *Handbook of crime correlates*. San Diego: Academic Press; pp. 208-210.

Xie, Y. (1999). Simplified Coping Style Questionnaire. In X. Wang, X. Wang & H. Ma (Eds.), *Rating scales for mental health (suppl)* (Vol. 12, pp. 122-124): Chinese Journal Mental Health.

Zhang, J. (2000). Gender differences in athletic performance and their implications in gender ratios of suicide: A comparison between the USA and China. *Omega: Journal of Death and Dying, 41*(2), 117-123. doi: 10.2190/K1K2-MUW6-9YGG-PVBK

Zhang, J. (2010). Marriage and Suicide among Chinese Rural Young Women. *Social Forces, 89*(1), 311-326. doi: 10.1353/sof.2010.0065

Zhang, J., Conwell, Y., Wieczorek, W. F., Jiang, C., Jia, S. H., & Zhou, L. (2003). Studying Chinese suicide with proxy-based data: Reliability and validity of the methodology and instruments in China. *The Journal of Nervous and Mental Disease, 191*(7), 450-457.

Zhang, J., & Gao, Q. (2012). Validation of the Trait Anxiety Scale for State-Trait Anxiety Inventory in Suicide Victims and Living Controls of Chinese Rural Youths. *Archives of Suicide Research, 16*(1), 85-94. doi: 10.1080/13811118.2012.641440

Zhang, J., Norvilitis, J. M., & Jin, S. H. (2001). Measuring gender orientation with the Bem Sex Role Inventory in Chinese culture. *Sex Roles: A Journal of Research, 44*(3/4), 237-251.

Zhang, J., Sun, W., Kong, Y., & Wang, C. (2012). Reliability and validity of the Center for Epidemiological Studies Depression Scale in 2 special adult samples from rural China. *Comprehensive Psychiatry, 53*(8), 1243-1251.

In: Neurobiology of Men's Mental Health
Editors: Leo Sher and Timothy R. Rice
ISBN: 978-1-63483-191-8
© 2015 Nova Science Publishers, Inc.

Chapter 15

MALE SUICIDALITY ACROSS THE LIFESPAN

N. Lubbad[1], MD, and G. Zalsman[1,2,3,*], MD, MHA
[1] Child and Adolescent Division, Geha Mental Health Center, Petah Tikva, Israel
[2] Sackler Faculty of Medicine, Tel Aviv University, Tel Aviv, Israel
[3] Molecular Imaging and Neuropathology Division, Psychiatry Department, Columbia University, NY, NY

ABSTRACT

Few studies have focused on the research of male suicidality. The World Health Organization estimated in 2012 the global male: female suicidality ratio to be 1:9. Some of the most often researched factors in relation to suicidality are mental health problems and especially depression. Other factors that are being suggested as possible contributors to male suicide are stress and stressful events. Psycho-social factors such as social relationships, impulsiveness and availability of suicide means have been shown to be significant, but have a different impact on male suicidality versus female. It was also found that suicidal ideation is seven times higher in the elderly.

Keywords: male, suicide, suicide means, depression

INTRODUCTION AND EPIDEMIOLOGY

Suicide continues to be the second cause of death. While there is an abundance of research on both completed and attempted suicide, few studies focus specifically on these typologies of suicide in males.

The World Health Organization (WHO)'s most recent epidemiological data on suicide, which dates from 2012, shows that in 2012 there were an estimated 804,000 deaths by suicide

[*] Corresponding author: Prof. Gil Zalsman, Geha Mental Health Center, P.O. Box 102, Petach Tikva 49100, Israel, E-mail zalsman@post.tau.ac.il

worldwide. This indicates an annual global suicide rate of 11.4 per 100,000 population (8.0 for females and 15.0 for males). The global Male: Female ratio of age-standardized suicide rate was 1.9 in that year [1].

According to the WHO, the major differences between high-income countries and low- and middle-income countries (LMICs) are that in the LMICs, suicide rates of young adults and elderly women are much higher than in high-income countries, while in middle-aged men the suicide rates are significantly higher in high-income countries than in the LMICs [1]. For example, in Sweden, suicide rates have been decreasing in all age groups except the 15-24 year group. It is the main cause of death for men aged 15-44 years, and the second for women in this age group [2].

Suicide rates vary with age and gender: 0% for girls and boys ages 5-9 years. For ages 10-14 the rate for boys is 2.4/100.000 and for girls 0.96/100.000. In adolescents aged 15-19 years it is 3.48/100.000 for girls and 18.25/100.000 for boys. These rates indicate that males are more inclined than females to complete suicide rather than only attempt it. Causes and risk factors are not yet well understood, and research has found both differences and similarities between the genders [3].

Many studies focus on the epidemiology of suicide and gender differences in attempted versus completed suicide. Few, however, have attempted to understand the reasons for choosing one over the other and the differences between the two types of suicidal behavior.

In this chapter we will discuss those two behaviors.

From Childhood to Puberty

A Finnish study followed a group of children from age 8 till adolescence or early adulthood. The authors of this study found that most males who committed suicide or made serious attempts (serious attempt, in this study, was defined as requiring patient admission to hospital) in adolescence were diagnosed with psychiatric problems by the age of 8, indicating a trajectory that persists through life. Among females however, suicidality is not predictable by psychopathologic disorders. This finding highlights the fact that suicidality has different trajectories in males vs. females [4].

Studies show that suicide occurs in children of parents suffering from mood disorders and impulsive aggression. Suicide in children also occurs in those who come from families with a history of suicide. Suicide becomes increasingly common after puberty, probably due to the association between puberty and the onset of depression as well as substance abuse. Evidence shows that the risk for suicide increases after stress periods. Moreover, cognitive immaturity, age related impulsivity and availability of suicide methods also play an important role [5].

Another Finnish study which compared the characteristics of suicide of male children and adolescents to that of their female counterparts, indicated that while violent suicide methods predominated in both genders the majority were in males. Furthermore symptoms of mental illness were common in persons who committed suicide but only a minority of males had a previous history of psychiatric hospitalization.

It has been suggested that the difference between male and female suicidality may be explained by the higher rates of emotional and behavioral problems among adolescent females and the cultural view of suicide as a primarily feminine action. On the other hand, the traditional view of the male role may discourage males from admitting to mental health problems and thus from seeking help [6].

Impulsivity is another factor that is associated with more suicide attempts in males than females [7].

Stress

It is known that life stress events are a risk factor for suicide in youth. Boys are more likely to respond to stressors with externalizing symptoms such as aggression [8].

Adolescence is a critical period for learning stress management. The second decade of life involves tremendous cognitive development that increases adolescents' ability for abstract thinking, for reasoning and for planning [9].

Although women attempt suicide two to three times more frequently than men, death by suicide occurs more frequently among men. This may be because of men's preference for highly lethal methods. Possible reasons for more lethal behavior in males could be a stronger intent to die, higher threshold for seeking help (to be discussed later) and more aggressive and impulsive personality traits. Other potential factors include the involvement of alcohol and the role of unemployment. Moreover many cases of suicide occur in the context of psychiatric disorders like depression and schizophrenia [10].

Depression

As mentioned earlier, many individuals who commit suicide appear to have a psychiatric disorder at the time of their death. Psychological autopsy found depression to be the most common of these disorders. The highest suicide risk was found in males [11]. This may be accounted for in part by men's tendency to irritability and anger as an expression of depressive affect [12].

It is important to take into account social factors such as work and employment situation as well as increased psychological demands and job insecurity, which are strongly associated with depression especially among men.

Social Relationships

Some studies have shown that socially isolated men had a greater incidence of death from a variety of reasons such as cardiovascular disease and others. For suicide, it is though important to consider social networks. Studies have shown that the loss of family and other social support through divorce constitutes a stressful life crisis that may lead to psychological distress and places the individual at a higher risk for suicide. Widowed and divorced men are at higher risk for suicide compared to women in the same situation. This may be due to men's tendency to shy away from discussing intimate topics even with people close to them and from seeking professional help, as women tend to do.

Difficulties with social relationships, absence of cooperation, lack of trust, shallow relationships and difficulty establishing new relationships may all partially explain why their risk to die of suicide is higher. A study that investigated immigrants additionally found that for people who immigrated to a new country at a relatively young age immigration presented a major disruption in their social support network and enhanced the risk for anxiety and depressive disorders that in some of them led to suicidal behavior.

This study also found impulsiveness and aggressiveness to present additional risk factors for suicidal behavior.

A longitudinal study that followed men over a 24 year period showed that social integration was associated with lower risk for suicide and so were elements of both family and religious integration [13].

AVAILABILITY

The importance of accessibility to firearms has also been investigated in the context of suicidal behavior. It has been found that access to firearms does not seem to increase the risk of committing or attempting suicide. But people who already have suicidal ideation or plan, and in addition have access to such fire arms, are 7 times more likely to have a plan involving firearms [14].

Fire arms are considered to be the more available and more "masculine" means by which to commit suicide. Thus restrictions in availability of fire arms can serve to prevent suicide [14].

PSYCHOLOGICAL EXPLANATION

There is ample evidence in professional literature that the traditional male gender role tends to discourage males from admitting to mental health problems and from seeking help.

Few studies have investigated the reasons for these differences between the genders, and few have explained the gender differences by the attitude toward seeking psychological help [15].

Some of the barriers to men seeking help from mental health services are disinclination to express emotions - also known as restrictive emotionality, embarrassment and anxiety [15]. Being strong, invincible, unemotional, independent and resilient are traditional norms of masculinity, and they are key to men not seeking psychological help. The cognitive conflict and the negative emotions that result from the struggle with these masculine norms are known as gender-role conflict.

It has been found that younger participants expressed even more negative attitudes toward seeking psychological help than their older counterparts. This is explained by the lack of experience and understanding of mental health issues in younger people as compared to older ones [15].

Thus the male adherence to masculine norms may increase the risk for death by suicide.

It is important to understand the difference between people who desire death and those who are capable of acting on it. Studies which have differentiated between desiring and having the capability have shown the importance of the interpersonal theory that comprises elements and feeling of being isolated socially and feeling of being a burden. According to the theory, the capability for suicide comprises fearlessness of death and the ability to tolerate the physical pain that may be involved in suicide. Thus the combination of all three factors can contribute to increased risk for suicide even though having the capability to commit the act of suicide is the most important one [19-16].

High levels of distress tolerance facilitate suicidal behavior in at-risk populations and suggest that the capacity to tolerate aversive physiological and affective arousal might be vital to engagement in serious or lethal suicidal behavior [20].

Another factor that is associated with suicide is violence. It has been indicated in another study that violence and suicidality mutually affect each other from adolescence into young adulthood. Evidence was found showing that the association between suicidality and violence was stronger in males than in females, particularly in early and young adulthood. This confirms previous findings by demonstrating that violence is a risk factor for future suicide [21].

Geriatrics

Some research on the risk factors for suicide attempts and completed suicide has been done in older people as well, but much more is needed.

Suicidal ideation is seven times higher in older adults with comorbid anxiety and depression disorders than it is in the general population. The group with the highest rate of suicidality consisted of, unmarried immigrant males, who by the nature of their existence, tend to have little social support.

A 10 year survey was conducted in older suicide completers. It showed that the prevalence of suicide among people 75 years and older is the highest of all age groups, while there is a predominance of male suicide completers. Hanging as a mean for suicide was the predominant method in the older group. Other means, like shooting and jumping off a high place were also frequently employed [22].

Research shows that old age and male gender are major risk factors for completed suicide when associated with low social support, positive psychiatric history and physical illness [22, 23]

Conclusion

Further research is needed in order to reveal the differences in suicide and suicidal behavior between the genders. Nonetheless, this chapter showed epidemiologically that men are more susceptible to committing suicide. Some studies argue that the reason for this phenomenon may be under-diagnosis of suicidality in men, due to their tendency to minimize symptom severity. We therefore, need more accurate tools for the evaluation of male suicidality, including investigating the effects of stress, social relations and availability of suicide means. It is particularly important to recognize and take into consideration men's difficulties in seeking help.

REFERENCES

[1] World Health Organization. Suicide Statistics: WHO; 2014 [cited 2014 November]. Available from: http://www.who.int

[2] Hadlaczky G, Stefenson A, Wasserman D. The state of psychiatry in Sweden. International review of psychiatry. 2012;24(4):356-62.

[3] Zalsman G, Oquendo MA, Greenhill L, Goldberg PH, Kamali M, Martin A, et al. Neurobiology of depression in children and adolescents. *Child and adolescent psychiatric clinics of North America*. 2006;15(4):843-68, vii-viii.

[4] Sourander A, Klomek AB, Niemela S ,Haavisto A, Gyllenberg D, Helenius H, et al. Childhood predictors of completed and severe suicide attempts: findings from the Finnish 1981 Birth Cohort Study. *Archives of general psychiatry*. 2009;66(4):398-406.

[5] Dervic K, Brent DA, Oquendo MA. Completed suicide in childhood. *The Psychiatric clinics of North America*. 2008;31(2):271-91.

[6] Lahti A, Harju A, Hakko H, Riala K, Rasanen P. Suicide in children and young adolescents: a 25-year database on suicides from Northern Finland. *Journal of psychiatric research*. 2014;58:123-8.

[7] Rueter MA, Holm KE, McGeorge CR, Conger RD. Adolescent suicidal ideation subgroups and their association with suicidal plans and attempts in young adulthood. *Suicide & life-threatening behavior*. 2008;38(5):564-75.

[8] Vo DX, Park MJ. Stress and stress management among youth and young men. *American journal of men's health*. 2008;2(4):353-66.

[9] Halpern-Felsher BL, Millstein SG, Irwin CE. Work group II: healthy adolescent psychosocial development. *The Journal of adolescent health : official publication of the Society for Adolescent Medicine*. 2002;31(6 Suppl):201-7.

[10] Cibis A, Mergl R, Bramesfeld A, Althaus D, Niklewski G, Schmidtke A, et al. Preference of lethal methods is not the only cause for higher suicide rates in males. *Journal of affective disorders*. 2012;136(1-2):9-16.

[11] Hawton K, Casanas ICC, Haw C, Saunders K. Risk factors for suicide in individuals with depression: a systematic review. *Journal of affective disorders*. 2013;147(1-3):17-28.

[12] Lemieux AM, Saman DM, Lutfiyya MN. Men and suicide in primary care. Disease-a-month : DM. 2014;60(4):155-61.

[13] Tsai AC, Lucas M, Sania A, Kim D, Kawachi I. Social integration and suicide mortality among men: 24-year cohort study of U.S. health professionals. *Annals of internal medicine*. 2014;161(2):85-95.

[14] Anglemyer A, Horvath T, Rutherford G. The accessibility of firearms and risk for suicide and homicide victimization among household members: a systematic review and meta-analysis. *Annals of internal medicine*. 2014;160 (2.10-101).

[15] Yousaf O, Popat A, Hunter MS. An investigation of masculinity attitudes, gender, and attitudes toward psychological help-seeking: American Psychological Association; 2014 [cited 2014 November]. Available from: http://psycnet.apa.org/psycinfo/2-014 ./001-12450

[16] Van Orden KA, Witte TK, Cukrowicz KC, Braithwaite SR, Selby EA, Joiner TE, Jr. The interpersonal theory of suicide. *Psychological review*. 2010;117(2):575-600.

[17] Granato SL, Smith PN, Selwyn CN. Acquired Capability and Masculine Gender Norm Adherence: Potential Pathways to Higher Rates of Male Suicide: *American Psychological Association*; 2014 [cited 2014 November]. Available from: http://psycnet.apa.org/psycinfo/2014-45104-001./

[18] Smith PN, Cukrowicz KC. Capable of suicide: a functional model of the acquired capability component of the Interpersonal-Psychological Theory of Suicide. *Suicide & life-threatening behavior*. 2010;40(3):266-75.

[19] Moller-Leimkuhler AM. Barriers to help-seeking by men: a review of sociocultural and clinical literature with particular reference to depression. *Journal of affective disorders*. 2002;71(1-3):1-9.

[20] Anestis MD, Knorr AC, Tull MT, Lavender JM, Gratz KL. The importance of high distress tolerance in the relationship between nonsuicidal self-injury and suicide potential. *Suicide & life-threatening behavior*. 2013;43(6):663-75.

[21] Van Dulmen M, Mata A, Claxton S, Klipfel K, Schinka K, Swahn M, et al. Longitudinal Associations between Violence and Suicidality from Adolescence into Adulthood. *Suicide and Life-Threatening Behavior*. 2013;43(5):523-31.

[22] Shelef A, Hiss J, Cherkashin G, Berger U, Aizenberg D, Baruch Y, et al. Psychosocial and medical aspects of older suicide completers in Israel: a 10-year survey. *International journal of geriatric psychiatry*. 2014;29(8):846-51.

[23] Oliffe JL, Han CSE, Ogrodniczuk JS, Phillips JC, Roy P. Suicide From the Perspectives of Older Men Who Experience Depression: *A Gender Analysis American journal of men's hea*lth. 2011;5:444-54.

In: Neurobiology of Men's Mental Health
Editors: Leo Sher and Timothy R. Rice ISBN: 978-1-63483-191-8
© 2015 Nova Science Publishers, Inc.

Chapter 16

TESTOSTERONE AND HOMICIDE AMONG YOUNG MEN

Allan Mazur, PhD*
Syracuse University, Syracuse, New York, US

ABSTRACT

High testosterone does not appear to contribute directly to violence among young men. However there is good evidence that high testosterone, possibly in combination with low cortisol, promotes assertive, leader-like, and antisocial behavior. Thus high testosterone seems to encourage young men to engage in "dominance contests" to determine superior rank in a group's status hierarchy. Face-to-face dominance contests are normally nonviolent, even friendly or conducted jokingly. However violence may occur when dominance contests escalate beyond their usual bounds. The likelihood of escalation to violence is increased by the influence of alcohol and the cost of losing status in front of an audience. Whether such violent encounters turn out to be fatal or merely injurious is largely a matter of chance and extenuating circumstances like the ready availability of medical care or having weapons at hand. "Honor subcultures" such as are found in our inner cities place inordinate importance on the enhancement of personal reputations and the humiliation of losing face. This atmosphere of persistent challenge may produce heightened testosterone in young black men of the inner city, raising the likelihood that they will engage in dominance competition, which in turn raises the likelihood of a violent, even fatal, outcome.

Keywords: homicide, testosterone, dominance, dominance contests, honor cultures

* Tel./Fax: +001 315 445 1970; E-mail: amazur@syr.edu.

INTRODUCTION

There are diverse reasons for murder including drug marketing, robbery, jealousy, mental derangement, religious or ideological commitment, cash payment, and familial obligation.

Here I focus on what criminologist Jack Katz calls the most common type of criminal homicide, the impassioned killing of someone for what the killer regards as a good moralistic reason, perhaps the defense of his family, his property, or his good name [1].

Usually these murders of passion occur without premeditation. Katz has in mind episodes that develop quickly, occurring without thought of legal consequences. Often the killers are surprised by the unintended fatality, regarding the outcome as an accident. Reflecting their lack of forethought, many killers do not attempt to escape, or do it ineptly. Police make arrests in roughly 80 percent of homicide cases, usually within a day, compared to arrests in about 25 percent of robberies and 15 percent in burglaries.

Death is not necessarily the desired end point of an impassioned attack so much as hurting or physically punishing the victim. Whether an attack is fatal or not may be incidental, depending on such chance factors as the time to reach an emergency room, the quality of medical service, whether a gun was used, whether the falling victim's head hit concrete, and so on. There is little reason to think such killings differ much from impassioned attacks whose victim survived. Murders are better documented than non-fatal assaults and therefore more amenable to analysis.

Victims are usually relatives, friends, or at least acquainted with their killers. It is well established that fatal aggression between men is usually precipitated by a trivial altercation, perhaps an insult, curse, or jostling. This is followed by an escalation of hostile verbal actions that may look no different than many nonfatal arguments.

DISTINGUISHING DOMINANCE FROM AGGRESSION

Numerous animal experiments, mostly on rodents, show that raising testosterone increases aggressiveness. It is important in interpreting this work to distinguish aggressive behavior from dominance behavior. An individual will be said to act *aggressively* if its apparent intent is to inflict physical injury on a member of its species. An individual acts *dominantly* if its apparent intent is to achieve or maintain high status – i.e., to obtain power, influence, or valued prerogatives – over a conspecific. Rodents typically dominate aggressively, but that is not true among the higher primates.

The difference between aggression and dominance is especially important for humans, because we normally assert our dominance without violence. Sports, spelling bees, elections, criticism, competitions for promotion, and academic debate all involve contests for domination without intending physical injury. We understand that there are different motivations for dominance and violent aggressions, though they may sometimes work together.

Earlier researchers thought that young men with high testosterone were especially aggressive, like male rodents. This stereotype persists among the public, with images of steroid-primed body builders having outbursts of violence, or "roid rages." Actually, empirical research shows little if any *direct* link between testosterone and physical aggressiveness. However, testosterone may play an *indirect* role in interpersonal violence [2].

Testosterone is related primarily to dominant behavior among men, not to aggression per se [3, 4]. Nearly all animal studies once interpreted as linking testosterone to aggression may as easily be interpreted as linking testosterone with dominance. Dominating tendencies, whether aggressive or nonaggressive in action, would confer an evolutionary advantage in helping an individual acquire valued resources, especially in competition for mates. This is not simply a matter of a dominant man taking what he wants; women regard men who *look* dominant or *act* dominantly as attractive.

TESTOSTERONE WORKS DIFFERENTLY IN UTERO, PUBERTY, AND ADULTHOOD

Testosterone affects human males importantly but differently at three stages of life: perinatally, during puberty, and in adulthood. The mammalian fetus of both XX and XY individuals begins with undifferentiated sexual parts. A gene on the Y chromosome causes the asexual gonads to develop as testes; lacking this gene the gonads become ovaries. The sex chromosomes have little more to do with sex differentiation which hereafter is driven by hormones produced in the now sex-specific gonads. The testes produce testosterone during gestation, and production peaks again a month or two after birth, then declines by 6 months of age to the low range seen in later childhood. Testosterone and other testicular secretions cause the external genitalia to form into penis and scrotum rather than clitoris and labia. Many perinatal hormone effects are regarded as *organizing* the architecture of the body and brain, and the distribution of hormone receptors, into a relatively male-like configuration. When male testosterone increases later in life, it *activates* these preexisting structures. Thus, behaviors derive from the interaction of long-term organizational *and* shorter-term activation effects.

At puberty the testes greatly increase production of testosterone, promoting growth of the penis, larynx (and deeper voice), muscles, beard and body hair, and sex interest. It is tempting to assume the testosterone surge in adolescent boys causes their seemingly heightened aggressiveness or pugnacity at that age. However, research has not verified a causal link between the hormonal and behavioral changes. First of all, physical aggression in boys does *not* generally rise during adolescence [5]. Actually, it is littler boys who are most prone to physically assault their peers, though they lack the muscle or weapons to do much damage. The well-documented rise in boys' antisocial behavior with puberty is due mainly to nonviolent delinquency such as vandalism and status violations [6].

A recent meta-analysis of puberty studies concludes that there are insufficient longitudinal data of high quality to confirm that changing testosterone per se causes changes in mood and behavior in adolescent boys [7]. These are difficult studies to conduct because investigators must untangle the direct effect testosterone from other physical changes in the boy's body at puberty, which affect how people respond to him. Also, there are important *social* changes during the early teen years – entry into high school, taking a job, prolonged absence from parents, more dependence on peer approval – any of which may affect behavior independently of hormonal effects.

Among American males, testosterone peaks in the late teens and early twenties, and usually declines slowly with age, more so if men become obese [8]. The hormone decline among middle-age men does not reliably occur in non-industrial societies. For unknown reasons, testosterone levels of American men have declined since the 1980s independently of increasing obesity and aging of the population [9].

By the end of teenage, the physical form of a boy has changed into that of a man so testosterone no longer influences behavior through major reorganization of the body. However, the level of testosterone circulating in the bloodstream at any moment may affect dominance behavior by activating receptors in organs or the nervous system. The remainder of this chapter focuses on such post-adolescent effects.

RECIPROCAL CAUSATION

Much evidence from a variety of settings shows that in men, circulating testosterone is correlated with dominant behavior [10-11]. Most impressive are recent studies linking, with some qualification, endogenous increases (or decreases) in testosterone with increased (decreased) readiness to compete further [12-13]. Cortisol, which reliably elevates under stressful conditions, may also play a role. According to the "dual hormone hypothesis," testosterone is associated with dominance or leadership especially when cortisol is low [14]. This hypothesis is intuitively appealing, suggesting that the people most likely to act dominantly, to break norms or otherwise scale stressful barriers, are those least bothered by the stressor, thus low in cortisol. The import of a testosterone-by-cortisol interaction for dominant behavior is now an active area of research.

Not only does testosterone affect dominance, but changes in dominance behavior or in social status cause changes in testosterone level. We have strong evidence on this "reverse" effect in humans. By now there are many reports of testosterone changes in young men during athletic events as well as other competitive situations that are not physically taxing. Again with some qualification, male testosterone varies in predicable ways both before and after competitive matches. First, testosterone rises shortly before the match, as if in anticipation of the competition. This pre-competition boost may promote dominant behavior, increasing the chance victory, as occurs in rodents but has not yet been demonstrated in humans [15]. Second, for one or two hours after the match, testosterone of winners is usually high relative to that of losers. The win/loss effect even works vicariously, with supporters of one side or another in an athletic contest or an election showing corresponding changes in testosterone depending on whether or not "their" side won or lost [16-18]. Thus, the pattern of testosterone fluctuations appears in nonphysical as well as physical competition, and in response to symbolic challenges and status changes among men. (Whether similar changes occur among women is unresolved.)

In theory, a man who has experienced a recent rise in testosterone, perhaps from a prior victory or a symbolic elevation in status, should be unusually assertive and may challenge someone of relatively high status in a contest for dominance. If the winner experiences rising testosterone as a result of his victory, this should sustain or increase his assertiveness and his display of dominant signs such as erect posture, sauntering or striding gait, and direct eye contact with others. Thus bolstered, the dominant actor may seek out new challenges and is primed to win them. The feedback between high testosterone and dominant demeanor may explain the momentum often associated with strings of triumphs. Success begets a high testosterone response, which begets more dominant behavior, which begets more success.

The loser experiences a drop in testosterone, reducing his assertiveness and increasing his display of deferential signs such as stooped posture, smiling, or eye aversion. Faced with a new dominance encounter, he is now at a psychic and physiological disadvantage. One defeat begets another because the loser becomes increasingly likely to retreat or submit. This may be an adaptive response, saving losers from further losses and possible damage.

Dominance Contests

As is common among primates, humans in face-to-face groups form themselves into fairly consistent dominance/status hierarchies so that higher-ranked members have more power, influence, and valued prerogatives than lower-ranked ones. Ranks are allocated either *cooperatively*, by consensus of those involved, or *competitively*, when there is disagreement over who should outrank whom.

To appreciate a person's decision to compete or cooperate, visualize two individuals (Ego and Alter) meeting for the first time. If their interaction is very brief or casual, the notion of ranking may never arise. However, in more extended or serious meetings, each will size up the other and gain some sense of their relative standings. If Ego thinks that Alter's status does or should exceed his own, he may defer to Alter without any competition. In human terms, Ego may believe that Alter belongs in the higher rank, that Alter deserves it, that Alter could easily take it if Ego resisted, or that Alter would be more competent in the duties of high rank. In any case, ranks are allocated quickly and cooperatively. If Ego and Alter do not agree on their relative standings, then they may either break off the interaction or vie for the contested rank.

Ego's decision to compete or to comply will also depend on his motivation to dominate, which seems related to his testosterone level (among other factors). A man who has experienced a recent rise in testosterone, perhaps from a victory or a symbolic elevation in status, will be unusually assertive and may challenge someone of relatively high status. If both Ego and Alter decide to compete, their relative ranks are then determined by the outcome of one or more short dominance contests between them.

Nonhuman primates are commonly observed to establish and maintain their status hierarchies through a series of short face-to-face competitions between members of the group. Some competitions involve fierce combat; others are mild, as when one animal is obviously the more powerful and assertive one or the other appears fearful. In such cases, a simple stare by the powerful animal, followed by the other animal's eye aversion or by yielding something of value (perhaps food or a sitting place), may suffice. Sometimes a single contest is all that is needed to allocate ranks or to verify a preexisting rank relationship, but often the outcome is settled only after a series of contests.

A psychophysiological mechanism operating across this range of competition is the manipulation of stress levels [19]. An exchange of threats or attacks is seen as an attempt by each individual to "outstress" or intimidate the other by inducing discomfort. Stress is experienced both psychologically and as a syndrome of neurological responses. The individual who outstresses his adversary is the winner.

The model becomes clearer if we consider a concrete example. Imagine two strangers, Ego and Alter, whose eyes meet, by chance, across a room. Let us say that one of them, Ego, decides to hold the stare. The chance eye contact now becomes a dominance encounter. Ego's stare makes Alter uncomfortable. Alter may then avert his eyes, thus relieving his discomfort while, in effect, surrendering, or he may stare back, making Ego uncomfortable in return. In the latter case, the stare-down would continue, with each individual attempting to outstress the other until finally one person succumbed to the discomfort (and the challenger) by averting his eyes. The matter thus settled, the yielder usually avoids further eye contact, though the winner may occasionally look at the loser as if to verify his victory.

In this example, Ego's stare is assumed to elicit feelings of stress in Alter. Alter's eye aversion is assumed to relieve his own felt stress. Staring – the stress-inducing behavior – is a dominant sign associated with high status. Eye aversion is a deferential sign associated with low status. In other words, a dominant act (staring) elicits stress in the recipient; a submissive act (eye aversion) relieves stress in the actor. It is a central assumption of this model that most dominant and deferential acts work this way, inducing or relieving stress, respectively. These acts are the means whereby the adversaries wage their stress contest, each aiming "darts" at the other. Finally, when the stress is too great for one, he switches from dominant to deferential actions, thereby relieving his stress and simultaneously signaling his acceptance of the lower rank.

Within minutes of this outcome, we assume Ego (the loser) experiences a drop in testosterone, reducing his assertiveness, diminishing his propensity to display the dominant actions associated with high status, and increasing his display of such submissive signs as stooped posture, smiling, or eye aversion. Faced with a new dominance encounter, Ego is more likely than before to retreat or submit. On the other side, Alter, the winner, experiences the opposite effects: rising testosterone, increased assertiveness, and a display of dominant signs such as erect posture, sauntering or striding gait, and direct eye contact with others. Alter may seek out new dominance encounters and is bolstered to win them. This feedback between high (or low) testosterone and dominant (or submissive) demeanor would help to explain the momentum often associated with strings of triumphs or defeats: success begets a high testosterone response which begets more dominant behavior which begets more success

As an aside, the phrase "dominance contest" was adopted from the literature on nonhuman primates, which commonly refers to "dominance hierarchies," whereas the equivalent term for face-to-face human groups would usually be "status" or "prestige" hierarchies. Unfortunately the word "dominance," when applied to humans, often evokes images of forceful, coercive, even physically punishing interaction, an inference especially likely in a chapter on homicide. In fact, among humans, dominance contests are usually gentle, barely if at all coercive, and very rarely physically damaging.

MURDER

In everyday life, dominance contests are based on the subtle manipulation of psychological and physiological stress, not on causing or even threatening physical harm to one's adversary. Most adult humans experience little violence, but occasionally face-to-face competition escalates to a violent stage not originally intended or foreseen.

While a killing is rarely the outcome of a violent dominance contest, I focus on murder because it represents an unambiguous endpoint, has good statistics, has been much studied by criminologists, and is so distressing a phenomenon. In the U.S., killers and their victims are disproportionately young adult males. There is a large racial disparity in murder rates. For example, in New York City from 2002 through 2005 there were 1,662 reported murders. Killers were 93% male, 76% between ages 18 and 40, and 61% black (compared to a black population of 25%). Victims were 82% male, 69% between 18 and 40, and 60% black [19]. The reasons for these patterns are not fully understood, but I propose that testosterone plays an indirect role.

Often the killer feels provoked by the victim, being the target of an insult, the butt of a joke, or an object of humiliation. "From the killer's perspective, the victim either teases, dares, defies, or pursues the killer... That the killer feels compelled to respond to a fundamental challenge to his worth is indicated as well by the frequent presence and the role of an audience" [1, 20].

Extreme anger (rage) heightens the potential lethality of competition by shifting the adversaries' intentions from dominating to damaging the opposition. The red face of anger, with its raised voice and universally recognized facial gestures and body postures, likely entails particular neurophysiological actions. Alcohol may intensify the normal neurophysiological effect of rage.

Nisbett and Cohen [21] attribute the historically high violence in the American South, compared to the North, to its "culture of honor" whereby Southern men, when challenged by insults to themselves or their families, are required to defend themselves as virtuous warriors or else lose face. Apparently as a result, Southern men are unusually alert to possible insults, reacting dominantly – sometimes violently – to speech or actions that might not be perceived as injurious in other cultures.

Leaving aside the particular historic roots of the South, there may be a general hypersensitivity to insult in any subculture that is (or once was) organized around young men who are unconstrained by traditional community agents of social control, as often occurs in frontier communities, gangs, among vagabonds or bohemians, and after breakdowns in the social fabric following wars or natural disasters. When young men place special emphasis on protecting their reputations, and they are not restrained from doing so, dominance contests become ubiquitous, the hallmark of male-to-male interaction. The leading student of street behavior in America's inner cities, sociologist Elijah Anderson [22, 88–89], vividly portrays the importance of dominance contests and their constant presence for poor young black men:

> (M)ost youths have...internalized the code of the streets..., which chiefly (has) to do with interpersonal communication..., (including) facial expressions, gait, and verbal expressions – all of which are geared mainly to deterring aggression...
>
> Even so, there are no guarantees against challenges, because there are always people looking for a fight to increase their share of respect – of "juice," as it is sometimes called on the street. Moreover, if a person is assaulted, it is important, not only in the eyes of his opponent but also in the eyes of his "running buddies," for him to avenge himself. Otherwise he risks being "tried" (challenged) or "moved on" by any number of others. To maintain his honor he must show he is not someone to be "messed with" or "dissed."

> ...The craving for respect that results gives people thin skins. Shows of deference by others can be highly soothing, contributing to a sense of security, comfort, self-confidence, and self-respect... Hence one must be ever vigilant against the transgressions of others or even appearing as if transgressions will be tolerated. Among young people, whose sense of self-esteem is particularly vulnerable, there is an especially heightened concern with being disrespected. Many inner-city young men in particular crave respect to such a degree that they will risk their lives to attain and maintain it.

We know that testosterone rises in men awaiting a contest, regardless of the eventual outcome of that contest. Generalizing to the street, hormone levels should be elevated in young men who are constantly vigilant against challenges to their reputations. Testosterone is also affected by the outcome of the contest, so persistent losers might be hormonally depressed, but most men – those with mixed outcomes or better – are expected to have elevated testosterone, thus hormonally primed to react assertively to real or perceived insult. But is it in fact true that testosterone levels are especially high among young black men living in poor neighborhoods?

The question has been addressed in two large epidemiological studies. One, based on 4,462 male U.S. Army veterans, answered in the affirmative. Younger black men who were poorly educated – those veterans most likely to be participants in inner-city honor cultures – did as a group have very high testosterone; among older men or well-educated men, race made no difference in hormone levels [19]. However a second study, based on different sample of 1,637 men drawn from the U.S. population, showed testosterone of young, poorly educated black men to be unexceptional, thus contradicting the first study [8]. Like so many questions about hormones and human behavior, we do not have a clear answer.

CONCLUSION

There is no indication that heightened testosterone in young men leads *directly* to violence. However in rare instances testosterone may contribute *indirectly* to violent behavior, even to homicide.

Sometimes one man's status relative to another is negotiated in one or more dominance contests in which each participant displays linguistic and nonverbal signs and signals that put stress on the other man. The usual outcome of one or more dominance contests is a settled ranking, one man having higher status than the other. While dominance contests among nonhuman primate species often take violent form, this is unusual among humans, who typically carry out their competitions for status in a friendly and polite manner without physical contact, often barely aware that they are competing or stressing one another. Apparently, high testosterone encourages a man to participate in dominance contests and to be assertive in his signaling of high rank. Conversely, low or declining testosterone encourages a man to avoid or withdraw from such contests, accepting low rank.

Occasionally dominance contests escalate beyond their usual boundary of friendly or at least nonviolent competition. The escalation might move from loud voices through angry insults, swearing and shouting, to placing hands on the other person. At this stage, physical violence is increasingly likely. This is a typical scenario for what criminologist Jack Katz calls "impassioned killing," the most common form of homicide. Usually there is no premeditation, no intent to kill. What distinguishes physical assault in such escalations from killing is often a matter of chance or bad luck. A blow, not intended to be lethal, may have that consequence. A weapon at hand, having the confrontation in front of an audience, using alcohol or other disinhibiting intoxicants -- all these increase the likelihood of escalation and a fatal outcome. Thus testosterone, while not a direct precursor to violence, can indirectly produce that outcome by increasing the tendency toward dominance competition. Honor cultures as described by Elijah Anderson in poor black neighborhoods, in which young men are hyper-vigilant about their personal reputations and persistently on alert for challenges, ought theoretically to be associated with high testosterone levels, which might reinforce this dysfunctional behavior. Whether or not testosterone is actually higher is uncertain, with one study suggesting yes, another no.

The biosocial model sketched here raises issues of scientific validity and policy ethics. The link between testosterone and dominance has primarily correlational support and will require experimental manipulation, comparing men treated with testosterone versus placebo, to establish causal connections. In prior decades it was considered incautious to dose healthy subjects with testosterone because of possibly negative effects, however as testosterone therapy has recently boomed, apparently without causing harm, we are on the verge of experimental studies of its behavioral effects. Biomarkers are increasingly included in large population surveys and may settle whether or not inner city men have unusually high testosterone, however any parsing of testosterone levels by race and social class must be sensitive to the potential for invidious comparisons of groups.

If high testosterone does play a causal role in the high homicide rate among young black men, there would be a troubling policy question of what, if anything, to do about it. Any notion of a medical or pharmaceutical fix, rather like prescribing Ritalin for hyperactivity, would reek of race-based chemical castration and should be regarded as outside the pale.

REFERENCES

[1] Katz J. Seductions of Crime. New York: Basic Books, 1988.

[2] Mazur A. Dominance, violence and the neurohormonal nexus. Handbook of Neurosociology. Edited by Franks D., Turner J. Springer, 2012.

[3] Mazur A., Booth A. Testosterone and dominance in men. *Behav. Brain Sci.*, 1998; 21: 353–363.

[4] Archer J. Testosterone and human behavior. *Neurosci. Biobehav. Rev.,* 2006; 30: 319–345.

[5] Tremblay R. The development of aggressive behaviour during childhood. *Int. J. Behav. Dev*., 2000; 24: 129–141.

[6] Rutter J., Giller H., Hagell A. Antisocial behavior by young people. New York: Cambridge University Press, 1998.

[7] Duke S. A., Balzer B. W., Steinbeck K. S. Testosterone and its effects on human male adolescent mood and behavior: a systematic review. *J. Adolesc. Health,* 2014; 55: 315-322.

[8] Mazur A. The age-testosterone relationship in black, white and Mexican-American men, and reasons for ethnic differences. *Aging Male,* 2009; 12(2): 66–76.

[9] Mazur A., Westerman R., Mueller U. Is rising obesity causing a secular (age-independent) decline in testosterone among American men? *PLoS ONE,* 2013; 8: e76178.

[10] Burnham T. High-testosterone men reject low ultimatum game offers. *Proc. Biol. Sci.,* 2007; 274: 2327–2330.

[11] Carré J. M., McCormick C. M., Hariri A. The social neuroendocrinology of human aggression. *Psychoneuroendocrinology,* 2011; 36: 935–944.

[12] Carré J. M., Putnam S. K., McCormick C. M. (2009). Testosterone responses to competition predict future aggressive behaviour at a cost to reward in men. *Psychoneuroendocrinology*, 2009; 34: 561–570.

[13] Mehta P. H., Josephs R. A. Testosterone change after losing predicts the decision to compete again. *Horm. Behav.*, 2006; 50: 684–692.

[14] Mehta P. H., Josephs R. A. Testosterone and cortisol jointly regulate dominance: Evidence for a dual-hormone hypothesis. *Horm. Behav.,* 2010; 58: 898-906.

[15] Gleason E. D., Fuxjager M. J., Oyegbile T. O., Marler C. A. Testosterone release and social context. *Front. Neuroendocrinol.*, 2009; 30: 460-469.

[16] Bernhardt P., Dabbs J., Fielden J., Lutter C. Testosterone changes during vicarious experiences of winning and losing among fans at sporting events. *Physiol. Behav.*, 1998; 65: 59–62.

[17] Stanton S., Beehner J., Saini E., Kuhn C., Labar K. Dominance, politics, and physiology: voters' testosterone changes on the night of the 2008 United States presidential election. *PLoS One*, 2009; 4(10): e7543.

[18] Carré J. M., Putnam S. K. Watching a previous victory produces an increase in testosterone among elite hockey players. *Psychoneuroendocrinology,* 2010; 35: 475–479.

[19] Mazur A. Biosociology of Dominance and Deference. New York: Rowman and Littlefield, 2005.

[20] McGinity J. New York killers, and those killed, by numbers. New York Times 2006, April 28.

[21] Nisbett R., Cohen D. Culture of Honor. Boulder CO: Westview Press, 1993.

[22] Anderson E. The code of the streets. *Atlantic Monthly*, 1994; 5: 81–94.

In: Neurobiology of Men's Mental Health
Editors: Leo Sher and Timothy R. Rice
ISBN: 978-1-63483-191-8
© 2015 Nova Science Publishers, Inc.

Chapter 17

SUICIDAL TERRORISM: PSYCHIATRIC ASPECTS

Leo Sher[1,2,] ***, M.D***
[1]Department of Psychiatry,
Icahn School of Medicine at Mount Sinai, New York, NY, US
[2]James J. Peters Veterans' Administration Medical Center,
Bronx, New York, US

ABSTRACT

Acts of terror, go back many hundreds, if not thousands, of years. A terrorist act is rarely an end in itself but is rather designed to instill fear in whole populations by targeting a small, representative group. Terrorism has achieved special prominence in the modern technological era. Suicide bombing looks like suicide, but in important aspects it is incomparable with suicide. It is part of a political and military strategy. There are divergent opinions whether suicide bombers are mentally healthy or not. It has been suggested that suicide bombers have no underlying psychopathology, and that it is, in fact, their mental stability helps them to endure pressure. Psychological strength, not weakness, may underlie the characters of the most effective terrorists. However, several lines of evidence suggest that suicide terrorists have classic suicide risk factors, such as depression, post-traumatic stress disorder, drug addictions, other psychiatric disorders, serious physical injuries or disabilities, or having suffered the sudden death of a loved one or from other personal crises.

Keywords: men's mental health, suicide, terrorism, politics

INTRODUCTION

The term terrorist was first used by Edmund Burke in 1790 [1]. Acts of terror however, go back many hundreds, if not thousands, of years [1, 2]. The last groups happy to call themselves terrorists were Russian anarchists in the late 1880s. In 1920, Lev Trotsky, the first military minister of the Soviet Russia, explained the rationale of terrorism in the following way: "War, like revolution is founded upon intimidation. A victorious war, generally speaking, destroys only a portion of the conquered army, intimidating the remainder and breaking their will … terror … kills individuals and intimidates thousands" [1]. A terrorist act is rarely an end in itself but is rather designed to instill fear in whole populations by targeting a small, representative group. Mao Zedong spoke of "killing one to move a thousand" [3].

Terrorism has achieved special prominence in the modern technological era, beginning in the1970s as international terrorism, continuing in the1980s and 1990s as American domestic terrorism, and apparently coming full circle in the 21st century [2]. Many terrorism experts believe that the worst is yet to come [2, 4].

The U.S. Federal Bureau of Investigation defines terrorism as "the unlawful use of force or violence against persons or property to intimidate or coerce a government, the civilian population, or any segment thereof, in furtherance of political or social objectives" [5]. According to the U.S. Department of Defense, "terrorism is the calculated use of violence or threat of violence to instill fear, intended to coerce or try to intimidate governments or societies in the pursuit of goals that are generally political, religious, or ideological" [5]. The operative terms in both of these definitions are coercion and intimidation and the ability to convert weakness of numbers into strength of impact.

It is important to note that some, usually totalitarian, governments around the world were terrorizing their populations under the guise of counter-terrorism. The worst state terror was employed by Stalin, Hitler and Pol Pot. But it has existed in many countries.

Suicide Terrorism

Various definitions of suicide attacks/suicide terrorism have been proposed [6-10]. Dr. Boaz Ganor, the Executive Director of the International Policy Institute for Counter-Terrorism in Israel defines suicide terrorism as "An operational method in which the very act of the attack is dependent upon the death of the perpetrator" [7].

A research paper on suicide terrorism in Sri Lanka by the Institute of Peace and Conflict Studies in India defines suicide terrorism as, "The readiness to sacrifice one's life in the process of destroying or attempting to destroy a target to advance a political goal. The aim of the psychologically and physically war-trained terrorist is to die while destroying the enemy target" [8]. The "Free Dictionary" defines suicide terrorism as "Suicide for a perceived greater cause, in which a person kills himself/herself—and many others—to advance a particular political or religious agenda" [9] "Wikipedia. The Free Encyclopedia" states that "A suicide attack is an attack upon a target in which an attacker intends to kill others and/or cause great damage, knowing that he or she will either certainly or most likely die in the process" [10].

There were 1200 suicide attacks around the world between 1981 and 2006 [11]. It constituted 4% of all terrorist attacks but 32% (about 15,000 people) of all terrorism-related deaths. Suicide bombing looks like suicide, but in important aspects it is incomparable with suicide [12]. In the Western mind, suicide is associated with despair, defeat, depression, and a disordered mind [13, 14]. Suicide terrorism is comparable to killing in a war. It is part of a political and military strategy.

In his speech before the U.S. House of Representatives on July 14, 2005, Rep. Ron Paul, M.D., said: "It is virtually impossible to stop determined suicide bombers. Understanding why they sacrifice themselves is crucial to ending what appears to be senseless and irrational" [15].

Forensic experts can usually distinguish a suicide bomber from other victims of violence. Analysis of injuries observed in the suicide bombers in Sri Lanka showed distinctive injury patterns consisting of detachment of the head and limbs, severe disruption of the trunk, burns at the transected tissue margins, presence of cyanide capsule in the neck, and absence of the shrapnel injuries [16]. These findings may help to recognize the suicide bomber.

PSYCHIATRIC ASPECTS

Education and Training

The ability to commit otherwise unspeakable acts, not to mention giving one's own life, necessitates a strong belief that these acts are in the cause of some absolute and worthy purpose. The preparatory work is typically done by others behind the scene, leaving the suicide attacker's mind clear to pray and reflect on the purpose of his/her mission.

Education of future suicidal terrorists involves indoctrination [17, 18]. Indoctrination utilizes personal, video and written testament by figures of power and authority, heroes, martyrs and the families of the organization.

The media play a vital role in the propagation of terror-related suicide [17]. Studies have shown that sensational media reporting induces "copycat" suicidal behaviors. It is particularly relevant to suicide bombing, where the attempt to inflict as many casualties as possible arouses tremendous media coverage.

Suicide Terrorists are Mentally Stable

There are divergent opinions whether suicide bombers are mentally healthy or not. Some authors propose that possibly, vulnerable members in a society may be seduced into acting for terrorists, but this runs the risk of exposure and failure so is unlikely [1]. It has been suggested that suicide bombers have no underlying psychopathology, and that it is, in fact, their mental stability helps them to endure pressure [12, 17, 19, 20]. Psychological strength, not weakness, may underlie the characters of the most effective terrorists [20]. Some terrorist organizations routinely exclude unstable persons as candidates for suicidal terrorism [12, 17]. Interestingly, some terrorist recruiters will not select candidates they deem to have suicidal tendencies [20]. In one study, thirty-four parents and siblings of suicide bombers were interviewed from 1993 to 1998 and no clear evidence of psychopathology and no common personality type were found [21]. However, about one-third of the terrorists had suicidal tendencies, although they did not display the main recognized risk factors for suicide, including depression, substance abuse, or a history of suicide attempts.

Suicide Terrorists Have Psychiatric Problems

Several lines of evidence suggest that suicide terrorists have classic suicide risk factors, such as depression, post-traumatic stress disorder, drug addictions, other psychiatric disorders, serious physical injuries or disabilities, or having suffered the sudden death of a loved one or from other personal crises [22, 23].

Some reports suggest that exposure to recent violence in traumatic situations, especially combined with childhood exposure to violence, may lead to posttraumatic stress disorder, which in turn may result in suicidal violence and suicidal terrorism [24, 25].

Lester [25] is referring to a CNN report aired in August 2007. In October 2001, a Palestinian girl was wounded by Israeli troops. S. was the first to reach her. She died in his arms. After that incident, he stopped laughing and talking and became silent and quiet. He prayed all the time. He was then recruited by terrorists for his mission. He opened fire at a busy intersection in the Israeli city of Hadera, killing and wounding many people before being killed by police. It is possible that S. suffered from posttraumatic stress disorder, or depression, or comorbid posttraumatic stress disorder and depression which can be conceptualized as posttraumatic mood disorder [26, 27].

R., a 22-yr.-old mother of two small children, blew herself up at a Gaza border crossing [28-30]. There were reports that she had arguments with her in-laws and husband and was having an extramarital affair. She was forced by her husband and her lover to sacrifice herself [29]. Israeli intelligence officials suggested that she had been depressed for a long time and had previously attempted suicide.

It has been suggested that there are two categories of suicide bombers: retaliation bargaining defined as arising from feelings of humiliation and powerlessness (which depicts male suicide terrorists) and negative inclusive fitness defined as arising from a feeling of being a burden to others (which depicts female suicide terrorists) [28].

Probably, some suicidal terrorists are mentally healthy individuals and some of them have psychiatric problems. It has been suggested that the causes of suicide terrorism lie not in individual psychopathology but in broader social conditions and that understanding and knowledge of these conditions is crucial for developing appropriate public policies and responses to protect the public [11]. Suicide terrorism is a very complex phenomenon.

CONCLUSION

Let's hope that the world will become more just, peaceful and humane. Every day, TV, radio, Internet and newspapers report on violence, terror and wars. Currently, 64 countries are involved in wars [31]. Possibly, human mental functions have differentiated unevenly: emotionally, the contemporary human functions almost like the Stone Age human, but human intellectual capacity has developed enormously [32, 33]. Emotions control human behavior, and humans employ their intellect to reach their goals. It is amazing to watch how people create more and more advanced methods to study brain receptors and, at the same time, they create more and more advanced weapons to kill each other.

Let's be kind to each other. People inflict pain on others in the selfish pursuit of their happiness or satisfaction. But true happiness comes from love and compassion.

Let's free our hearts from hatred. Let's give more. Let's expect less. Let's remember that life is not primarily a quest for pleasure, or a quest for power, but a quest for meaning.

REFERENCES

[1] Palmer I. Terrorism, suicide bombing, fear and mental health. *Int. Rev. Psychiatry,* 2007; 19(3):289-96.

[2] Miller L. The terrorist mind: I. A psychological and political analysis. *Int. J. Offender Ther. Comp. Criminol.*, 2006; 50(2):121-38.

[3] Bolz F., Dudonis K. J., Schultz D. P. The counter-terrorism handbook: Tactics, procedures, and techniques. Boca Raton, FL: CRC Press, 1996.

[4] Savitch H. V. Does 9-11 portend a new paradigm for cities? *Urban Affairs Review,* 2003; 39:103-127.

[5] Seger K. A. Deterring terrorists. In A. Silke (Ed.). Terrorists, victims and society: Psychological perspectives on terrorism and its consequences. Chichester, UK: Wiley, 2003, pp. 257-269.

[6] DCSINT Handbook No. 1.03, Suicide Bombing in the COE. Fort Leavenworth, Kansas, 15 August 2005.

[7] Ganor B. The first Iraqi suicide bombing: A hint of things to come? International Policy Institute for Counter-Terrorism Article, Herzliya, Israel, 30 March 2003.

[8] Gunaratna R. Suicide terrorism: a global threat. In: “Suicide Terrorism in Sri Lanka” IPCS. Research Papers, no. 5, R. Ramasubramanian, August 2004.

[9] The Free Dictionary by Farlex. Suicide terrorism. http://medical-dictionary.thefreedictionary.com/Suicide+Terrorism Accessed: November 2, 2014.

[10] Wikipedia. The Free Encyclopedia. Suicide attack. http:// en.wikipedia.org/wiki/Suicide_attack Accessed: November 2, 2014.

[11] Hassan R. What motivates the suicide bombers? Yale Global, 3 September 2009. http://yaleglobal.yale.edu/content/what-motivates-suicide-bombers-0 Accessed: October 27, 2014.

[12] Grimland M., Apter A., Kerkhof A. The phenomenon of suicide bombing: a review of psychological and nonpsychological factors. *Crisis,* 2006; 27(3):107-18.

[13] Sher L. Preventing suicide. *QJM*, 2004; 97(10):677-80.

[14] Maris R. W. Suicide. *Lancet*, 2002; 360(9329):319-26.

[15] Paul R. Suicide terrorism. Before the US House of Representatives, July 14, 2005. URL: http://archive.lewrockwell.com/paul/paul262.html Accessed October 31, 2014.

[16] Ruwanpura R., Kumara K. S., Jayawardane H., de Alwis L. B. Suicidal bomb explosions in Sri Lanka. *Am. J. Disaster Med.*, 2008; 3(1):47-51.

[17] Merari A. The readiness to kill and die: Suicidal terrorism in the Middle East. In: Reich W. (Ed.). *Origins of Terrorism*. Washington DC: Woodrow Wilson Center Press, 1990, pp. 192–207.

[18] Lelyveld J. All suicide bombers are not alike. *The New York Times,* 2001, October 28.

[19] Miller L. The terrorist mind: II. Typologies, psychopathologies, and practical guidelines for investigation. *Int. J. Offender Ther. Comp. Criminol.*, 2006; 50(3):255-68.

[20] Moghadam A. Palestinian suicide bombing in the Second Intifada: Motivations and organizational aspects. *Studies in Conflict and Terrorism,* 2003; 26:65–92.
[21] Merari A. Political, organizational, and psychological factors in suicide terrorism. Paper presented at the NATO Advanced Research Workshop on Suicide Terrorism: Strategic Importance and Counterstrategies, Lisbon, Portugal, 2004.
[22] Lankford A. (2013). The Myth of Martyrdom: What Really Drives Suicide Bombers, Rampage Shooters, and Other Self-Destructive Killers. New York: Palgrave Macmillan, 2013.
[23] Merari A. Driven to Death: Psychological and Social Aspects of Suicide Terrorism. New York: Oxford University Press, 2010.
[24] Hoffman B. Inside terrorism. New York: Columbia University Press, 1998.
[25] Lester D. Possible role of posttraumatic stress disorder in suicide terrorists: an hypothesis. *Psychol. Rep.,* 2008; 102(2):614-5.
[26] Sher L. The concept of post-traumatic mood disorder. *Med. Hypotheses*, 2005; 65(2): 205-10.
[27] Sher L. A model of suicidal behavior in war veterans with posttraumatic mood disorder. *Med. Hypotheses*, 2009; 73(2):215-9.
[28] Lester D. Female suicide bombers and burdensomeness. *Psychol. Rep.,* 2010; 106(1):160-2.
[29] Bloom M. Mother, daughter, sister, bomber. *Bulletin of the Atomic Scientists,* 2005; 61(6);54-62.
[30] McGreal C. Palestinians shocked at use of suicide mother. The Guardian, 26 January 2004. URL: www.guardian.co.uk/world/2004/ jan/27/israel/print Accessed: November 2, 2014.
[31] Wars in the World. http://www.warsintheworld.com/?page=static1258254223 Accessed: November 4, 2014.
[32] Gottfries C. G., Roos B. E. Man's emotional capacity--an unexplored and unexploited possibility. *Med. Hypotheses*, 1994; 43(5):347-55.
[33] Sher L. Behavioral genetic research: achievements and concerns. *Med. Hypotheses,* 1998; 50(3):265-6.

In: Neurobiology of Men's Mental Health
Editors: Leo Sher and Timothy R. Rice

ISBN: 978-1-63483-191-8
© 2015 Nova Science Publishers, Inc.

Chapter 18

HOMICIDAL TERRORISM: PSYCHIATRIC ASPECTS

Zebulon Taintor, MD

Adjunct Professor of Psychiatry, New York University School of Medicine; Organizer, World Psychiatric Association Section on Terrorism, US

ABSTRACT

Technological changes have enabled terrorist recruitment to become global. More attention is paid to women and terrorism, but often in their traditional roles or as human rights activists, less so as murderers. Men are the homicidal terrorists. In Western countries recent shootings have been carried out by men whose histories have recurring themes: a sense of powerlessness, frustration over lack of personal achievement, identity diffusion with seeking to define one's self through a cause and memorable action. They often have been living on society's margin, engaging in petty crime, and are not particularly religious. In terrorism their identities often are the opposite of what they were before. They draw such different meanings from events that their reality testing must be suspect, so psychiatric considerations have become more relevant than when terrorist organizations could be expected to screen out impaired would-be recruits.

Keywords: terrorism, homicide, crime, identity, culture, internet, social media, Pakistan, Taliban, Sabaoon

INTRODUCTION

Terrorism is a threat not only to society overall but to psychiatry and mental health clinicians. No other medical specialty is challenged by individuals and groups who aim to induce a state that leads to disorders that it treats: terror, existential threat, stress, anxiety, frustration, anger, and helplessness.Although other specialties deal with specific methods terrorists use, such as radiation, chemical and biological agents, etc., psychiatrists and other mental health clinicians are left to deal with the feelings produced by terrorist actions. Usually attention is focused on the victims, who are both more numerous and more damaged. Most victims would prefer that terrorists be killed, or at least locked up and the key thrown away. Terrorists seek no care for themselves and acknowledge no vulnerabilities. Except for those who are sympathetic to their causes, the terrorists are condemned.

Terrorism is an "-ism," i.e., it can be believed in and practiced. Terrorism has been described since ancient times and studied before the World Trade Center attacks of 9/11, but had not produced the mass casualties that brought about great efforts to understand and counter it. Now there are more than 100 journals on terrorism [1] and many books. Academics have worked to evolve a definition that will be more useful than the observation that one person's terrorist is another person's freedom fighter. For this paper the definition used by the National Consortium for the Study of Terrorism and Responses to Terrorism (START; www.start.umd.edu) is used: The act is a) political (including ethnic), economic, religious goal-directed; b) intended to coerce, intimidate, or convey some other message to larger audience; c) involves such heinous harm to innocent noncombatants that the act itself, even if committed during war, would be regarded as a crime.

There is also a need for a typology of terrorism. Post distinguished criminal terrorism and pathological terrorism from political terrorism, further dividing the last into sub state, state-supported or a regime or state terrorism.Within sub state terrorism there are social revolutionary terrorism (from the left), right wing terrorism, nationalist-separatist terrorism,single issue terrorism andreligious extremist terrorist, within which are religious fundamentalism terrorism and new religion terrorism [2]. Setting pathological terrorism apart no longer serves, as dismissing some terrorists as pathological defines those remaining as not having psychopathology. This leaves us considering psychology and psychodynamics (both useful), but excluding psychiatry, illness, and impairment. Even if clinicians do not agree, the general public reads the newspaper biographies of the most recent terrorists and concludes there's something abnormal about these murderers and clinicians should get back to work. This paper will argue that the current terrorism problem in the Western world requires some use of a psychiatric frame of reference to explain how messages that revolt and unify the majority against terrorism at the same time serve to recruit and inspire a frustrated, powerless, marginalized minority to participate in a powerful, invincible, community.

METHODS

Studies on terrorists in different cultures associated with different causes were reviewed. Newspaper reports provide detailed biographies of terrorists, especially Westerners who turn against their cultures. The author has visited Pakistan frequently since 1968 and is involved with programs for child soldiers (particularly Sabaoon, which, for security reasons, does not have a web site, but has been described in the press [3]), street children, and psychiatric patients. Criminal justice experience was acquired through several years of working with men and women in the New York State prison system.

FINDINGS

Internet and social media: While some doubt that terrorism has changed much over the centuries and point to historical continuities [4], newspaper reports show increased use of social media for rapid communication, increased use of web sites (especially for recruitment), dissemination of methods for producing explosives, and changing symbolism in target choice. Web sites bridge geographical spaces and are found by those who are looking for answers, even though oceans away [5]. Rid and Hecker note that "increased efficiencies in the distribution, manufacturing, and marketing of Salafist ideas and tactics, the demand and support for – and thus the feasibility of – violent extremism has become more independent of mainstream popular backing" [6]. They also note that terrorism's goals can be left vague so that the recruit fill in the blanks as s/he wants. Radicalization can still be said to require involvement with a group, but the group need not be face-to-face or nearby. Recent lone-wolf killers in Canada and the United States found their radicalizing groups on the Internet. Freedom of travel and evading diminished border controls is a recent change that governments are seeking to thwart, but smart phones and updates on safe routes have made it easy for would-be militants to attract little attention before their surprising disappearances in their Western homelands and reemergence in combat zones weeks or months later, such as the five Washington area college students detained in Pakistan after making a "disturbing" video [7]. The Charlie Hebdo shooting videos, horrifying to the mainstream, were given jihadist anthem overlays and put on the web within a day as a recruiting tool, demonstrating the power to punish those who disrespected the prophet [8].

These qualities of homicidal terrorism show how an image of invincible power is filled in:

Increasing frequency: Homicidal terrorism is the oldest form of terrorism [9] and by far still the most frequent, despite the rise of suicidal terrorism and increased use of kidnapping/hostage taking (which can lead to continuing publicity, then murder or profit, as ransom is paid). According to START, terrorism reached new records in 2013: 9,707 attacks killing nearly 17,800 people, in 93 countries, a steady increase in the past decade except for slight dips in 2004 and 2009. There were 32,500 injuries and 2900 kidnappings. The Afghan Taliban were the worst perpetrators, carrying out 525 attacks that killed 1842 people. By country, Iraq, Afghanistan, and Pakistan accounted for 57% of the attacks and 73% of fatalities. Statistics for 2014 are predicted to exceed 2013 in all categories, which terrorists say shows the futility of efforts to stop it.

Challenging locus: Homicidal terrorism usually is practiced amid a population unlikely to be won over to the cause by persuasion. Gandhi was successful in getting the British to leave India using nonviolent passive resistance because he got so many of his countrymen to practice it. Terrorists try to be threatening, intimidating, coercive, and exhausting/burdensome. Often they hope to provoke an authoritarian response that will radicalize parts of the population who then will become sympathetic to the cause. Suicidal terrorism often is more effective in such settings in that sacrifice is appreciated. Murderers inspire more fear and little sympathy. They are sent to penetrate secure locations, demonstrating their invincibility and the target's weakness. Terrorist groups try hard to manage the publicity after an incident to highlight the cause, generate sympathy for the attackers, and emphasize (sometimes fabricating) brutality of the authorities. Prolonged public anguish often confers the bonus of copycat events, giving an impression of coordination, communication and command.

Powerful methods: The more macho the better: bigger guns, more destructive bombs show power and demonstrate ability to overcome resistance. Video recording has led to more use of beheading, long used as an instrument of power and final justice [10]. Beheadings are staged to be a prolonged, colorful, powerful execution, echoing the guillotine Robespierre made the instrument of the first Terror, saying: "Terror is nothing other than justice, prompt, severe, inflexible. It is therefore an emanation of virtue." [11].

Invincibility: Homicidal terrorism requires intelligence and planning, in that an escape from the scene must be carried out. But suicide bombers can be used only once, while murderers live to kill another day. They can be drawn from among criminals who know how. In the summer of 2014 a decrease in suicide bombing in Pakistan and the capture of some terrorist assassins revealed that terrorism had been contracted out to criminals, working more for money than martyrdom and thus preferring to be captured rather than killed. The detailed narratives collected at Sabaoon from more than 200 youths who were in terrorist training in Pakistan report that martyrdom was extolled (and preferences for it honored) as a form of invincibility, conferring the heroic status of martyrdom and sure admission to Paradise. However, weapons training and survival were seen as demanding more skills, so selection for that track conferred higher status. Weapons training can be accomplished in weeks. Suicide bombing training mechanistically is much simpler: how to get to the spot undetected and how to set off the device, usually a vest. More important for suicide bombing is developing, assessing, and verifying motivation and resolve. Two boys came to Sabaoon after a compassionate off-duty policeman worshipping at a mosque in Islamabad noticed their anxiety and agitation. He engaged them in conversation. They decided to show him, rather than detonate, the explosives in which they were wrapped. Their stories are an embarrassment to terrorism.

Claimed efficacy: The toll of stress, anxiety, depression, PTSD, etc. has, of course, been described at such length as to require no elaboration here. Terrorists know and gloat over the damage they cause, but these effects rarely are statedas goals beyond "vengeance."Were the terrorists' specific goals achieved? They claim so, as evidenced by claims of responsibility (sometimes multiple and competing) after attacks. However, reality creeps in as reported efficacy of terrorist acts varies widely [12], with better results in areas of low education and poor communication. Generally, the goals are not only not achieved but become less attainable as reaction to terrorism sets in. For example, Germany's Baader-Meinhof Gang killed 34 people and carried out many bombings in an effort to remake society, but turned society against it and is remembered only as a group of dangerous criminals (those captured having passed sanity tests). The Peshawar school shootings of 2014 have so far united Pakistani society to stop distinguishing between "good and bad" Taliban, mute would-be negotiators, end a paralyzing political standoff, create a military court system, step up cleaning out terrorist strongholds, etc. But the murderers mistakenly believe they will be effective, a cognitive distortion that raises questions about who they are and how they came to think as they did.

Promise vs. Reality

According to Emad Shahin, on the 2015 Charlie Hebdo shooters, "Islam now provides an effective way to appeal to feelings of identity, community, justice, freedom, and nationalism all at once ….a 'just war'" [13]. One could infer homicidal terrorists lack these things and are defined by seeking them as well as by the cognitive distortions that prevent them from distinguishing the gap between fantasy and reality, as below:

Islam: Religious terrorism appeals on the basis of promising the one true religion. In reality Islam consists of many sects and schools, numbered by some scholars up to 95, members of which often see members of other sects as the worst sort of infidel, meriting slaughter. Most militants are recruited by Salafis (Wahabis), Sunnis who often see killing Shia Muslims as part of purifying the world. Militants call for implementation of Sharia law, not specifying which of its major, conflicting schools would be chosen [14] . But religion often is not explored deeply. Karen Armstrong, who argues that religion is a convenient banner under which wars are fought for other purposes [15], notes that the Boston marathon bombers had not progressed beyond "Islam for Dummies." Yet Islam continues to offer simple demands and a simple narrative, elements of which have been incorporated into the Al Qaeda narrative, per Vlahos: 1) a heroic journey and a mythic figure, 2) the rhythm of history captures as epic struggle and story, 3) the commanded charge of renewal, 4) history revealed through and enjoined through mystic literary form [16].

Identity: Russell Razzaque's superb study of the 9/11, London transport, and other terrorists [17] shows how choices (including his own) were made at each step of the way, starting with reactions to family dynamics (his subjects shared distant, emotionally cold, but authoritarian fathers), peer group acceptance and rejection, work, status, and frustration with results leading to escapist, grandiose choices. Whoever or whatever one was before matters little compared to what one will become, known through the ages for a terrorist deed. One's creative juices turn from the everyday humdrum to what will be most memorable. Thus the more vulnerable are the goal-less with shaky identities, self-dissatisfaction, few social satisfactions and supporting relationships.The total identity recruitment into terrorism offers a total identity replacement from powerlessness powerful from goalless to having a specific grandiose goal. Post cites the German government study of the social revolutionaries in the 1970s. All could be described as "a great deal had gone wrong in their lives.They were comforted by joining a group whose credo was, 'It's not us it's them.They are responsible for our problems.'" [18].

Community: Thus the individual identity problems can be eased by getting involved with a group sharing one's views. But this turns the problem into a community identityproblem. Schwartz, Dunkel, and Waterman dissect these strands, adding cultural identity [19]. Western societies have learned how not to be provoked and to allow expression of all shades of opinion. In some communities in the developing world violence is a norm. Farzana Parveen, a 25 year-old pregnant woman was bludgeoned to death by her immediate family members on a busy city street in front of the courthouse where her case for having married the man she loved was to be heard. Her husband had murdered his first wife five years before to be with Farzana. He had been freed from that charge using an Islamic provision of Pakistani law [20]. The history of the Pashtuns in northwestern Pakistan and Afghanistan [21] shows how some communities develop around a sense of oppression and revenge. Groups within communities coalesce around grievances, so as some young Pakistanis have chosen to wage jihad against a Pakistani government seen as corrupt and incompetent individual problems disappear into group and community dysfunction [22]. So when a large enough group decides jihadism is OK, a person who doesn't go along is not OK. The "Jihadi Cool" may be recruit able [23]. The vulnerable group in Western countries is the second generation of immigrants, as seen in several country studies in Ruggiero's *Perspectives on Immigration and Terrorism* [24]. After all, it was not their decision to come to a strange country in which they have to form their identities under stress. Post describes how terrorism arises from hatred bred in the bone over generations, with different types depending on the psychodynamics between children and parents, parents and the state [25].

Justice: Terrorism offers the chance to take a law into one's own hands, appealing if one perceives the law as unjust, contrary to the overwhelming majority of people living in Western countries. Pakistan is reacting to the Peshawar shootings by creating military courts, a tacit admission that the civil justice system is not able to do the job. A recent graduate from an American law school returned to Pakistan to help his country. A law firm offered him a job, explaining that the first step in a case is to pay 5000 Rupees to guarantee delaying the case for at least a year, during which time witnesses can be bribed. Criminals already have different ideas of justice and often evolve into terrorists, seeking a higher morality. Almedy Coulibaly, a 2015 Paris shooter, got involved by helping to supply weapons he learned to obtain as a criminal [26].

Freedom: Overall, the only freedom exercised is to submit to an authority that will take control of one's life. But there are choices along the way that give the illusion of freedom, starting with going to Internet sites or getting involved with others that preach the cause.Then there is the freedom to leave behind one's family, country and life up to then. Sounds good, but may not work well in practice, as disillusioned youths returning from ISIS testify [27].

Nationalism: Taking up arms against one's government can be seen as true nationalism. ISIS promises a caliphate, an Islamic state, so one can be both a nationalist and devout. Existing Muslim states, many of which have Islam in their names, e.g., "The Islamic Republic/Kingdom of …" are dismissed as corrupt and contaminated by decadent Western influences.

PSYCHOPATHOLGY

Excessive time on the Internet and social media has been described as an addiction [28, 29] and can ease isolation from family and local community, leading to new contacts, new groups and a new community that provide support and validation. John Horgan's superb studies of the Irish Republican Army and other terrorists and their organizations [30] noted that terrorist organizations screen out recruits who show psychopathology because they can't be expected to function well. The cause should not suffer from the stigma of mental illness. He provides many examples of rational choices, stable interpersonal relationships, etc. He notes radical beliefs being acquired over time by people whose identities have formed, only in part around radical beliefs. He describes disengagement, rather than deradicalization, as the process by which IRA militants turned to peaceful activities. The problem of diagnosing psychopathology is best illustrated by the Fort Hood shooter, Maj. Nidal Malik Hasan, MD, born in the USA, where he had graduated from medical school, psychiatry residency, and a two year fellowship in preventive and disaster psychiatry, then earning a MPH. He and no one near him diagnosed or prevented anything before he killed 13 people. A turning point in his life seemed to be isolating himself socially and delving into books on Islam following his parents' deaths in 1998 and 2001 [31]. So far he has passed all sanity tests and is providing his own defense at trial. Whatever his colleagues may opine, Thomas Friedman probably spoke for many when he wrote Dr. Hasan "may have been mentally unbalanced – I assume anyone who shoots up innocent people is. But the more you read … the more it seems that Major Hasan was just another angry jihadist spurred to action by "The Narrative" [America's war on Terror is a war on Islam] [32]." So the psychopathology is at least believing "The Narrative" despite abundant evidence to the contrary.

Aaron Beck described the steps of cognitive distortion involved in externalizing and adopting a view of the Enemy: a) overgeneralization - complaints about specific individuals are merged and individual opponents become the Enemy group; b) dichotomous thinking in which groups are seen as all good or all bad; c) tunnel vision in focusing everything on destruction of the target [33]. Bandura has described the moral distortions justifying murdering innocents: cognitive restructuring has to include morality. "People with strong convictions generally are influenced by strong convictions. But most moralities are opposed to violence. All religions oppose killing other people. To justify violence, morals have to be disengaged, compartmentalized, and rationalized.There are many variations on "the end justifies the means" - the greater good that will result from doing this one bad thing." [34]. These and 322 other reports on the psychology of terrorism as of 2004 can be found in the University of South Florida's excellent annotated bibliography [35].

Diagnoses perforce are made after a person has come to the attention of someone who can diagnose, and an early diagnose may not be alarming. Almedy Coulibaly had a psychiatric evaluation in December 2012: "immature and psychopathic personality" [36]. Now screening roles are reversed in that terrorists welcome self-starting lone wolves who can cause mayhem without being sufficiently involved in an organization to learn its secrets while governments are casting about for ways of detecting terrorists before they act. There are no hard data. This won't stop efforts at trying, such as the United Kingdom's charge to teachers to detect warning signs in their pupils [37]. A lot is known about youths entering gangs and group deviant behavior, some of whom develop diagnosable antisocial personality disorder [38]. Sitting with street children in Karachi this author was struck by how the described being involved in crime rings (learning how to stop a bus and rob everyone on it) led to vulnerability to the relative higher morality of the jihad, especially as they were offered steady meals and income and training to fight for a good cause. Their personalities seemed more opportunistic and fluid than presenting fixed diagnoses. As the NIMH has decided that Research Diagnostic Criteria will be used for research rather than DSM-5, perhaps the best diagnosis to use for now is "homicidal terrorist."

RESEARCH SUGGESTIONS

Male-female differences: In a study of 119 lone-actor terrorists convicted of engaging or planning to engage in terrorism in the USA only four were female. Of those four only two had been violent (one stabbing, one shooting, no homicides) [39]. Mia Bloom's historical multinational survey includes detailed biographical information on some women who chose involvement with terrorism [40]. Some have murdered, in effect saying a woman can do a man's job, the same argument used for combat roles in the military. But women have chosen soldier-terrorist roles for decades. In 1970 Ulrike Meinhoff, seen as a pioneering feminist in social revolutionary terrorism, visited Lebanon and found Palestinians training women as terrorists [41]. But most of the women in Bloom's account have chosen or submitted to being assigned more traditional roles. While crime committed by women has increased more rapidly than crimes committed by men, women commit only a tenth the number of murders compared to men. Recorded murders in the USA by both sexes have been decreasing since 1976 [42]. There is increased attention to the status of women in all parts of life, including the rise in suicide bombings by women. But there are practical reasons for using women as suicide bombers: a) they usually are not objects of suspicion. Police in Pakistan will stop two men on a motorbike, but not a man and a woman; b) they cover themselves and are not searched by men, easily concealing explosives. A 10 year old girl used to set off a time bomb killing 20 in a Nigerian market may not even known she was carrying a bomb [43]. While some women have taken a leading role for human rights in countering Islamic terrorists' commands that they submit [44, 45], others have traveled to ISIS to take more traditional roles, having been promised strong and good husbands [46]. Research should continue to focus on sex differences. In particular, men are angrier; terrorist acts show marked sadistic elements. Sex and gender roles merit continued monitoring and investigation.

Stress biomarkers: Research funds are still likely to be directed more to victims of terrorism than to terrorists themselves. But the research will be relevant as there is a focus on the common denominator of stress and reactions to it. While testosterone may play a biological role, the psychological determinant between depression and anger is often just a question of whether one feels at fault versus blaming others. Biomarkers of stress may lead to biomarkers for terrorism.

Psychodynamic defenses: The cognitivedistortions described above involve denial, reaction formation, and projection, among others. Razzaque has incorporated them and family issues in the questionnaire with which he closes his book.

Predisposing factors: A study of 608 men and women of Pakistani or Bangladeshi Muslim origin found vulnerability to radicalization was characterized by depression but resistance to radicalization showed a different pattern of health and psychosocial variables, with the paradoxical role of social capital warranting further investigation [47]. The lone-wolf study in the USA found many were social isolated [39].

Intolerance: In one study [48] intolerance of ambiguity was related to paranoid severity, as some subjects jumped to incorrect conclusions rather than accepting they couldn't identify ambiguous drawings. Intolerance of ambiguity is a measure of vulnerability to conspiracy theories. Tolerance of ambiguity has been described as a desirable quality to develop in leaders [49] Development of critical thinking has been used to counter radicalization in Kenya [50].

CONCLUSION

Homicidal terrorism is committed almost exclusively by men, unlike suicide bombing which is also committed by women (albeit sometimes under orders from men). It presents unique opportunities for understanding how some marginal men who feel frustrated and powerless choose to become terrorists, with power over life and death.

REFERENCES

[1] Tinnes J., (2013). 100 core and periphery journals for terrorism research. Perspectives on Terrorism Vol 7(2). http://www. terrorism analysts.com/pt/index.php/pot/ article/view/258/html, accessed 1/8/15.

[2] Post J: When hatred is bred in the bone: psycho-cultural foundations of contemporary terrorism. *Political Psychology* 26(4): 615-636, (table p617), 2005.

[3] http://www.ndtv.com/aticle/world/pak-school-to-rehabilitate-boys-recruited-by-Taliban. Associated Press. August 10, 2011, accessed January 12, 2011. Also Lakhani K: Pakistan's child soldiers. Foreign Policy, January 12, 2011. http://afpak. foreignpolicy.com/posts/2010/03/29/ pakistan's_child_soldiers, accessed January 12, 2011.

[4] Duyvesteyn I: How new is the new terrorism? Studies in Conflict and Terrorism, Volume 27, Issue 5, 2004, www.tanfonline.com/doi/abs/10. 1080/10576100490483750, accessed January 5, 2015.

[5] Lieberman J, Beckner C: The homegrown jihadist threat grows. *Wall Street Journal*, October 24, 2014, pA13.

[6] Rid T, Hecker M: The terror fringe. *Policy Review* December 2009 – January 2010. Washington, The Hoover Institution, 2010, http://www. hoover.org/research/terror-fringe, accessed January 7, 2015.

[7] Shane S: Pakistan detains five U.S. Muslims in raid by police. *New York Times* December 10, 2009, pA1

[8] Hubbard B: Jihadists and supporters take to social media to praise attack on Charlie Hebdo. *New York Times,* January 11, 2015, pA1.

[9] Chailand G, Blin A: The History of Terrorism: From Antiquity to Al Qaeda. Oakland, CA, University of California Press, 2007.

[10] Larson F: Severed: A History of Heads Lost and Found. New York, Liveright, 2014.

[11] Robespierre as quoted in Psychology of Terrorism, Bonga B, Brown LM, Beuler LE, Breckinridge JN, Zimbardo, PG, Oxford University Press, New York, 2013.

[12] Merari A: Terrorism as a strategy of insurgency. *Terrorism and Political Violence* 5(4): 213-251, 1193.Reports of efficacy.

[13] Kirkpatrick D: Raising questions about Islam. *New York Times* January 9, 2015, p1.

[14] Kadri S: *Heaven on Earth: A Journey Through Shari'a Law from the Deserts of Ancient Arabia to the Streets of the Modern Muslim World.* New York, Farrar, Straus and Giroux, 2012.

[15] Armstrong K: *Fields of Blood: Religion and the History of Violence*. New York, Knopf, 2014.

[16] Vlahos M: Terror's mask: insurgency within Islam. Occasional Paper, Joint Warfare Analysis Department, Applied Physics Laboratory, John Hopkins University, May 2002.

[17] Razzaque R: *Human Being to Human Bomb*. Cambridge, Icon Books, 2008.

[18] Post, ibid, p618.

[19] Schwartz S, Dunkel CS, Waterman AS: Terrorism: an identity theory perspective. *Studies in Conflict and Terrorism*, 32: 537-559, 2009.

[20] Walsh D: One man, two wives and many accepted forms of violence in Pakistan. *New York Times*. May 31, 2014, pA8.
[21] Siqqique A: *The Pashtun Question*. London, Random House Group Ltd., 2014
[22] Tavernise S, Gillani W: Frustrated generation wages jihad on Pakistan. *New York Times*, February 28, 2010 Sec I, p1.
[23] McCaul M: Islamic state is recruiting America's "Jihadi Cool" crowd. *Wall Street Journal*, September 12, 2014, pA13.
[24] Ruggiero GM: Perspectives on Terrorism. *IOS Press*, Amsterdam, 2011.
[25] Post, ibid, p418.
[26] Higgins A: French police say suspect in attacks evolved from petty criminal to terrorist. *New York Times*, January 11, 2015, pA1.
[27] Abi Habib M: The child soldiers of Syria. *Wall Street Journal*, December 27-28, 2014, pA1.
[28] Weintraub P, Dunn TM, Yager J, Taintor Z: Internet Addiction pp 407-418 in Ruiz P and Strain E: *Lowinson & Ruiz's Substance Abuse: A Comprehensive Textbook.* Fifth edition. Philadelphia, Lippincott Williams & Wilkins, 2011.
[29] Rosenthal R, Taintor Z: Microprocessor abuse and Internet addiction. pp 980-990 in *The ASAM Principles of Addiction Medicine*, eds Ries RK, Fiellin DA, Miller, SC, Saitz, R., Philadelphia, Wolters-Kluwer, 2014.
[30] Horgan J: *The Psychology of Terrorism.* Routledge, New York, 2005.
[31] McKinley JC, Dao J: After years of growing tensions, 7 minutes of bloodshed. *New York Times*, November 9, 2009, pA1.
[32] Friedman T: America vs. the narrative. *New York Times*, November 29, 20009, Sec. IV, p9.
[33] Beck AT: Prisoners of hate. *Behavior Research and Therapy* 40(3): 209-216, 2002.
[34] Bandura A: The consequences of moral disengagement: A social learning perspective. in *Understanding Terrorism: Psychosocial Roots, Consequences, and Interventions*. American Psychological Association, Washington, DC, 2003.
[35] Borum R: *The Psychology of Terrorism*. University of South Florida, Tampa, 2004. https://www.ncjrs.gov/pdffiles1/nij/grants/208552.pdf?q= psychology-of-terrorism.
[36] Callimachi R, Higgins A: Video shows a Paris gunman declaring his loyalty to the Islamic state. *New York Times*, January 12, 2015, pA6.
[37] Connolly J: Terror threat: teachers become spies? *BBC newsbeat*, November 26, 2014. http://www.bbc.co.uk/newsbeat/30216768, accessed January 10, 2015.
[38] Black DW: *Bad, Boys, Bad Men: Confronting Antisocial Personality Disorder (Sociopathy)*. New York, Oxford University Press, 2013.
[39] Gill P, Horgan J, Deckert P: Bombing alone: Tracing the motivations and antecedent behaviors of lone-actor terrorists. *Forensic Sciences* 59(2): 425-435.
[40] Bloom M: *Dying to Kill: The Allure of Suicide Terror*. New York, Columbia University Press, 2005.
[41] Noe D: The Baader-Meinhof Gang. http://www.crimelibrary.com/ terro rists_spies/ terrorists/meinhof/12.html, 2005, accessed January 10, 2015.
[42] Greenfield LA, Snell TL: Women Offenders, Special Report, December 1999 (as revised 10/3/2000) http://www.crimelibrary.com/ terrorists_ spies/terrorists/meinhof/ 12.html,accessed January 10,2015.

[43] Nossiter A: Child bombs Nigeria market, killing 20. *New York Times*, January 11, 2015, pA1.

[44] Yousakzai M: *I Am Malala*. New York, Little Brown, 2013.

[45] Bennoune K: *Your Fatwa Does Not Apply Here*. New York, Norton, 2013.

[46] Daley S, de la Baume M: A French town reels as teenage girls disappears, apparently to join jihadists. *New York Times*, December 1, 2014, http://www.nytimes.com/2014/12/02/world/europe/a-french-town-reels-after-teenage-girl-vanishes-apparently-to-join-jihadists.html?src= xps, accessed December 1, 2014.

[47] Bhui K, Everitt B, Jones E: Might depression, psychosocial adversity, and limited social assets explain vulnerability to and resistance against violent radicalization? *PLOS ONE* 9(9): 1-10, September 2014, e105918.

[48] Abroms G., Taintor Z, Lhamon W.: Percept assimilation and paranoid severity. *Arch. Gen. Psychiat.* 14:491-96, 1966.

[49] Wilkerson D: *The Ambiguity Advantage. What Great Leaders Are Great At*. New York, Palgrave Macmillan, 2006.

[50] Savage S, Khan A, Liht J: Preventing violent extremism in Kenya through value complexitydevelopment: assessment of being Kenyan being Muslim. *Journal of Strategic Security* 7(3): 1-26, 2014.

SECTION THREE: MAJOR MENTAL ILLNESS IN MEN

In: Neurobiology of Men's Mental Health
Editors: Leo Sher and Timothy R. Rice

ISBN: 978-1-63483-191-8
© 2015 Nova Science Publishers, Inc.

Chapter 19

POSTTRAUMATIC STRESS DISORDER IN MEN: PSYCHOSOCIAL, CULTURAL AND NEUROBIOLOGICAL FACTORS

Lisa J. Norelli*[1,*], *MD, MPH,
and John J. O'Connor*[2,†], *PhD
[1]Capital District Psychiatric Center, Albany, NY, US
Associate Professor, Albany Medical College,
Department of Psychiatry, Albany, NY, US
Assistant Professor, SUNY Albany School
of Public Health, Renssalaer, NY, US
[2]Capital District Psychiatric Center, Albany, NY, US
Assistant Professor, Albany Medical College,
Department of Psychiatry, Albany, NY, US

ABSTRACT

Gender is an important determinant of health across the human lifespan. Different gender-specific prevalence rates of certain disorders have been well established. For example, autoimmune diseases, eating disorders, depression, anxiety disorders, and post-traumatic stress disorder (PTSD) are more common in women, and hypertension, alcohol abuse, substance abuse, impulsive and aggressive disorders are more common in men. PTSD is prevalent in both civilian and military populations. Numerous epidemiological studies have shown that exposure to pathological stress, such as war, sexual assault, physical violence, and disasters result in higher lifetime prevalence rates of PTSD in woman than men even though men experience more traumatic events over their lifetime.

* Corresponding author: Lisa J. Norelli, MD, MPH. Capital District Psychiatric Center, 75 New Scotland Avenue, Albany, NY 12208 US. Associate Professor, Albany Medical College, Department of Psychiatry, Albany, NY 12208 US. Assistant Professor, SUNY Albany School of Public Health, Renssalaer, NY 12144 US. E-mail: NorelliMD@me.com.

† John J. O'Connor, PhD: Capital District Psychiatric Center, 75 New Scotland Avenue, Albany, NY 12208 US. Assistant Professor, Albany Medical College, Department of Psychiatry, Albany, NY 12208 US.

However, other research has shown that female gender is of limited use as a predictor of stress-related disorders. Gender-related susceptibility (or resilience) to stress-related disorders such as PTSD has been attributed to differences in a variety of psychosocial, cultural, and neurobiological factors. In humans, studies of sex and other hormone levels, structural and functional brain imaging, and gene expression examine the neurobiological gender differences that may increase or mitigate the risk of developing PTSD.

Psychosocial and cultural differences such as lifetime exposure to trauma, individual and trauma-related characteristics, coping style, supports, and cognitive appraisal styles are also hypothesized to contribute to observed gender differences in PTSD. By controlling for sociocultural factors, animal models of PTSD are postulated to better elucidate the gender-specific neurobiological determinants of PTSD.

In contrast to epidemiological studies of human subjects, a recent review of animal study data show that male animals are significantly more vulnerable to acute and chronic stress and female animals are more resilient. These contradictory gender-related differences demand further study to refine our diagnostic conceptualization of PTSD, help identify individuals at risk, and provide more effective prophylaxis and treatment. In this chapter we review the current gender-related knowledge regarding the sociocultural and neurobiological effects of PTSD.

Keywords: posttraumatic stress disorder, PTSD, men, gender differences, risk factors, resilience

INTRODUCTION

Post-traumatic stress disorder (PTSD) is a complex and prevalent disorder that can significantly disrupt interpersonal relationships, quality of life, and overall functioning. The disorder can occur at any age. PTSD is associated with a higher burden of comorbid physical and psychiatric health disorders, including cardiovascular disease, migraine, gastrointestinal disorders, substance abuse, anxiety disorders, and major depression [1-4].

The majority of humans experience at least one traumatic event over their lifetime, yet the majority of people exposed to a traumatic event do not develop PTSD [5, 6]. It is clear that exposure to a traumatic event alone is not sufficient cause for the development of the disorder. However, PTSD is prevalent in both civilian and military populations worldwide, is generally found at double the rate in women, even though men are exposed to more traumatic events during their lifetime [6, 7]. The reasons why women appear more likely than men to develop PTSD (or why men appear more resilient to traumatic stressors) are exceedingly complex, and remain the subject of ongoing research and debate.

Broad areas of research include gender differences in pre-morbid factors, peri- and post-traumatic factors, and the interaction and mediation of psychological, social, and cultural variables. Some of the increased prevalence in women is attributable to social factors and higher rates of exposure to childhood abuse and interpersonal violence, particularly potent causes of PTSD. Other potential factors include, differences in coping strategies, loss or lack of resources, and neurobiological differences [8].

Research methodology (e.g., sample population, reporting biases, self-report versus clinician interview), terminology revisions, discrepancies, and lack of consensus regarding the diagnostic conceptualization of PTSD often make comparisons between studies difficult. Concerning gender differences, the data are contradictory. A recent meta-analysis of civilian and military studies found limited predictive value of gender in the development of PTSD in adults [9].

Several studies have shown no significant gender-related differences on the development of the disorder, associated symptoms, or outcome [10, 11]. Interestingly, a recent review of animal studies found male animals were significantly more vulnerable to acute and chronic stress and female animals were more resilient [12].

With respect to prevention, diagnosis, and treatment of PTSD, it is crucial to understand any potential gender differences in neurobiological, social, psychological, cultural, event-specific, and other factors that confer risk and/or resilience in the development of PTSD.

DEFINING POSTTRAUMATIC STRESS DISORDER

The American Psychiatric Association first defined PTSD in the Diagnostic and Statistical Manual of Mental Disorders in 1980 and has included the disorder in subsequent revisions, most recently DSM-5, published in 2013 [13-15]. Since 1994, the World Health Organization has concurrently developed the International Classification of Diseases (ICD) to provide standard international diagnostic classification [16]. The most recent revision, ICD-10, includes similar diagnostic criteria for PTSD, but there are differences in some details between classification systems that may impact study results, including gender diagnosis rates [17]. However, both ICD and DSM conceptualize PTSD similarly: That is, in some individuals, a traumatic event or events are etiologically linked to a specific syndrome. Therefore, and this point cannot be overstated, the development of a serious psychiatric disorder can be prevented by decreasing exposure of individuals to traumatic stressors when possible, including, for example, child abuse and other interpersonal violence.

A traumatic event is defined as an extreme stressor or catastrophic event whereby an individual directly encounters actual or threatened death, serious injury, or sexual violation to them or others. In addition to the traumatic event experience criterion, the PTSD syndrome consists of a combination of symptoms from the following categories: 1) intrusive re-experiencing of the event (e.g., intrusive memories, dreams, severe psychological or physical distress when exposed to an event reminder), 2) avoidance (e.g., efforts to avoid thoughts, feelings, activities, or places that may arouse recollections of the event, inability to recall important aspects of the event), 3) emotional numbing (e.g., decreased interest, diminished emotions, detachment from others), 4) hyperarousal (e.g., hypervigilance, increased startle reflex, irritability, sleep difficulties), and in DSM-5, 5) negative alterations in mood and cognition [15, 16]. DSM requires that the symptoms have to persist for at least one month and cause clinically significant distress or functional impairment [15]. Whereas ICD specifies that onset symptoms must be within six months of the stressor, characterizing it as ‘a delayed or protracted response to a catastrophic or exceptionally threatening event likely to cause pervasive distress in almost anyone’ [16]. Longitudinal studies have shown that after a stressor, PTSD symptoms usually appear quickly, and often subside over weeks to months. However, in some trauma survivors PTSD becomes chronic when symptoms persist and continue to interfere with functioning beyond three months.

A minority of survivors may have delayed onset of PTSD, first manifesting symptoms six months or longer after the stressor [18, 19].

Complex PTSD is a disorder that can affect survivors of prolonged, repeated and severe trauma lasting months or years, and/or often beginning in childhood. Examples of this type of trauma include concentration camps, political torture, child trafficking, prostitution, long-term domestic violence, certain religious cults, and long-term physical, emotional, and sexual abuse and neglect in childhood. Judith Herman first described Complex PTSD in 1992, but the disorder has not been included in the diagnostic classification system editions of DSM-IV, DSM-V, or ICD-10 [20]. It is currently being considered for inclusion in ICD-11, due for publication in 2017. Individuals with Complex PTSD have, in addition to customary PTSD symptoms, pervasive disturbances such as: disordered attachments, interpersonal difficulties, impaired self-identity, dissociation, somatization, affect dysregulation, and emotional instability [20].

Neurobiological Factors

The neurobiological correlates of PTSD have been under intense scrutiny, but few studies to date demonstrate clinically useful data regarding gender differences. As pointed out by Selye in 1936 and currently by McEwen, there is an inherent paradox in the human stress response system (e.g., the autonomic nervous system and the adrenocorticoid system). That is, in the short term in response to acute threat or stress, the responses are adaptive and protective. However, when activated over prolonged periods of time, these systems cause physical and psychological damage, and accelerate disease [21, 22]. Chronic dysregulation of the neural networks that modulate stress-responsivity, learning, and memory is associated with PTSD and other psychiatric disorders, such as depression [4, 22-24]. In general, people with PTSD have chronic activation of the sympathetic nervous system (SNS) and renin-angiotensin-aldosterone system (RAAS), underactivity of the parasympathetic and GABAergic systems, changes in neuropeptide Y (NPY), hypothalamic-pituitary-adrenal (HPA) system, and brain-derived neurotrophic factor (BNDF). The immune system is suppressed, a proinflammatory state is created, and there is increased allostatic load [22-25].

Gender differences in the response of the HPA axis to stress may account for some of the gender differences observed in PTSD. Most studies of HPA axis function show high levels of cortisol associated with acute exposure to traumatic events, but low levels in those who have already developed PTSD [25, 26]. PTSD trauma survivors have been shown to have both hypocortisolism and reduced responsiveness to glucocorticoids.

One study showed that the HPA axis response in men associated with a more robust adrenocorticotropic hormone (ACTH) response and higher levels of cortisol release compared to women. A small study comparing cortisol levels in survivors of the 9/11 attack showed that in men with PTSD, higher cortisol was associated with more severe re-experiencing symptoms and less severe avoidance symptoms. The women with PTSD showed no correlation [25-27]. Chronic hypocortisolism is associated with pain, fatigue, enhanced stress sensitivity, and may contribute to changes in insulin sensitivity.

Hypocortisolism and SNS hyperactivity may also contribute to proinflammatory states and increase production of proinflammatory markers and cytokines [28, 29]. PTSD is associated with increased levels of proinflammatory cytokines including tumor necrosis factor (TNF)-alpha, interleukin (IL)-1beta and IL-6 [30, 31]. Diminished activation of genes associated with immune cells has been reported in subjects with PTSD. Men with PTSD showed decreased expression of monocyte genes pertaining to cytokine/chemokine signaling, platelet functioning and histone activity. A pilot sample of women PTSD subjects showed a more balanced pattern of monocyte gene expression, suggestive of a gender-mediated dimorphism in monocyte gene activation in PTSD [32].

Neuropeptide Y (NPY), a neurotransmitter involved in the modulation of stress responses, has anxiolytic and neuroprotective properties. Usually stress-induced NPY release promotes cellular recovery. Chronic NPY release amplifies metabolic changes leading to hypertension, obesity and metabolic syndrome. In studies of male veterans with PTSD, NPY levels are reduced both at rest and during stress, and appear to increase sensitivity to stress [33]. Chronic SNS hyperarousal (and hypoactive parasympathetic activity) is reflected by elevated resting heart rates and low heart rate variability in people with PTSD. Low heart rate variability is a robust predictor of cardiac mortality [34, 35]. The long-term activation of stress pathways and proinflammatory responses may contribute to the increased rates of obesity, insulin resistance, diabetes mellitus, hypertension, cardiovascular and neurovascular diseases observed in PTSD individuals [36, 37].

Structural and functional neuroimaging studies have demonstrated some gender differences. Magnetic resonance imaging (MRI) of the brains of children with PTSD show decreased cerebral volumes compared to controls. The male children had a significantly greater reduction in the corpus callosum and a trend for greater cerebral volume reduction compared to female subjects.

The male children with PTSD also had larger prefrontal lobe cerebrospinal fluid volumes and smaller splenium compared to controls, and greater lateral ventricular volume compared to female children with PTSD [38]. Functional MRI (fMRI) has shown men with PTSD have enhanced brainstem activity, and greater hippocampal activity to fear compared to women with PTSD [39, 40]. These studies may reflect normal gender differences and do not as yet inform us about gender differences in PTSD risk or neuropathology.

Animal models of PTSD are useful to help illuminate underlying gender-dependent biological differences, symptomatology, and coping mechanisms. Such research also holds the promise of guiding the development of general and gender-dependent targets for prevention and treatment. The limitations of animal models include the validity of the PTSD paradigm, study design regarding gender (e.g., using only male animals, using both sexes, or ovariectomized and/or hormone replaced females).

When interpreting the results of animal studies it is important to keep in mind that study conditions do not have the ability to simulate the complex impact of gender-dependent social and cultural factors encountered by humans. However, by controlling for sociocultural factors, animal studies lead to a clearer understanding of gender-dependent and non gender-dependent neurobiological factors concerning PTSD.

In contrast to epidemiological studies of human subjects, a recent review of animal study data show that male animals are significantly more vulnerable to acute and chronic stress and female animals are more resilient [11]. A recent comprehensive review of animal studies by Cohen and Yehuda show gender related differences on behavior (phenotype) and neurobiology (endophenotype) at baseline conditions and in response to stress. In adult female rodents higher plasma corticosteroid levels are observed both at rest and during stress. This difference may reflect effects of gonadal steroids, and gender differences may have underlying genetic and environmental components.

Overall, the animal data suggests that higher basal cortisol levels may decrease the risk of PTSD [11]. In humans there is some evidence that lower baseline cortisol levels are a pretraumatic risk factor for PTSD [40, 41].

Other Psychosocial and Cultural Factors

Some of the observed gender difference may be related to social and culture-bound attributes and expectations. Some of the initial responses to traumatic events may be related to differences in the meaning of the event or gender-specific attributes such as increased guilt, diminished rights and resources that burden women in many societies. Men experience traumatic events more often than women, women and men often differ in the type of trauma they experience, and women more often develop PTSD after a traumatic event especially if there has been a higher prevalence of past child sexual abuse [8]. Breslau found that in a US community sample, 36% of women experienced assault compared to 6% of men [42, 43]. The international literature on prevalence rates of childhood sexual abuse show wide variations in rates but generally confirm higher prevalence rates in girls compared to boys. Finkelhor reviewed child sexual abuse surveys in over 20 countries and rates of 7-36% were reported for women and 3-29% for men [44]. Perada et al., updated the review including new data from 21 countries and 39 prevalence studies. Prevalence rates ranged from 0-53% for women and 0-60% for men. The persistence of high rates of childhood sexual abuse (as well as emotional abuse, physical abuse and neglect) highlights the inadequacy of the professional, cultural and social response towards a preventable cause of serious psychological morbidity in children and adults [45].

Research has consistently shown that for both men and women, social isolation, avoidance, disengagement, rumination, self-destructive behaviors, and self-blame are maladaptive coping strategies in the aftermath of extreme stress, possibly increasing PTSD risk [46]. Women tend to use avoidant and emotionally-mediated coping styles compared to men, who report fewer emotion-based symptoms such as intrusive thoughts, remorse, and avoidance. Men may also tend to express their psychological distress via somatic symptoms [46]. Bryant et al., identified two PTSD modes of coping: emotional numbing (superior for men, no difference with women), and dysphoria [47]. In terms of coping, one study found increased resilience in women against the potential impact of PTSD symptoms associated with their ability to continue to provide intimacy behaviors [48]. Women and girls are more likely to have peritraumatic dissociation, shown to be a predictor of PTSD in many studies [2, 40, 42]. Other recent research has found gender differences concerning greater acquisitions of conditioned fear, in vulnerability to terror (speculated to be related to lower self-efficacy), and a higher sense of threat in women who developed PTSD [2, 40, 42].

Cognitive appraisal is how people evaluate events and experiences, reflecting subjective perception, interpretation, and evaluation of their experience. Cognitive appraisal of the traumatic event may also account for some of the observed gender differences. Studies suggest that men and women have different levels of control appraisal, perceived control, and tendency towards self-blame. In general, men have higher measures of control appraisal, perceived control, and rely less on self-blame; factors which are shown to predict lower risk of developing PTSD [40, 49]. Gender differences have been shown in how people process strong emotional memories.

Some of these variances may contribute to the observed differences in gender sensitivity or response to stress. Our knowledge of the gender-specific psychosocial and cultural roles and influences mediating the risk of developing PTSD and the longitudinal outcomes remains limited.

Peritraumatic dissociation is a state of limited or distorted awareness at or around the time of the traumatic event. Women are known to have higher rates of peritraumatic dissociation compared to men, and this phenomenon has been shown to predict development of PTSD in many studies.

The underlying biological mechanisms associated with peritraumatic dissociation appear to be related to dihydroepiandrosterone (DHEA), dihydroepiandrosterone sulfate (DHEA-S), and HPA axis modulation but data remain inconclusive [40, 50].

COMORBIDITY

Comorbidity of PTSD and other psychiatric disorders is common and associated with increased symptom severity, chronicity, higher health care utilization, homelessness, and increased functional impairment. The European Study of the Epidemiology of Mental Disorders (ESEMeD) found significant gender differences in the 12-month prevalence comorbidity in broad diagnostic categories. The Odds Ratio (OR) of PTSD and panic disorder and agoraphobia were higher in women, whereas alcohol abuse was higher in men [6, 51]. Comorbid symptoms and diagnoses may mask the existence of PTSD (e.g., substance and alcohol abuse, traumatic brain injury, depression, anxiety, eating disorders, personality disorders) which some postulate frequently leads to PTSD being underdiagnosed. Additional studies have confirmed comorbidity between PTSD and other psychiatric disorders particularly alcohol and substance abuse, major depression, eating disorders, personality disorders, and anxiety disorders [6, 52].

People with serious mental illness (SMI) (e.g., schizophrenia spectrum disorders, mood disorders) often have higher rates of childhood adversity, including physical abuse, sexual abuse and neglect compared to the general population. Estimates of lifetime prevalence rates of PTSD in this population range from 14-53%, much higher in comparison to the general population. Elevated PTSD prevalence rates are found in both men and women with SMI, with women at higher risk. In addition, people with SMI are more vulnerable to trauma and victimization in adulthood. In contrast to general population studies, women with SMI have higher rates of exposure to traumatic events compared to men with SMI [53].

Similar to general population studies, women with SMI report more sexual assault and men have higher rates of physical assault. This group has higher overall rates of homelessness, suicidal behavior, non-suicidal self-harm, more severe symptoms, more impaired cognitive function, overall poorer health and quality of life, impairment in functioning, and higher rates of alcohol and substance use [53].

Independent of gender, PTSD is associated with an increase in somatic health disorders and premature mortality. Individuals with PTSD have higher rates of obesity, metabolic syndrome, cardiovascular and cerebrovascular disease, hypertension, diabetes mellitus, and non-cirrhotic liver disease. Individuals with PTSD can have impaired immune function with higher rates of stomach ulcers, infection, and increased autoimmune diseases such as thyroid disorders and rheumatoid arthritis. Chronic pain syndromes such as fibromyalgia, chronic fatigue syndrome, migraine headaches, chronic musculoskeletal pain, osteoarthritis and osteoporosis are also more prevalent in those with PTSD.

Individuals with PTSD have premature all-cause mortality, and have increased risk of death following acute coronary events and surgery compared to the general population [36, 22, 54-56].

EXPOSURE TO TRAUMATIC EVENTS

Over the lifetime, trauma exposure is believed to occur in the majority of the general population. In the United States (US) the National Comorbidity Survey (NCS) reported that a community sample of 18-54 year old adults, 61% of men and 51% of women had at least one lifetime trauma [6]. In the Detroit Area Survey of Trauma of 18-45 years old adults, 92.2% of men and 87.1% of women reported at least one traumatic event [43]. A study of a community sample of young Germans (14-24 years old) found 26% of male subjects and 17.7% of female subjects reported at least one traumatic event [7].

TYPE OF TRAUMA EXPOSURE

The type of trauma predicts likelihood of developing PTSD, with certain events more noxious than others. Although men experience more traumatic events over their lifetime compared to women, there are gender differences in the types of trauma exposure. Women are more likely to experience sexual abuse and rape, while men are more likely to experience physical assault and serious accidents. Mass trauma events such as natural or man-made disasters generally are comparatively less noxious determinants of PTSD [6, 7, 42, 43]. Among traumatic events in both community and military samples, rape has the highest conditional risk for PTSD for both men and women. The research to date shows the risk for developing PTSD from rape is as high or higher compared to combat exposure [6, 57, 58]. In the NCS, the most likely trauma events that lead to PTSD in men are rape (65%), combat (35%), and sexual assault (15%). About 45% of women develop PTSD after rape, a lower rate compared to men, but the difference was not statistically significant. Women are significantly more likely than men to develop PTSD after a physical assault (20% versus 2%), threat with a weapon (33% versus 2%), and sexual assault (27% versus 15%) [6].

Earlier age of trauma exposure, and ongoing exposure as a child or adult are important predictors of PTSD and other psychopathology. During childhood, women are more likely to experience sexual abuse compared to men (27% versus 8%) [43-45]. In children and adolescents living in areas of war and ongoing political conflict, studies have shown that boys are more likely to directly experience shooting, shelling, the wounding of a friend or others, living in concentration camps, and to be physically injured. Girls reported more maltreatment (e.g., sexual and physical abuse), loss, and wounding of close friends and relatives. A review of the effect of a wide variety of traumatic events in child and adolescent populations showed the most common psychiatric disorder was PTSD (even surpassing physical injury), followed by mood disorders, anxiety disorders, learning difficulties, and conduct disorders [59]. A review of studies in children and adolescents exposed to armed conflict in the Middle East show high rates of PTSD including: 5-8% in Israel, 23-70% in Palestine, and 10-30% in Iraq [60].

These rates are comparable to other studies of children in countries under armed conflict. The strongest predictors of PTSD in children are reported as amount and type of trauma exposure, age (older children have more posttraumatic stress symptoms), gender, socioeconomic adversity, poor social support, and religiosity. In terms of gender differences, boys had more objective or direct exposure to traumatic events, more behavioral problems, aggression and hyperactivity. Girls had greater subjective exposure (defined as "cognitive and emotional experiences following a traumatic event"), reported more posttraumatic stress symptoms, depression, and separation anxiety than boys [60].

Prevalence in the General Population

Reported lifetime prevalence rates of PTSD in the general population vary widely by country, from 0.3% in China to 6.8% in the United States [6, 7]. Most general population studies have found that women are about twice as likely to develop PTSD over their lifetime compared to men. In the United States, the lifetime prevalence of PTSD in women versus to men is estimated to be 10.4% versus 5.0% [6]. The ESEMeD study, conducted in community samples of six European countries, reported lifetime prevalence rates of PTSD of 0.9% in men and 2.9% in women [61]. Overall, countries have found similar gender differences and comorbidity. However, compared to US, community samples in other countries report fewer trauma events and lower PTSD prevalence, but the conditional probability for PTSD after trauma was similar [6, 7, 61]. The wide differences in prevalence rates have been attributed to a variety of factors including: changes in the diagnostic criteria over time, classification system used, heterogeneous study methodology (e.g., self-report, clinician interview, diagnostic instrument), recall bias, differences in cultural symptom expression, comorbid disorders, social and cultural mediation of symptoms contributing to underreporting and underdiagnosis. Peters et al., found significant gender differences in the prevalence of PTSD for ICD-10 (with women endorsing more symptoms than men), however no gender differences were found using DSM-IV criteria. The investigators raised the issue that gender differences could be attributable to different patterns of symptom endorsement leading to over- or underdiagnosis of PTSD [17]. Community samples of civilians exposed to war or other violent conflict show elevated PTSD prevalence rates ranging from 16-37%; people in countries with higher rates positively correlated to more frequent trauma events or longer conflict duration [62]. Broadly speaking, US and European epidemiological studies show age-related declines in the overall prevalence of mental disorders. However, a significant part of the older population continues to manifest or develop PTSD, and as the population ages, prevalence rates of dementia and other neurodegenerative conditions rise and can complicate the clinical presentation [6, 61]. Age-related changes in PTSD symptoms need further investigation. In studies that used both self-report and clinician-rated measures of PTSD in older and younger veterans, older men self-reported significantly lower numbing symptoms when compared to younger men. Older men also had lower clinician-rated PTSD severity, possibly leading to underdetection and undertreatment [62]. Longitudinal studies of PTSD offer further insight into risk factors and the course of the disorder.

However, aside from confirming the finding of increased prevalence in women, most studies did not examine or were not designed to study other gender-dependent variables.

Groups at higher risk of PTSD include military personnel, first responders, people with serious mental illness (SMI), victims of war, torture, violence, child abuse, rape, Holocaust survivors, and survivors of severe accidents, industrial disasters and natural disasters. Research in the community is also exploring the risk of PTSD following certain events that can be experienced as traumatic, such as: accidents, (e.g., motor vehicle accidents) and being diagnosed with a life-threatening illness, such as cancer. A prospective study of motor vehicle accident survivors showed no gender differences in development or recovery from PTSD [10].

Populations Exposed to Mass Conflict

Many countries worldwide are engaged in ongoing pervasive conflict that subjects their civilian population to frequent traumatic events, displacement, multiple losses and deprivation, disruption of community supports, and an environment of ongoing threat and instability. The World Health Organization estimates the prevalence of mental health disorders as 20% in the population of countries with ongoing humanitarian emergencies [64]. Steel et al., conducted a systematic review and meta-analysis of adult civilian populations exposed to mass conflict, displacements and torture included data from 161 articles and 181 surveys in 40 countries. The weighted average of PTSD prevalence was estimated to be between 13%-25%. Torture was reported in 29 of the 40 countries. In 84 surveys 21% of subjects reported personal experience of torture. Torture was the strongest predictor of PTSD, followed by exposure to repeated traumatic events. Gender was not a predictor of either PTSD or depression. PTSD was higher in populations that were displaced within or outside their country or in refugee camps compared to those permanently resettled in other countries [65].

Natural or Man-Made Disasters

Each year, a large proportion of the population is exposed to trauma through natural disasters (e.g., floods, earthquakes) or man-made disasters (e.g., industrial accidents, terrorism). In the NCS, 18.9% of men and 15.2% of women reported a lifetime exposure to a disaster [6]. International data are limited but ongoing conflicts, community displacements, terrorist attacks, and other violence expose many worldwide communities to high rates of mass trauma [66].

Shalev et al., comprehensively reviews the complexities regarding the psychiatric epidemiology of mass disasters [67]. The current empirical evidence suggests that survivors with direct exposure to disasters have a 30-40% 12-month prevalence of PTSD. Overall, disaster studies reflect the general population finding that women are at higher risk than men.

Research indicates that that after a mass trauma event women exhibited initially higher PTSD symptoms, but the gender difference disappeared with a high rate of spontaneous remission, yielding low prevalence 18 months later. Other groups at high risk of PTSD include children, rescue workers, and those with pre-existing psychiatric disorders.

Psychosocial correlates of PTSD after natural or man-made disasters include neuroticism, guilt, difficulty concentrating, obsessive traits, maladaptive coping strategies, low social support, and greater degree of trauma exposure [2, 68-70].

Military Personnel

Early studies of PTSD focused mainly on Vietnam veterans and showed high prevalence rates. A recent re-examination of PTSD prevalence in Vietnam veterans with combat exposure reported a lifetime prevalence of 18.7% and a point prevalence of 11.1% [71]. Data from a Vietnam veteran twin study revealed 16.8% of monozygotic (MZ) twins who served in Vietnam had PTSD compared to 5.0% of their MZ twin siblings who had not served. Twins with high combat exposure had about 9 times the prevalence of PTSD compared to their MZ twin sibling [72-74]. The reported prevalence rates of PTSD amongst current military personnel and veterans vary widely. Some of this marked variation is attributed to heterogeneous study methodology, especially the administration of self-report versus clinician-administered instruments. However, United States combat veterans are estimated to have a two- to four-fold increase in PTSD prevalence compared to the US civilian population. PTSD prevalence in VA primary care clinics is 11.5% [75, 76].

Toomey et al., conducted self-report and in-person psychological evaluations of Gulf war veterans (Operation Desert Storm) about ten years post-deployment. This study examined 1061 deployed veterans and 1128 non-deployed veterans.

Both groups had 78% male subjects. They found 6.3% of deployed troops had war-onset PTSD compared to 1.1% of non-deployed Gulf war veterans. Their analysis, which included gender differences, showed that deployment was the only significant independent variable predictive of PTSD. They also found the prevalence of PTSD decreased over time [77]. This is in contrast to other studies in which the prevalence of PTSD increased from 3-7% within 5 days of return from deployment to 8-16% at 18-24 month follow-up [57].

More recent studies of military personnel deployed to Iraq and Afghanistan also show wide variations, but elevated prevalence rates of PTSD. New onset PTSD prevalence rates based on self-report instruments were 7.6-8.7% in deployed, combat exposed troops, compared to 2.3-3.0% in non-deployed troops [78].

The military forces in the US and other countries are predominantly male, and historically many have limited the combat service roles of women. Recent trends over the past few decades have expanded the role of women in the military exposing more female soldiers to direct combat. A recent review of Iraq and Afghanistan soldiers found gender was not a reliable predictor of PTSD, but acknowledged the methodology of the available studies limited analysis of gender differences [78].

A recent study by Kline et al., examined risk and protective factors for PTSD in National Guard troops deployed to Iraq. Despite the same level of combat exposure, female gender remained a small, but significant predictor of higher post-deployment posttraumatic stress symptoms. In their analysis, the authors attribute the differences to female soldiers endorsing lower unit cohesion and lower military preparedness compared to male soldiers. The authors point out that both of these are modifiable risk factors that can be addressed to decrease vulnerability to PTSD [79].

Military Sexual Trauma

Studies of gender differences and PTSD in the military have also focused on military sexual trauma (MST). The US Veteran Health Administration (VHA) defines MST as sexual assault and/or severe and threatening sexual harassment that occurred during military service. MST has serious consequences for both men and women [80, 81]. MST is also estimated to be much higher in women compared to men service personnel. About 22% of women and 1% of men under VHA care report MST [81]. In one investigation that used a self-report instrument with soldiers deployed to Iraq and Afghanistan 42% of women endorsed MST compared to 12.5% of men. In this study, the men with MST had higher rates of multiple war related stressors, however no significant gender differences were found in prevalence of symptoms, PTSD, or readjustment to civilian life [82]. A study of homeless veterans reported MST in 39.7% of women and 3.3% of men. Male homeless veterans with MST had significantly higher rates of other psychiatric disorders, suicide attempts, non-suicidal self-harm, and more mental health clinic visits compared to male homeless veterans without MST [83]. A review of MST in US men estimated a lifetime prevalence of 1.1%, however this is likely an underestimate. Survey nonresponse rates varied from 3-92% and there are many barriers to disclosure of MST for both men and women soldiers. It is estimated that male servicemen do not report 66% of MST incidents [84, 85]. In active servicemen and male veterans, MST is likely to be a highly noxious predictor of PTSD, but a large proportion of these incidents may go unreported, and symptoms of PTSD may be overlooked or overshadowed by other physical and psychopathological problems.

First Responders and Rescue Workers

First responders, including emergency medical personnel, police, firefighters, rescue and other disaster responders, are considered to be at higher risk for development of PTSD or posttraumatic stress symptoms because the nature of their occupation exposes them to traumatic stressors. This group includes both professional and voluntary first responders. Emergency responders are a self-selected group and often undergo pre-employment screening and extensive training. These factors along with a strong sense of group cohesion, sound organizational environment, systematic training, support and debriefing after an event increase resilience towards stressors [70]. Studies of this group have shown PTSD prevalence rates ranging from 7-10% [86-89]. A systematic review and meta-analysis of an international sample of first responders showed no gender-related differences and a pooled prevalence rate of PTSD of 10%. The review confirmed previous research findings that in this group ambulance workers tend to have highest rates of PTSD and police have lowest rates [89].

Studies of police officers have generally shown no significant gender differences and PTSD prevalence rates less than 10% [90-92]. Some suggest that the lack of gender differences in PTSD observed in police (and in many military studies) is that female officers may use characteristically male ways of suppressing, inhibiting, or otherwise controlling their emotional distress, including increased alcohol use and somatic expression [90]. In a study of police and other first responders, Marmar et al., found five variables associated with development of PTSD: greater peritraumatic distress, greater peritraumatic dissociation, greater problem-solving coping, greater routine work environment stress, and lower levels of social support [93]. In a prospective study of traumatized police officers, 7% developed PTSD, and 34% had PTSD symptoms but did not meet diagnostic criteria. Trauma severity was the only predictor of persistent PTSD symptoms at 3 and 12 month follow up. There were no differences in gender rates. At 12 months, low job satisfaction, social isolation outside of work, lack of hobbies, acute hyperarousal, and subsequent traumatic event exposure were the factors that predicted symptom persistence [87].

Studies of firefighters, who often have additional duties as paramedics, have shown PTSD prevalence rates approaching 20% [94, 95]. The available studies include all or mostly male subjects. A 2-year prospective study reported psychological characteristics associated with increased PTSD risk in firefighters were: high levels of hostility and low self-efficacy, depression, anxiety and alexithymia (difficulty understanding, processing or describing emotions) [96]. Pre-morbid factors such as higher neuroticism, history of psychiatric disorders, fear of emotions, negative social interactions, and fewer social supports were better predictors of PTSD risk than severity of exposure or property loss [69, 97].

In summary, current research of this heterogeneous, and predominantly male, group suggest that individual pre-trauma event factors that increase resilience to stressors include: older age, more experience, higher self-efficacy, lower levels of hostility, hobbies and social supports, and a negative history of psychiatric disorders. Peri-traumatic factors include less peri-traumatic dissociation and distress; and post-trauma factors that mitigate the risk of PTSD are the use of positive coping strategies such as the use of social supports, active problem solving and analysis of the situation, and positive reappraisal style that fosters creativity and personal growth.

Conversely, detrimental coping styles may increase the risk of PTSD include avoidance, distancing, confrontational style (e.g., hostility, risk-taking), accepting responsibility, and efforts at emotional self-control [70, 88]. At an organizational level, pre-employment selection, training in stress management, early intervention, and a supportive work environment are factors that generally contribute to resilience [70].

CONCLUSION

PTSD is a prevalent and complex clinical disorder that exacts a high cost to families and society through severe human suffering, intergenerational effects, functional and role impairment, high health care needs, and other economic impacts. It can become chronic, and is associated with poor quality of life, significant occupational/role dysfunction, interpersonal difficulties, suicide, comorbid psychiatric and physical health disorders, and premature mortality.

The observed, and sometimes contradictory differences in gender-related prevalence rates are not fully understood. There is a paucity of research in many aspects of gender-differences in PTSD. Gender-related susceptibility or resilience to stress-related disorders has been attributed to differences in a variety of psychosocial, cultural, and neurobiological factors.

In humans, studies of sex and other hormone levels, structural and functional brain imaging, and gene expression continue to explore the neurobiological gender differences that may increase or mitigate the risk of developing PTSD.

Heterogeneity and lack of agreement regarding diagnostic classification, comorbidities, and study methodology have contributed to problems interpreting the research literature [17, 98]. Certain biological, psychosocial and cultural factors and mediators such as lifetime exposure to trauma, childhood exposure, repetitive exposures, individual coping methods, social supports, and cognitive appraisal styles, and trauma-event related characteristics, have been associated with increased risk of PTSD.

Important areas requiring further research include: gender-related issues, mediating and modulating factors, manifestation over the lifespan, social, cultural and neurobiological factors that contribute to risk and resilience. Although not all traumatic events exposures are preventable, many humanitarian disasters are, such as war and other forms of interpersonal violence. As a society we must commit resources addressing the social determinants and inequities that contribute to violence, provide individual and community interventions that decrease the risk of childhood abuse, rape and other forms of interpersonal violence – preventable traumatic stressors known to be potent determinants of PTSD and other disorders. Finally, research towards a more comprehensive understanding of this disorder will aid in identifying people at risk regardless of gender, and help develop more effective prevention and treatment strategies for both men and women.

REFERENCES

[1] Schnurr, P. P., Friedman, M. J., Sengupta, A., Jankowski, M. K., Holmes, T.: PTSD and utilization of medical treatment services among male Vietnam veterans. *J. Nerv. Ment. Dis*. 2000 Aug; 188:496-504.

[2] Shalev, A. Y.: Posttraumatic stress disorder (PTSD) and stress related disorders. *Psychiatr. Clin. North Am*. 2009; 32:687-704.

[3] Kulka, R. A., Schlenger, W. E., Fairbank, J. A., Hough, R. L., Jordan, B. K., Marmar, C. R., Weiss, D. S.: *Trauma and the Vietnam War generation: Report on the findings from the National Vietnam Veterans Readjustment Study*. New York: Brunner/Mazel; 1990.

[4] Ouimette, P., Cronkite, R., Henson, B. R., Prins, A., Gima, K., Moos, R. H.: Posttraumatic stress disorder and health status among female and male medical patients. *J. Trauma Stress* 2004; 1:1-9.

[5] Vermetten, E., Lanius, R.: Biological and clinical framework for posttraumatic stress disorder. *Handb. Clin. Neurol*. 2012; 106:291-342.

[6] Kessler, R. C., Sonnega, A., Bromet, E., Hughes, M., Nelson, C. B.: Posttraumatic stress disorder in the National Comorbidity Survey. *Arch. Gen. Psychiatry* 1995; 52: 1048-1060.

[7] Perkonigg, A., Kessler, R. C., Storz, S., Wittchen, H. U.: Traumatic events and post-traumatic stress disorder in the community: prevalence, risk factors and comorbidity. *Acta Psychiatr. Scand.* 2000 Jan; 101:46-59.

[8] Gavranidou, M., Rosner, R.: The weaker sex? *Gender and post-traumatic stress disorder: Depress Anxiety* 2003; 17:130-139.

[9] Brewin, C. R., Andrews, B., Valentine, J. D.: Meta-analysis of risk factors for posttraumatic stress in trauma-exposed adults. *J. Consult. Clin. Psychol.* 2000; 68:748-766.

[10] Freedman, S. A., Gluck, N., Tuval-Mashiach, R., Brandes, D., Peri, T., Shalev, A. Y.: Gender differences in responses to traumatic events: A prospective study. *J. Trauma Stress* 2002; 15:407-413.

[11] Yasan, A., Saka, G., Ozkan, M., Ertem, M.: Trauma type, gender, and risk of PTSD in a region within an area of conflict. *J. Trauma Stress* 2009; 22:663-666.

[12] Cohen, H., Yehuda, N. Gender differences in animal models of posttraumatic stress disorder. *Disease Markers* 2011; 30:141-150.

[13] *American Psychiatric Association Diagnostic and Statistical Manual of Mental Disorders (DSM-III)*, 3rd Edition. Washington, DC. American Psychiatric Press, 1980.

[14] *American Psychiatric Association Diagnostic and Statistical Manual of Mental Disorders (DSM-IV-TR)*, revised 4th Edition. Washington, DC. American Psychiatric Press, 2000.

[15] *American Psychiatric Association Diagnostic and Statistical Manual of Mental Disorders (DSM-5)*, 5th Edition. Arlington, Virginia. American Psychiatric Press, 2013.

[16] *The International Classification of Diseases 10th edition (ICD-10)*, Classification for Mental and Behavioral Disorders. World Health Organization, Geneva, 2010, pp. 120-121.

[17] Peters, L., Issakidis, C., Slade, T., Andrews, G.: Gender differences in the relevance of DSM-IV and ICD-10 PTSD. *Psychol. Med.* 2006; 36: 81-89.

[18] Peleg, T., Shalev, A. Y.: Longitudinal studies of PTSD: Overview of findings and methods. *CNS Spectr.* 2006; 11:589-602.

[19] Weisæth, L.: The stressors and post-traumatic stress syndrome after and industrial disaster. *Acta Psychiatr. Scand.* Suppl. 1989; 355:25-37.

[20] Herman, J. L.: Complex PTSD: A syndrome in survivors of prolonged and repeated trauma. *J. Trauma Stress* 1992; 5:377-391.

[21] Selye, H.: A syndrome produced by diverse nocuous agents: *Nature* 1936; 138:32.

[22] McEwen, B. S.: The neurobiology of stress: from serendipity to clinical relevance: *Brain Res.* 2000; 886:172-189.

[23] Bracha, H.: Human brain evolution and the "Neuroevolutionary Time-Depth Principle": implications for the reclassification of fear-circuitry-related traits in DSM-V and for studying resilience to warzone-related posttraumatic stress disorder. *Prog. Neuropsychopharmacol. Biol. Psychiatry* 2006; 30:827-853.

[24] Vermetten, E., Bremner, J.: Circuits and systems in stress. *I. Preclinical studies: Depress Anxiety* 2002; 15:126-147.

[25] Yehuda, R.: The biology of posttraumatic stress disorder. *J. Clin. Psychiatry.* 2001;62 (suppl. 17):41-46.

[26] Elzinga, B. M., Schmahl, C. G., Vermetten, E., van Dyck, R., Bremner, J. D.: Higher cortisol levels in response to traumatic reminders in abuse-related PTSD. *Neuropsychopharmacol.* 2003; 28:1656-1665.

[27] Dekel, S., Ein-Dor, T., Gordon, K. M., Rosen, J. B., Bonanno, G. A.: Cortisol and PTSD symptoms among male and female high-exposure 9/11 survivors. *J. Truama Stress* 2013; 26:621-625.

[28] Raison, C. L., Miller, A. H.: When not enough is too much: the role of insufficient glucocorticoid signaling in the pathophysiology of stress-related disorders. *Am. J. Psychiatry* 2003; 160:1554-1565.

[29] Fries, E., Hesse, J., Hellhammer, J., Hellhammer, D. H.: A new view on hypocortisolism. *Psychoneuroendocrinol.* 2005; 30:1010-1016.

[30] Gill, J., Vythilingam, M., Page, G. G.: Low cortisol, high DHEA, and high levels of stimulated TNF-alpha, and IL-6 in women with PTSD. *J. Trauma Stress* 2008; 21:530-539.

[31] Sarchielli, P., Alberti, A., Baldi, A., Coppola, F., Rossi, C., Pierguidi, L., Floridi, A., Calabresi, P.: Proinflammatory cytokines, adhesion molecules, and lymphycyte integrin expression in the internal jugular blood of migraine patients without aura assessed ictally. *Headache* 2006; 46:200-207.

[32] Neylan, T. C., Sun, B., Rempel, H., Ross, J., Lenoci, M., O'Donovan, A., Pulliam, L.: Suppressed monocyte gene expression profile in med versus women with PTSD. *Brain Behav. Immun.* 2011; 25:524-531.

[33] Rasmusson, A. M., Schnurr, P. P., Zukowska, Z., Scioli, E., Forman, D. E.: Adaptation to extreme stress: post-traumatic stress disorder, neuropeptide Y and metabolic syndrome. *Exp. Biol. Med.* (Maywood). 2010; 235:1150-1162.

[34] Villareal, R. P., Lui, B. C., Massumi, A.: Heart rate variability and cardiovascular mortality. *Curr. Atheroscler. Rep.* 2002; 4:120-127.

[35] Dennis, P. A., Watkins, L. L., Calhoun, P. S., Oddone, A., Sherwood, A., Dennis, M. F., Rissling, M. B., Beckham, J. C.: Posttraumatic stress, heart rate variability, and the mediating role of behavioral health risks. *Psychosom. Med.* 2014; 76:629-637.

[36] Levine, A. B., Levine, L. M., Levine, T. B.: Posttraumatic stress disorder and cardiometabolic disease. *Cardiology* 2014; 127:1-19.

[37] Seeman, T. E., Singer, B. H., Rowe, J. W., Horwitz, R. I., McEwen, B. S.: Price of adaptation—allostatic load and its health consequences: MacArthur studies of successful aging. *Arch. Intern. Med.* 1997; 157:2259-2268.

[38] De Bellis, M. D., Keshavan, M. S., Shifflett, H., Iyengar, S., Beers, S. R., Hall, J., Moritz, G.: Brain structures in pediatric maltreatment-related posttraumatic stress disorder: a sociodemographically matched study. *Biol. Psychiatry* 2002; 52:1066-1078.

[39] Felmingham, K., Williams, L., Kemp, A., Liddell, B., Falconer, E., Peduto, A., Bryant, R.: Neural responses to masked fear faces: sex differences and trauma exposure in posttraumatic stress disorder. *J. Abnormal Psychol.* 2010; 119:241-247.

[40] Olff, M., Langeland, W., Draijer, N., Gersons, B. P. R.: Gender differences in posttraumatic stress disorder. *Psychol. Bull.* 2007; 133:183-204.

[41] Yehuda, R., Flory, J. D., Pratchett, L. C., Buxbaum, J., Ising, M., Holsboer, F.: Putative biological mechanisms for the association between early life adversity and the subsequent development of PTSD. *Psychopharmacol.* (Berl.) 2010; 212:405-417.

[42] Breslau, N.: Gender differences in trauma and posttraumatic stress disorder. *J. Gend. Specif. Med.* 2002; 5:34-40.

[43] Breslau, N., Kessler, R. C., Chilcoat, H. D., Schultz, L. R., Davis, G. C., Andreski, P. Trauma and posttraumatic stress disorder in the community: the 1996 Detroit area survey of trauma. *Arch. Gen. Psychiatry.* 1998; 55:626-632.

[44] Finkelhor, D.: The international epidemiology of child sexual abuse. *Child Abuse Negl.* 1994; 18:409-417.

[45] Pereda, N., Guilera, G., Forns, M., Gomez-Benito, J.: *The international epidemiology of child sexual abuse: A continuation of Finkelhor.* 2009; 33:331-342.

[46] Norris, F. H., Friedman, M. J., Watson, P. J., Byrne, C. M., Diaz, E., Kaniasty, K.: 60,000 disaster victims speak: Part I. An empirical review of the empirical literature 1981-2001. *Psychiatry* 2002; 65:207-239.

[47] Bryant, R. A., Creamer, M., O'Donnell, M., Silove, D., McFarlane, A. C.: The capacity of acute stress disorder to predict posttraumatic stress disorders. *J. Psychiatr. Res.* 2012; 46:168-173.

[48] Hanley, K. E., Leifker, F. R., Blandon, A. Y., Marshall, A. D.: Gender differences in the impact of posttraumatic stress disorder symptoms on community couples' intimacy behaviors. *J. Fam. Psychol.* 2013; 27:525-530.

[49] Dickerson, S. S., Kemeny, M. E.: Acute stressors and cortisol responses: a theoretical integration of laboratory research. *Psychol. Bull.* 2004; 130:355-391.

[50] Rasmusson, A. M., Vythilingam, M., Morgan, C. A. 3rd: The neuroendocrinology of posttraumatic stress disorders: new directions. *CNS Spectr.* 2003; 8:651-666.

[51] ESEMeD/MHEDEA 2000 Investigators: 12-Month comorbidity patterns and associated factors in Europe: results from the European Study of the Epidemiology of Mental Disorders (ESEMeD) project. *Acta Psychiatr. Scand.* 2004; 109:28-37.

[52] Pietrzak, R., Goldstein, R., Southwick, S., Grant, B.: Psychiatric comorbidity of full and partial posttraumatic stress disorder among older adults in the United States: results from wave 2 of the National Epidemiologic Survey on Alcohol and Related Conditions. *Am. J. Geriatr. Psychiatry* 2012; 20:380-390.

[53] Shack, A. V., Averill, P. M., Kopecky, C., Krajewski, K., Gummattira, P. Prior history of physical and sexual abuse among the psychiatric inpatient population: A comparison of males and females. *Psychiatric Quarterly* 2004; 75:343-359.

[54] Boscarino, J.: Posttraumatic stress disorder and physical illness: results from clinical and epidemiologic studies. *Ann. NY Acad. Sci.* 2004; 1032:141-153.

[55] McEwen, B.: Protection and damage from acute and chronic stress: allostasis and allostatic overload and relevance to the pathophysiology of psychiatric disorders. *Ann. NY Acad. Sci.* 2004; 1032:1-7.

[56] Rohleder, N., Karl, A.: Role of endocrine and inflammatory alterations in comorbid somatic diseases of post-traumatic stress disorder. *Minerva Endocrinol.* 2006; 31:273-288.

[57] Wolfe, J., Proctor, S. P., Erickson, D. J., Heeren, T., Friedman, M. J., Huang, M. T., Sutker, P. B., Vasterling, J. J., White, R. F.: Relationship of psychiatric status to Gulf War veterans' health problems. *Pyschosom. Med.* 1999; 61:532-540.

[58] Kimmerling, R., Gima, K., Smith, M., Street, A., Frayne, S.: The Veterans Health Administration and military sexual trauma. *Am. J. Public Health* 2007; 97:2160-2166.

[59] Nooner, K. B., Linares, L. O., Batinjane, J., Kramer, R. A., Silva, R., Cliotre, M.: Factors related to posttraumatic stress disorder in adolescence. *Trauma Viol. Abuse* 2012; 13:153-166.

[60] Dimitry, L.: A systematic review on the mental health of children and adolescents in areas of armed conflict in the Middle East. *Child Care Health Dev*. 2012; 38:153-161.

[61] ESEMeD/MHEDEA 2000 Investigators: Prevalence of mental disorders in Europe: results from the European Study of the Epidemiology of Mental Disorders (ESEMeD) project. *Acta Psychiatr. Scand*. 2004; 109:21-27.

[62] Ben-Ezra, M., Palgi, Y., Hamama-Raz, Y., Shrira, A.: Factors predicting risk for PTSD among highly exposed civilians: Two longitudinal studies. *Psychiatr. Q* 2014; Sep. 30 [Epub. ahead of print.]

[63] Lunney, C. A., Schnurr, P. P., Cook, J. M.: Comparison of clinician- and self-assessments of posttraumatic stress symptoms in older versus younger veterans. *J. Trauma Stress* 2014; 27:1-8.

[64] *Inter-Agency Standing Committee. IASC Guidelines on Mental Health and Psychosocial Support in Emergency Settings*. Geneva, Switzerland: Inter-Agency Standing Committee; 2007.

[65] Steel, Z., Chey, T., Silove, D., Marnane, C., Bryant, R. A., van Ommeren, M.: Association of torture and other potentially traumatic events with mental health outcomes among populations exposed to mass conflict and displacement. A systematic review and meta-analysis. *JAMA* 2009; 302:537-549.

[66] Kessler, R. C. Posttraumatic stress disorder: the burden to the individual and to society. *J. Clin. Psychiatry* 2000; 61(suppl. 5):4-12.

[67] Shalev, A. Y., Tuval-Mashiach, R., Hadar, H.: Posttraumatic stress disorder as a result of mass trauma. *J. Clin. Psychiatry* 2004; 65(suppl. 1):4-10.

[68] Breslau, N.: The epidemiology of trauma, PTSD, and other posttraumatic disorders. *Trauma Violence Abuse* 2009; 10:198-210.

[69] McFarlane, A. C. The longitudinal course of posttraumatic morbidity. *J. Nerv. Ment. Dis*. 1988; 176:30-39.

[70] Skogstad, M., Skorstad, M., Lie, A., Conradi, H. S., Heir, T., Weisæth, L.: Work-related post-traumatic stress disorder. *Occup. Med*. 2013; 63:175-182.

[71] Dohrenwend, B. P., Turner, J. B., Turse, N. A., Adams, B. G., Koenen, K. C., Marshall, R.: The psychological risks of Vietnam for US veterans: a revisit with new data and methods. *Science* 2006; 313:979-982.

[72] Goldberg, J., Eisen, S. A., True, W. R., Henderson, W. G.: A twin study on the effects of the Vietnam conflict on alcohol drinking patterns. *Am. J. Public Health* 1990; 80: 570-574.

[73] True, W. R., Rice, J., Eisen, S. A., Heath, A. C., Goldberg, J., Lyons, M. J., Nowak, J.: A twin study of genetic and environmental contributions to liability for posttraumatic stress symptoms. *Arch. Gen. Psychiatry* 1993; 50:257-264.

[74] Chantarujikapong, S. I., Scherrer, J. F., Xian, H., Eisen, S. A., Lyons, M. J., Goldberg, J., Tsuang, M., True, W. R.: A twin study of generalized anxiety disorder symptoms, panic disorder symptoms, posttraumatic stress disorder in men. *Psychiatry Res*. 2001; 103:133-145.

[75] Magruder, K. M., Frueh, B. C., Knapp, R. G., Davis, L., Hamner, L. B., Martin, R. H., Gold, P. B., Arana, G. W.: Prevalence of posttraumatic stress disorder in Veterans Affairs primary care clinics. *Gen. Hosp. Psychiatry* 2005; 27:169-179.

[76] Rosenheck, R. A., Fontana, A. F.: Recent trends in VA treatment of post-traumatic stress disorder and other mental disorders. *Health Aff.* 2007; 26:1720-1727.

[77] Toomey, R., Kang, H. K., Karlinsky, J., Baker, D. G., Vasterling, J. J., Alpern, R., Reda, D. J., Henderson, W. G., Murphy, F. M., Eisen, S. A.: Mental health of US Gulf War veterans 10 years after the war. *Br. J. Psychiatry* 2007; 190:385-393.

[78] Crum-Cianflone, N. F., Jacobson, I.: Gender differences of postdeployment post-traumatic stress disorder among service members and veterans of the Iraq and Afghanistan conflicts. *Epidemiol. Rev.* 2014; 36:5-18.

[79] Kline, A., Ciccone, D. S., Weiner, M., Interian, A., St Hill, L., Falca-Dodson, M., Black, C. M., Losonczy, M.: Gender differences in the risk and protective factors associated with PTSD: a prospective study of National Guard troops deployed to Iraq. *Psychiatry* 2013; 76:256-272.

[80] Magley, V., Waldo, C., Drasgow, F., Fitzgerald, L.: The impact of sexual harassment on military personnel: Is it the same for men and women? *Mil. Psychol.* 1999; 11:283-302.

[81] Kimmerling, R., Street, A. E., Pavao, J., et al. Military-related sexual trauma among Veterans Health Administration patients returning from Afghanistan and Iraq. *Am. J. Public Health.* 2010; 100:1409-12.

[82] Katz, L. S., Cojucar, G., Beheshti, S., Nakamura, E., Murray, M.: Military sexual trauma during deployment to Iraq and Afghanistan: prevalence, readjustment, and gender differences. *Violence Vict.* 2012; 27:487-499.

[83] Pavao, J., Turchik, J. A., Hyun, J. K., Karpenko, J., Saweikis, M., McCutcheon, S., Kane, V., Kimmerling, R.: Military sexual trauma among homeless veterans. *J. Gen. Intern. Med.* 2013; 28:S536-S541.

[84] *Naval Inspector General: NAVINSGEN sexual assault study.* 2004 Washington, DC, Department of the Navy.

[85] Hoyt, T., Klosterman Rielge, J., Williams, L. F.: Military sexual trauma in men: A review of reported rates. *J. Trauma Dissoc.* 2011; 12:244-260.

[86] Marmar, C. R., Weiss, D. S., Metzler, T. J., Ronfelt, H. M., Foreman, C.: Stress responses of emergency service personnel to the Loma Prieta earthquake interstate 880 freeway collapse and control traumatic incidents. *J. Traum Stress* 1996; 9:63-85.

[87] Carlier, I. V. E., Lamberts, R. D., Gersons, B. P. R.: Risk factors for posttraumatic stress symptomatology in police officers: A prospective analysis. *J. Nerv. Ment. Dis.* 1997; 185:498-506.

[88] Holland, M.: The dangers of detrimental coping in emergency medical services. *Prehosp. Emerg. Care* 2011; 15:331-337.

[89] Berger, W., Freire Coutinho, E. S., Figueira, I., Marques-Portella, C., Pires Luz, M., Neylan, T., Marmar, C. R., Mendlowicz, M. V.: Rescuers at risk: a systematic review and meta-regression analysis of the worldwide current prevalence and correlates of PTSD in rescue workers. *Soc. Psychiatry Psychiatr. Epidemiol.* 2012; 47:1001-1010.

[90] Lilly, M. M., Pole, N., Best, S. R., Metzler, T., Marmar, C. R.: Gender and PTSD: What can we learn from female police officers? *J. Anxiety Disord.* 2009; 23:767-774.

[91] Pole, N., Best, S. R., Weiss, D. S., Metzler, T., Liberman, A. M., Fagan, J., et al. Effects of gender and ethnicity on duty-related posttraumatic stress symptoms among urban police officers. *Journal of Nervous and Mental Disease*. 2001; 189:442-448.

[92] Martin, M., Marchand, A., Boyer, R.: Traumatic events in the workplace: impact on psychopathology and healthcare use of police officers. *Int. J. Emerg. Ment. Health* 2009; 11:165-176.

[93] Marmar, C. R., McCaslin, S. E., Metzler, T. J., Best, S., Weiss, D. S., Fagan, J., Liberman, A., Pole, N., Otte, C., Yehuda, R., Mohr, D., Neylan, T. Predictors of posttraumatic stress in police and other first responders. *Ann. N Y Acad. Sci.* 2006; 1071:1-18.

[94] Beaton, R. D., Murphy, S. A. Sources of occupational stress among firefighter/EMTs and firefighter/paramedics and correlations with job-related outcomes. *Prehosp. Disaster. Med.* 1993; 8:140-150.

[95] Corneil, W., Beaton, R., Murphy, S., Johnson, C., Pike, K.: Exposure to traumatic incidents and prevalence of posttraumatic stress symptomatology in urban firefighters in two countries. *J. Occup. Health Psychol.* 1999; 4:131–141.

[96] Heinrichs, M., Wagner, D., Schoch, W., Soravia, L. M., Hellhammer, D. H., Ehlert, U.: Predicting posttraumatic stress symptoms from pretraumatic risk factors: a 2-year prospective follow-up study in firefighters. *Am. J. Psychiatry* 2005; 162:2276-2286.

[97] Farnsworth, J. K., Sewell, K. W.: Fear of emotion as a moderator between PTSD and firefighter social interactions. *J. Trauma Stress* 2011; 24:444-450.

[98] Yehuda, R., McFarlane, A. C.: Conflict between current knowledge about posttraumatic stress disorder and its original conceptual basis. *Am. J. Psychiatry* 1995; 152:1705-1713.

In: Neurobiology of Men's Mental Health
Editors: Leo Sher and Timothy R. Rice
ISBN: 978-1-63483-191-8
© 2015 Nova Science Publishers, Inc.

Chapter 20

TRAUMATIC BRAIN INJURY

Danielle B. Kushner, MD*
NYU School of Medicine, New York, NY, US

ABSTRACT

Traumatic brain injury (TBI) is one of the most under-recognized neuropsychiatric conditions despite its prevalence in the male population. Causes of TBI include falls, blunt trauma, and recurrent head injury. Initial injuries come from the direct physical injury of the brain, such as airway disturbance or impaired consciousness. Despite survival from acute injury, more evidence is showing the seriousness of the long-term neuropsychological sequelae from TBI. Such pathology ranges from memory loss, depression, psychosis, and substance use disorders, among others. Each subgroup of symptoms has developing modalities of treatment, including pharmacology. Overall, due to more recognition of the prevalence and seriousness of the condition, TBI is a growing area of research.

Keywords: men's mental health, traumatic brain injury

I. EPIDEMOLOGY/ETIOLOGY

Traumatic brain injury (TBI) is defined as a physiological disruption of the brain resulting from external application of force. Although the exact prevalence of TBI is difficult to predict due to poor reporting, according to the Centers for Disease Control and Prevention (CDC) in 2010 about 2.5 million emergency room visits, hospitalizations, or deaths in the United States were related to TBI [1]. Peak incidence is between the ages of 15 and 24 and older than 64 years [2]. Males are affected twice as often as females, though the rate has narrowed in the United States in recent years.

* Corresponding author: Danielle B. Kushner, MD. Department of Psychiatry, New York University School of Medicine. E-mail: Danielle.Kushner@nyumc.org.

From 2006-2010, falls were the leading cause of TBI, accounting for 40% of all TBIs in the United States. Falls led to 55% of TBIs among children and 81% of TBIs among the elderly. Unintentional blunt trauma was the second leading cause overall among all age groups, followed by motor vehicle accidents. Yet, motor vehicle crashes were the second leading cause mode of death and leading cause of injury in children and young adults [1]. In the civilian population, alcohol is involved in more than half of cases.

Recurrent head injury is common in patients with alcohol abuse and individuals playing contact sports. People from lower socioeconomic groups are also significantly more likely to be affected [2]. In war-torn areas, blast injuries from grenades, bombs, and improvised explosive devices are the most common cause of TBI, with other causes including object impact, burns, or inhalation of toxic vapors [3]. In the incarcerated population, 87% of people self-reported a lifetime history of head injury and 36.2% reported head injury in the preceding year [4]. Since young men have a higher incidence of TBI and incarceration, the prevalence of TBI in the incarcerated groups may not be higher than the general population [5].

II. Pathophysiology

The neuropathophysiology of TBI can be divided into primary and secondary injury. Primary injury results from an immediate acute insult to the brain. There are two basic types of brain injury: direct trauma and diffuse axonal injury. Direct trauma to the brain, causing contusions, hemorrhages, and lacerations, can result in a loss of the function served by a given brain area. Diffuse axonal injury from stretch injuries to the deep white matter can result in interruption of information being transmitted to various areas of brain [1]. In the most severe cases, the primary acute injury may be deadly.

Secondary, non-acute, injuries are the result of biomolecular and physiological changes that follow the initial insults. The inflammatory cascade causes further damage through transient cell membrane disruptions that alter ions and neurotransmitters, changes in cerebral glucose metabolism, and increased generation of free radicals [6]. These disturbances contribute to short and long term neuronal dysfunction and necrosis. Such injuries include seizures, hypoxia, infection, edema, ischemia, and metabolic abnormalities, among others [2]. Further research is developing in this area as this metabolic response proves to be both helpful and deleterious to an injured brain.

III. Course of Illness

Emergency management in the acute phase of TBI includes the initial administration and serial monitoring of Glasgow Coma Score classification, the monitoring fluid balance, and the maintenance of airway and oxygenation. Pharmacological, surgical, and invasive interventions are all included as part of acute TBI management [2]. Following acute stabilization, patients go through a state of impaired consciousness in the subacute phase. Risk factors for a more severe more prolonged delirium include older age, hypoalbuminemia, and prior neurological abnormalities.

The duration of delirium is also an indicator of severity of TBI. Once the patient is beyond the acute and subacute phases, many are at a higher risk for a variety of secondary sequelae including seizures, tremors, loss of taste or smell, and impaired balance [2]. Furthermore, TBI can predispose individuals to a variety of neuropsychiatric sequelae that is outlined below.

IV. Neuropsychiatric Sequelae

Cognitive

Cognitive dysfunction is characterized by impairment of attention, memory, language, or executive function. Post-traumatic cognitive disturbances depend both on the degree of diffuse axonal injury and the volume and location of focal injuries. Diffuse axonal injury is more associated with generalized cognitive dysfunction compared to focal injuries that cause specific deficits [2].

The prevalence of long-term cognitive impairment is approximately 65% in patients with moderate to severe TBI [7]. Following the period of acute delirium is usually a 6-12 month period of rapid recovery that concludes within the 12-24 months subsequent to the injury. Recovery after this time period is slow and limited. The remaining cognitive function is characterized as permanent cognitive sequelae [8]. The extent of deficits is usually correlated with TBI severity as manifested by impaired consciousness or posttraumatic amnesia, but more studies support cognitive defects even without such findings [2]. In adults with mild TBI, there is good recovery of initial cognitive symptoms by 1-3 months. However, a limited number of studies suggest that certain cognitive deficits may last longer in certain populations, but studies have not shown consistency among findings [9].

Treatment of the initial delirium involves treatment of any underlying disorders including metabolic or infectious processes. Atypical neuroleptics, except for clozapine, which lowers the seizure threshold and has significant anticholinergic properties, are the most useful pharmacologic agents. Yet there is evidence that neuroleptics may slow cognitive recovery and in some instances worsen the behavioral effects [10]. In regards to long-term cognitive effects, small studies have shown cholinesterase inhibitors, stimulants, L-dopa, tricyclic antidepressants, and lamotrigene to be helpful in treating cognitive deficits [11-15].

Motivation

Motivation refers to how behavior is started, energized, and sustained. A lack of motivation is common following significant head injury. Overall, there has been much debate about the conceptualization of the concept. Marin and Wilkosz proposed a definition of "disorders of diminished motivation" ranging in intensity from apathy, abulia, and to akinetic mutism [16]. Meanwhile the definition of apathy has also ranged from solely a lack of emotion or feeling in the absence of dysphoria [2], to a term that defines the disorders of overall diminished goal oriented behaviors and cognitions [17].

The reported frequency of generalized apathy ranges due to the comorbidity with other psychiatric diagnoses such as depression and cognitive impairment and lack of standardized assessments. Studies have shown increased apathy from injury to dorsolateral prefrontal lesions [18], anterior cingulate, insula, corpus callosum [19], and thalamic segments projecting to prefrontal and premotor areas [20]. The dysfunction of arousal systems such as glutamatergic and cholinergic neurons may play an additional role [21].

Treatment of amotivational symdromes involves both pharmacological treatment and behavioral interventions. Stimulants and dopaminergic medications, as well as activating antidepressants such as buproprion, monoamine oxidase inhbitors, or acetylcholinesterase inhibitors may increase motivation [22]. Selective serotonin reuptake inhibitors and typical neuroleptics may worsen amotivational syndromes. Nonpharmacological treatment includes rewarding goal-directed behavior, cognitive mediation, behavioral modification, and family interventions [2]. There is a lack of properly controlled trials for apathy in TBI at this time.

Depression

Depression is the most commonly reported psychiatric result of TBI with reports ranging from 25% to 50% of patients with TBI within 12 months of injury and 26% to 64% of patients lifetime. Development of depression has been linked to poorer treatment outcomes, including cognitive deficits as well as psychosocial and functional outcomes. The risk of developing depression is highest in the first year post injury, but the risk remains increased decades later. Depressive disorders are highly correlated with anxiety, substance misuse, and behavioral alterations like impulsivity and aggression [23]. Studies have shown that those with a history of TBI with loss of consciousness are four times more likely to attempt suicide than those without TBI [24].

Differential diagnosis of depressive symptoms includes delirium, substances, adjustment disorder, posttraumatic stress disorder, posttraumatic apathy, personality change due to a general medical condition, and depressive disorders [23]. Pre-injury risk factors for developing depression include past psychiatric history, substance abuse, risk-taking behaviors, younger age, and female gender [25]. Risk among the elderly may be confounded with cognitive defects. Injury risk factors include frontal lesions, left sided brain injury, and atrophy [2]. Other hypotheses suggest that the neural pathways involving the raphe nucleus, involved with serotonergic transmission, or amine-containing neurons are disrupted [8].

Post-injury risk factors include sleep complaints, family dysfunction, and lower socioeconomic status [25]. Therefore, mood disturbances in the early period following TBI may reflect the effects of the neurotrauma while those later in post-injury period reflect the psychological and social factors.

Selective serotonin reputake inhibitors (SSRIs) are preferred due to their safety profile. Sertraline and citalopram are favored due to their shorter half lives and limited side effects, as opposed to fluoxetine and paroxetine with their greater potential for side effects and drug-drug interactions [23]. Tricyclic antidepressants (TCAs) may have benefit in treating anxiety and depression, but should be limited to agents with the least anticholinergic activity, sedation, and effect on seizure threshold [2]. Treatments with other antidepressants have not been well established.

Monoamine oxidase inhibitors (MAOIs) and buproprion are discouraged due to their dietary restrictions and lowering seizure threshold, respectively. Stimulants are commonly used to augment partial responses to SSRIs, especially with ongoing cognitive problems and fatigue. Amantidine and anticholinesterase drugs have also been shown to be helpful in some studies [26, 27]. Electroconvulsive therapy may be used for those who fail other treatments. Repetitive transcranial magnetic stimulation (rTMS), transcranial direct current stimulation (tDCS), or vagal nerve stimulation has not been rigorously studied in TBI populations. However, due to the risk of seizure in TBI population, tDCS or rTMS appear to be more suitable options to treat depression in the TBI population. In addition to pharmacotherapy and brain stimulation, psychotherapy is a necessary part of all treatment plans. Psychoeducation, cognitive behavioral therapy, behavioral interventions, group and family therapy have each been found to be helpful [23].

Mania, Hypomania, and Mixed Disorder

Bipolar and related disorders are relatively uncommon following brain injury. Rates of secondary mania unequivocally related to trauma ranges from 1.7 to 9.0%. Studies have shown that mania postTBI was associated with injury severity and presence of posttraumatic epilepsy [23]. Anterior temporal lesions have shown a greater risk for mania [28], along with lesions of the right hemisphere [29].

Overdiagnosis of mania in this population is common due to misattribution of symptoms of affect dysregulation, impulsive or disinhibited behaviors, and alterations in sleep and appetite symptoms. Differential diagnosis overlaps with the differential of depressive disorders due to TBI, but also includes medication effect and pseudobulbar affect. In this patient population, additional neurophysiological studies such as EEG and neuroendocrine studies are especially important to rule out underlying causes [23].

Clonidine, carbamazepine, and divalproex are the drugs of choice for post-traumatic mania, with lithium being reserved for those with a prior history of mania, as lithium lowers the seizure threshold and its cognitive side effects are better tolerated in those with primary mania [23]. Newer anticonvulsants (lamotrigene and oxcarbaxepine) and atypical neuroleptics can be useful, but there are few published reports of their use. In additional to pharmacology treatment, psychotherapy, ECT, and brain-stimulation techniques can also be helpful in this population.

Psychosis

Approximately 0.9 to 8.5% of TBI patients develop psychotic symptoms. There are two general categories of postTBI psychosis: delusional disorders and schizophrenic-like psychosis. The most common delusional disorders are Capgras syndrome, the delusion that loved ones are imposters, and reduplicative paramnesia, the delusions that a familiar place is duplicated in another location, each of which occurs in approximately 32% of individuals with TBI-induced psychosis [30]. Schizophrenic-like psychosis involves the occurrence of hallucinations and delusions.

Delusional disorders tend to develop during the first year after injury whereas schizophrenic-like psychoses often appear three to four years after injury. Negative symptoms are present in only 43% of schizophrenic- like psychosis. Prodromal symptoms are not universally present, but one study shows behavioral changes included bizarre behavior (50%), affective instability (39%), antisocial behavior (36%), scholastic or work deterioration (33%), and social withdrawal (31%) [31]. Additional psychosis due to TBI is associated with cognitive impairments, most commonly in memory and executive functioning.

Differential diagnosis of post-TBI psychosis includes primary psychotic disorders, substance induced psychosis, post-traumatic disorder, delirium, and psychosis due to general medical condition, such as seizures. TBI psychoses are distinguished from primary psychotic disorders due to the lack of disordered thought process, later age of onset, less premorbid dysfunction, briefer duration, less common family history, better response to neuroleptics, less need for maintenance medication, and a better prognosis [32]. Yet studies have shown higher incidence rates of TBI in persons with schizophrenia before onset of illness [33].

There does not seem to be a correlation with severity of TBI and onset of psychosis. There are additional factors that can increase the risk of psychosis including male gender, previous psychological disturbance, family history of psychotic illness, and prior neurologic or neurodevelopmental conditions [33]. Proposed pathophysiologic mechanisms of psychosis include the dysregulation of dopamine in the temporal limbic and frontal areas of the brain [34]. Psychotic disorders after TBI are associated with focal lesions on imaging and slowing in EEG in the frontal and temporal lobes. Whereas in schizophrenia, neuroimaging shows enlarged ventricles with generalized atrophy, functional neuroimaging shows abnormal activities in several areas in TBI.

Patients with TBI-induced psychosis are more susceptible to adverse effects of medications, such as sedation, anti-cholinergic, and extrapyramidal effects than those with primary psychotic disorders. Treatment for post-TBI psychosis should be started at one-third to one-half the standard dosages due to potential adverse reactions. Yet, there is some evidence that neuroleptics may impede cognitive recovery by decreasing synaptic plasticity [35]. Atypical neuroleptics except for clozapine, which is strongly anti-cholinergc and lowers the seizure threshold, are the drugs of choice. Benzodiazepines should be used sparingly, if at all. In addition to pharmacological interventions, evidence-based psychosocial treatments for schizophrenia maybe useful in this population, such as assertive community training, supported employment, cognitive behavioral therapy, family-based services, skills training, and substance treatment, among others [30].

Anxiety Disorders

Anxiety following TBI includes all variants of anxiety disorders, including generalized anxiety disorder, panic disorder, phobias, obsessive compulsive disorder, and posttraumatic stress disorder. Prevalence of anxiety disorders ranges from 10%-70% [3]. Anxiety disorders frequently coexist with depression. Risk factors for developing acute stress symptoms include increased age, history of posttraumatic stress disorder, and an avoidant coping style. Acute stress disorder predisposes to the development of PTSD. Recall of the traumatic event has been identified as a predictor of developing PTSD [36, 37].

Civilian estimates of PTSD range from 12% to 30% (mild TBI), 15% to 27% (moderate TBI), and 3% to 23% (severe TBI). Among military samples with TBI, prevalence ranges from 12% to 89% [38]. PTSD is less common in patients who lost consciousness following head trauma. Proposed mechanisms for the development of anxiety include increased activity of the aminergic system and decreased activity of the gamma-aminobutyric acid (GABA) inhibitory network. Injury to amygdala and locus ceruleus may also play a role [34]. Right hemisphere lesions are more often associated with anxiety disorders than left-sided lesions [39].

Treatment of anxiety includes pharmacologic treatments in addition to behavioral and family therapy. SSRIs and TCAs maybe helpful, but their respective side effects should be monitored. The 2010 VA/Department of Defense Clinical Practice Guideline for PTSD recommends the SSRI antidepressants, sertraline and paroxetine, and the serotonin norepinephrine reuptake inhibitor (SNRI) antidepressant, venlafaxine, as the first line medication choices for treating PTSD [40]. Citalopram should also be considered due to its low side effect profile and less drug interactions. In general, long-term benzodiazepine treatment is avoided due to adverse reactions and potential paradoxical agitation in this population, but a short course of a short acting benzodiazepine maybe helpful in an acute stress situation. Neuroleptics should be minimized due to side effect of delaying cognitive recovery. Anecdotal reports have reported effectiveness of naltrexone and buspirone [41]. For PTSD, the effectiveness of current evidence based treatments for PTSD including cognitive processing therapy, prolonged exposure, and prazosin for nightmares have yet to be substantiated in those with TBI [38].

Aggressive Disorders

Some authors have divided post-TBI behavioral issues into two variants: *major variant* where the symptoms are chronic, persistent, and severe, and *minor variant* where symptoms are acute and transient [41]. In various studies, the incidence of aggression post-TBI ranges from 35% to 90%. Agitation commonly occurs in the delirious phase of recovery. Chronic irritability and aggression is seen in 40% to 70% of patients [42]. Aggressive behavior can become an obstacle to psychosocial recovery.

Different mechanisms of aggressive behavior after TBI have been described. The first involves a loss of inhibitory control where angry outbursts escalate from underlying irritability that is most likely due to loss of frontal modulatory control. The second process relates to episodic aggression with no evidence of underlying irritability or impulsivity between episodes. This process is assumed to be potentially epileptogenic in nature [43]. Overall, damage to the limbic system causes aggression, orbital-frontal area causes disinhibition, dorsal convexity of frontal lobes causes dysexecutive symptoms, and temporal lobes cause emotional lability and memory problems [8]. Studies have additionally found increased amounts of CSF norephinephrine and decreased amounts of serotonin [2]. Other risk factors for aggression include premorbid affective disorder, low IQ, low socioeconomic status, male gender, personality disorder, and alcohol or substance abuse [44].

In regards to treatment of aggressive symptoms, a multidisciplinary approach should be emphasized in addition to medications; including environmental modification, cognitive behavioral therapy, behavioral therapy, vocational training, supportive psychotherapy, and family therapy [8].

Neuroleptic medications do not help with chronic, non-psychotic aggression, but may be used for sedation in acute aggressive episodes. The side effect of akathisia may actually increase violent behavior and, as noted previously, neuroleptics have been shown to slow recovery. The use of benzodiazepine and buspirone may also cause a paradoxical reaction. Alternatively, the use of quetiapine in low dosages has shown good promise [45]. Carbamazepine, valproic acid, gabapentin, and oxcarbazepine have been used successfully in some patients, most likely those with episodic aggression [46]. The best quality evidence supports propranolol and pindolol [47] in higher dosages for both agitation and aggression. Methylphenidate may be effective on anger [48] and amantadine has also been shown to be helpful with impulsivity and aggression [49]. Current studies have not shown SSRIs to be effective.

Substance Use Disorders

Rates of pre-injury substance use and substance use disorders are greater in those with TBI than the general population. Among various studies, 44% and 79% of individuals with TBI had a pre-injury history of alcohol abuse, while 21% to 40% of illicit drug abuse. Substance abuse is common before injury and intoxication at time of injury is common. Premorbid substance use has been found to be associated with post-TBI drug usage.

A history of substance abuse predicts increased disability, poorer prognosis, and delayed recovery [50]. Chronic substance abuse can lead to cognitive deficits similar to those found in TBI itself. Studies suggest that mesolimbic dopaminergic neural pathways responsible for reward may be disrupted in TBI. Alternative explanations include disruption of incentive and motivation tracks, but will need further investigation [34]. Successful substance use disorder treatment helps reduce secondary illness, lowers mortality rates, and overall health care costs.

Conclusion

Traumatic brain injury is a growing cause of chronic illness and health care costs in the United States, especially in the male population. In addition to the serious medical issues associated with the initial trauma, increasing amounts of patients develop neuropsychological sequelae from their trauma. More awareness is needed in the community for ongoing psychiatric help and rehabilitation for these patients as they transition back to life in the community following injury.

References

[1] Division of Injury and Disability Outcomes and Programs. National Center for Injury Prevention and Control. Centers for Disease Control and Prevention, Department of Health and Human Services. *Traumatic Brain Injury in the United States: emergency*

Department visits, hospitalizations, and deaths. February 2014. Available at: http://www.cdc.gov/traumaticbraininjury/data/rates.html. Accessed November 2014.

[2] Nicholl, J., LaFrance, W. C. Neuropsychiatric Sequelae of Traumatic Brain Injury. *Semin. Neurol*. 2009;29:247-255.

[3] Vaishnavi, S., Rao, V., Fann, J. R. Neuropsychiatric Problems After Traumatic Brain Injury: Unraveling the Silent Epidemic. *Psychosomatics* 2009; 50: 198-205.

[4] Slaughter, B., Fann, J. R., Ehde, D. Traumatic brain injury in a county jail population:prevalence, neuropsychological functioning and psychiatric disorders. *Brain Inj*. 2003;17(9):731-41.

[5] Farrer, T. J., Hedges, D. W. Prevalence of traumatic brain injury in incarcerated groups compared to the general population: A meta-analysis. *Progress in Neuro-Psychopharmacology and Biological Psychiatry* 2011; 35: 390-394.

[6] Prins, M., Greco, T., et al. The Pathophysiology of Traumatic Brain Injury at a Glance. *Disease Models and Mechanisms* 2013; 6: 1307-1315.

[7] Rabinowitz, A. R., Levin, H. S. Cognitive Sequelae of Traumatic Brain Injury. *Psychiatr. Clin. N. Am*. 2014; 37: 1-11.

[8] Rao, V., Lyketsos, C. Neuropsyhciatric Sequelae of Traumatic Brain Injury. *Psychosomatics* 2000; 41: 95-103.

[9] Carroll, L. J., Cassidy, J. D., et al. Systematic Review of the Prognosis After Mild Traumatic Brain Injury in Adults: Cognitive, Psychiatric, and Mortality Outcomes: Results of the International Collaboration on Mild Traumatic Brain Injury Prognosis. *Archives of Physical Medicine and Rehabilitaion* 2014; 95: 152-73.

[10] Hoffmann, A. N., Cheng, J. P., Zafonte, R. D., Kline, A. E. Administration of haloperidol and risperidone after neurobehavioral testing hinders the recovery of traumatic brain injury-induced deficits. *Life Sciences* 2008; 83: 602-607.

[11] Zhang, L., Plotkin, R. C., Wang, G., Sandel, M. E., Lee, S. Cholinergic augmentation with donepezil enhances recovery in short-term memory and sustained attention after traumatic brain injury. *Arch. Phys. Med. Rehabil*. 2004;85(7):1050-1055.

[12] Whyte, J., Vaccaro, M., Grieb-Neff, P., Hart, T. Psychostimulant use in the rehabilitation of individuals with traumatic brain injury. *J. Head Trauma Rehabil*. 2002; 17(4):284-299.

[13] Lal, S., Merbtiz, C. P., Grip, J. C. Modification of function in head-injured patients with Sinemet. *Brain Inj*. 1988;2(3):225-233.

[14] Reinhard, D. L., Whyte, J., Sandel, M. E. Improved arousal and initiation following tricyclic antidepressant use in severe brain injury. *Arch. Phys. Med. Rehabil*. 1996;77 (1):80-83.

[15] Pachet, A., Friesen, S., Winkelaar, D., Gray, S. Beneficial behavioural effects of lamotrigine in traumatic brain injury. *Brain Inj*. 2003;17(8): 715-722.

[16] Marin, R. S., Wilkosz, P. A. Disorders of diminished motivation. *J. Head Trauma Rehabil*. 2005;20:377-88.

[17] Starkstein, S., Leentjens, A. F. The nosological position of apathy in clinical practice. *J. Neurol. Neurosurg. Psychiatr*. 2008;79:1088-92.

[18] Paradiso, S., Chemerinski, E., Yazici, K. M., et al. Frontal lobe syndrome reassessed: comparison of patients with lateral or medial frontal brain damage. *J. Neurol. Neurosurg. Psychiatr*. 1999;67:664-7.

[19] Knutson, K. M., Monte, O. D., Raymont, V., et al. Neural correlates of apathy revealed by lesion mapping in participants with traumatic brain injuries. *Hum. Brain Mapp.* 2013.

[20] Spalletta, G., Fagioli, S., Caltagirone, C., et al. Brain microstructure of subclinical apathy phenomenology in healthy individuals. *Hum. Brain Mapp.* 2013;34: 3193-203.

[21] Goldfine, A. M., Schiff, N. D. What is the role of brain mechanisms underlying arousal in recovery of motor function after structural brain injuries? *Curr. Opin. Neurol.* 2011; 24:564-9.

[22] Lane-Brown, A., Tate, R. Interventions for apathy after traumatic brain injury. *Cochrane Database Syst. Rev*. 2009;(2).

[23] Jorge, R. E., Arciniegas, D. B. Mood Disorders After TBI. *Psychiatr. Clin. N. Am.* 2014: 37: 13-29.

[24] Silver, J. M., Kramer, R., Greenwald, S., et al. The association between head injuries and psychiatric disorders; findings from the New Haven NIMH Epidemiologic Catchment Area Study. *Brain Inj.* 2001; 15; 935-945.

[25] Rapoport, M. J. Depression Following Traumatic Brain Injury. *CNS Drugs* 2012; 26(2): 111-121.

[26] Giacino, J. T., Whyte, J., Bagiella, E., et al. Placebo-controlled trial of amantadine for severe traumatic brain injury. *N. Engl. J. Med.* 2012;366 (9):819-26.

[27] Tenovuo, O. Central acetylcholinesterase inhibitors in the treatment of chronic traumatic brain injury-clinical experience in 111 patients. *Prog. Neuropsychopharmacol. Biol. Psychiatry* 2005;29(1):61-67.

[28] Jorge, R. E., Robinson, R. G., Starkstein, S. E., et al.: Manic syndromes following traumatic brain injury. *Am. J. Psychiatry* 150:916-921, 1993.

[29] Robinson, R. G., Boston, J. D., Starkstein, S. E., et al.: Comparison of mania with depression following brain injury: Causal factors. *Am. J. Psychiatry* 145:172-178, 1988.

[30] Fujii, D. E., Ahmed, I. Psychotic Disorder Caused by Traumatic Brain Injury. *Psychiatr. Clin. N. Am.* 2014; 37: 113-124.

[31] Sachdev, P., Smith, J. S., Cathcart, S. Schizophrenia-like psychosis following traumatic brain injury: a chart-based descriptive and case-control study. *Psychol. Med.* 2001; 31: 231-9.

[32] Sachdev, P., Smith, J. S., Cathcart, S. Schizophrenia-like psychosis following traumatic brain injury: a chart-based descriptive and case-control study. *Psychol. Med.* 2001; 31 (2): 231-239.

[33] Fujii, D., Fkii, D. C. Psychotic Disorder Due to Traumatic Brain Injury: Analysis of Case Studies in the Literature. *The Journal of Neuropsychiatry and Clinical Neurosciences* 2012; 24: 278-289.

[34] Bhalerao, S. U., Geurtjens, C., et al. Understanding the neuropsychiatric consequences associated with significant traumatic brain injury. *Brain Inj.* 2013; 27(7-8): 767-774.

[35] Hoffman, A. N., Cheng, J. P., Zafonte, R. D., Kline, A. E. Administration of haloperidol and risperidone after neurobehavioral testing hindYet, ers the recovery of traumatic brain injury-induced deficits. *Life Sci.* 2008;83(17-18):602-607.

[36] Gil, S., Caspi, Y., Ben-Ari, I. Z., et al. Does memory of a traumatic event increase the risk of posttraumatic stress disorder in patients with traumatic brain injury? A prospective study. *Am. J. Psychiatry* 2005;162: 963-9.

[37] Turnbull, S. J., Campbell, E. A., Swann, I. J. Post-traumatic stress disorder symptoms following a head injury: does amnesia for the event influence the development of symptoms? *Brain Inj.* 2001;15:775-85.

[38] Bahraini, N. H., Breshears, R. E., et al. Traumatic Brain Injury and Posttraumatic Stress Disorder. *Psychiatr. Clin. N. Am.* 2014; 37: 55-75.

[39] Jorge, R. E., Robinson, R. G., Starkstein, S. E., et al.: Depression and anxiety following traumatic brain injury. *J. Neuropsychiatry Clin. Neurosci.* 5:369-374, 1993.

[40] Management of Post-Traumatic Stress Working Group [Internet]. VA/DoD clinical practice guideline for management of post-traumatic stress. Washington (DC): Department of Veterans Affairs, Department of Defense; 2010. Available from: ⍰http://www.healthquality.va.gov/ptsd/ptsd_full.pdfT.

[41] Rao, V., Lyketsos, C. G. Psychiatric Aspects of Traumatic Brain Injury. *Psychiatry in the Medically Ill.* 2002; 25(1): 43-69.

[42] Tateno, A., Jorge, R. E., Robinson, R. G. Clinical correlates of aggressive behavior after traumatic brain injury. *J. Neuropsychiatry Clin. Neurosci.* 2003;15(2):155-160.

[43] Wood, R. L. Recognising and assessing neurodisability after traumatic brain injury. *NeuroRehabilitation* 2013; 32: 699-706.

[44] Wood, R., Liossi, C. Neuropsychological and neurobehavioral correlates of aggression following traumatic brain injury. *Journal of Neuropsychiatry and Clinical Neuroscience* 2006;18:3.

[45] Kim, E., Bijlani, M. A pilot study of quetiapine treatment of aggression due to traumatic brain injury. *Journal of Neuropsychiatry and Clinical Neurosciences* 2006; 18: 547-549.

[46] Huband, N., Ferriter, M., Nathan, R., Jones, H. Antiepileptics for aggression and associated impulsivity. *Cochrane Database of Systematic Reviews* 2010;17(2).

[47] Fleminger, S., Greenwood, R. J., Oliver, D. L.: Pharmacological management for agitation and aggression in people with acquired brain injury. *Cochrane Database of Systematic Reviews* 2006; 4.

[48] Mooney, G. F., Haas, L. J. Effect of methylphenidate on brain injury-related anger. *Archives of Physical Medicine and Rehabilitation* 1993; 74:153-160.

[49] Gualtieri, C. T., Chandler, M., Brown, L., et al.: Amantadine: A new clinical profile for traumatic brain injury. *Clin. Neuropharmacol.* 12: 258-270, 1989.

[50] West, S. L. Substance Use among persons with tramatic brain injury; A review. *NeuroRehabilation* 2011; 29:1-8.

In: Neurobiology of Men's Mental Health
Editors: Leo Sher and Timothy R. Rice

ISBN: 978-1-63483-191-8
© 2015 Nova Science Publishers, Inc.

Chapter 21

SUICIDAL BEHAVIOR IN MEN WITH COMORBID DEPRESSION AND SUBSTANCE USE DISORDERS: ROLES OF ACTIVE SUBSTANCE USE, EXPOSURE TO COMBAT IN THE MILITARY, AND ALTERATIONS IN TESTOSTERONE LEVELS

Drew D. Kiraly[1,2,*]***, MD, PhD, and Leo Sher***[1,2]***, MD***

[1]Department of Psychiatry, Icahn School of Medicine at Mount Sinai, New York, NY, US

[2]James J. Peters Veterans' Administration Medical Center, New York, NY, US

ABSTRACT

Suicide is a leading cause of death worldwide and, as such, remains an important public health issue. This is a particularly important topic of discussion in thinking of men's mental health as men are far more likely than women to complete suicide. Studies attempting to determine the factors that lead to suicide have been able to formulate numerous risk factors for suicide, but efficient tools for early targeting and intervention remain elusive. In this chapter we discuss two cases of men with serious suicidal behaviors that presented to our psychiatric units. Using these cases as a guide for discussion, we then evaluate the literature with respect to male-specific risk factors for suicide. Included in our discussion will be how military service and trauma may put men at higher risk for suicide, how alterations in alcohol and substance consumption may trend men more specifically towards suicidal behavior, and how alterations in serum testosterone may present a male specific risk factor for worsening depression and suicidal behavior. By considering these important factors and how they apply to specific populations, we hope to drive the process forward of developing more effective early interventions for at risk populations.

---

* Corresponding author: Drew D. Kiraly, M.D., Ph.D., Icahn School of Medicine at Mount Sinai Box 1230, One Gustave L. Levy Place, New York, NY 10029, USA, Email: drew. kiraly@mssm.edu.

Keywords: depression, suicide, mood disorder, substance abuse, alcohol, dual diagnosis, testosterone

INTRODUCTION

Suicide is one of the leading causes of death of young men worldwide and, as such, represents an important public health issue. Overall, major depression is the most significant risk factor for suicidal behavior – and a diagnosis is present in well over half of completed suicides [1, 2]. However, when looking specifically at the issue of how this pertains to men's health there is something of an epidemiological paradox. While men are only about half as likely as women to experience a major depressive episode, they are 3-10 times as likely to complete suicide [2-5]. Clearly clarifying the reason for this discrepancy and finding interventions to treat vulnerable men are both important tasks. In this chapter we will posit that differences in substance use and consumption, exposure to combat in the military, and alterations in testosterone levels may represent key factors in the gender differences seen with completed suicide. We will begin our examination of this issue with two case reports from our clinical practice. One report is of a young combat veteran with a highly lethal suicide attempt that occurred in the context of depression and alcohol use. The other of an older man and formal medical professional who had spent many years grappling with both depression and polysubstance use. These cases are both demonstrative of some of the struggles that are experienced by men with co-occurring mood and substance issues. It should be noted that full versions of each of these case reports are published elsewhere as well [6, 7]. We will then review the literature related to these issues and how specific risk factors present particular challenges in men's health.

Case #1 – A Young Military Veteran

This patient was a 29 year old man with a past psychiatric history of recurrent major depressive disorder (MDD), post-traumatic stress disorder (PTSD) brought on by events experienced in combat, and alcohol use disorder which was active with an intermittent binge pattern of use. This patient had one previous suicide attempt by overdose at age fifteen by overdose on unspecified pills. While he reported intermittently depressed mood since age fifteen, he did not seek psychiatric care until after his military service. During his military service his patrol was ambushed and an explosion caused the death of a close friend who was standing next to him, and left the patient with significant back, neck and shoulder pain.

When we met this patient it was in the context of a recent serious suicide attempt by hanging that had occurred while intoxicated. He reports that he had been experiencing an ongoing depression for approximately six months prior to admission. Along with depressed mood he reported difficulties with sleep initiation and maintenance, increased appetite, and difficulties concentrating. He denies having any thoughts of suicide prior to the attempt. In addition to his depressive symptoms he reported feeling increasingly frequent flashbacks to events of combat, and feelings of hypervigilance while out in public. He denied any symptoms consistent with mania or any of psychosis. He had continued to see his outpatient psychiatrist and was being treated with citalopram 60mg daily with minimal improvement in his symptoms. As his symptoms worsened he also began to escalate his alcohol intake to a level of 12 beers per night approximately four times per week, and had begun to smoke marijuana "at least every day."

On the night of admission the patient had drunk approximately twelve beers and became involved in an argument with his significant other. Shortly after the argument ended, he went into another room and attempted suicide by asphyxiation. He was found several minutes later unconscious and an ambulance was called. At the hospital the patient denied that this was a planned attempt, expressed regret at what he had done, and requested discharge as soon as possible. All admission bloodwork including infectious disease screening was normal, with the exception of his serum testosterone level which was found to be low – 194 ng/dl (ref. range: 270-1100 ng/dl).

Case #2 – A Former Nurse Practitioner

The second patient was a 57-year-old man who presented to the hospital requesting help for severely depressed mood, and detoxification from alcohol and cocaine use. The patient reported that he had been fully sober for more than six months, but had recently relapsed to substance use due to feelings of "overwhelming self-pity." The patient had previously been a practicing nurse practitioner, but had spent the past several years working as a dishwasher in a restaurant as a part of a government-sponsored program for recovering substance users. The patient reported a very long history of substance use and recurrent depression. He reported significant drinking of >10 drinks per day starting in his early twenties, and continuing to present day with only brief periods of sobriety. While working as a nurse practitioner he developed a dependence on prescription opiate medications which he obtained from his primary care doctor, emergency rooms, and dentists. His drinking and opiate use became so severe that he began to miss work frequently and made mistakes in caring for patients. He was eventually discovered and stripped of his license to practice after several failed rehabilitation attempts. After losing his license the patient continued drinking and developed a crack cocaine habit as well.

In addition to his serious substance abuse history the patient also reported repeated depressive episodes that usually remitted after starting antidepressant medications, but recurred whenever he came off of his medications. He had three previous suicide attempts – all by overdose. The most serious of which was approximately ten years prior to this admission which involved taking an entire bottle of nortriptyline pills and caused the patient to be in the intensive care unit for over one week. Prior to this admission the patient reports two weeks of worsening depressed mood, lack of interest in normal activities, decreased energy and poor sleep – which he attributes to the aforementioned self-pity. He also reports that during those weeks he relapsed to crack cocaine and alcohol use and had been drinking approximately ten beers daily and smoking cocaine at least daily. While he denied any suicidal thoughts, plans or intent on this admission he reported coming to the hospital for this help because "I have been down this road too many times before."

Discussion and Conclusions

These two cases are demonstrative of the struggles faced by men who have co-occurring depression and substance abuse. Depression and substance are significant causes of disability and mortality worldwide [8, 9]. Based on international epidemiological samples it is estimated that approximately 5% of the world's population will experience a major depressive episode at some point in their life, and difficulties with alcohol and substance use will affect as much as 20% [9]. While each of these problems is a significant public health issue when examined independently, the problem is confounded because the two conditions are frequently comorbid. Patients who are depressed are more likely to abuse substances and vice versa [8, 9]. These populations of "dual diagnosis" patients are very important as they are at particularly at high risk for suicide. Patients who have co-occurring substance and mood disorders are at higher risk for suicidal behavior, and have a higher lethality associated with this behavior than those that are not actively using [8, 10]. In addition, patients who are actively abusing substances are more reactive to acute psychosocial stressors, and are more likely to make an impulsive suicide attempt in response to an acute stressor [11], as was seen in our first case report above.

Importantly for men's health, there are multiple studies that show that increased alcohol consumption is an important risk factor for suicidal behavior. Epidemiological studies in the United States showed that men were more likely to have an elevated blood alcohol level at the time of completed suicide [12]. Other retrospective population level studies have demonstrated that population wide increase in alcohol consumption leads to an increase in the number of suicides in males much more so than in females [13, 14]. Another study that performed psychological autopsy analysis of completed suicide victims showed that 68% of male suicide victims met criteria for alcohol dependence while only 29% of female suicide completers did [15]. One possible explanation for the increased risk for suicide in males with high levels of alcohol consumption may be gender differences in the way alcohol is consumed. In many cultures excessive alcohol consumption is considered an acceptable male behavior, and a symbolic practice that is important for demonstration of masculinity or virility. In addition to these social factors, examination of other neurobiological or neuroendocrine factors remains an area open for further investigation.

In addition to the elevated risk for suicide seen in men with co-occurring substance and mood disorders it is important to consider psychosocial stressors that may drive men towards suicidal behavior. One extreme stressor that has traditionally been the bailiwick of young men is that of exposure to military combat. Soldiers who have seen active combat, or have underlying mood disorders show significantly increased risk for suicide when compared to the general civilian population [16, 17]. Veterans who have been in active combat are particularly at risk for post-traumatic stress disorder (PTSD), depression, and substance use - all of which are important risk factors for suicide [18, 19]. While in recent years female soldiers have played increasing roles in militaries and in combat situations, this is still a situation that applies primarily to men worldwide.

In considering the case report of the young veteran it is also important to consider the low serum testosterone level that was noted on his admission. Testosterone is the primary sex hormone produced in post-pubertal males and has been shown to have important effects on behavior and mood [20]. With respect to substance use heavy alcohol consumption has been shown to both suppress production of testosterone and lower overall serum levels [20-22]. In addition to the effect of alcohol on testosterone frequent cannabis use also decreases serum testosterone in some studies [23, 24].

At this time the role of testosterone in depression and suicide remains unclear. There have been multiple studies that have shown low testosterone leading to worse and more frequent depression in post-pubertal male patient populations [25, 26]. Studies looking at the role of testosterone and suicide have been somewhat mixed in their results. There are several studies that suggest that alterations in serum testosterone is an important risk factor and significantly correlated with suicidal behavior [27-30]. However, other studies have looked at this same question and found a lack of effect of serum testosterone level on risk for suicide [31, 32]. There is some data connecting alterations in serum testosterone, decreased cerebrospinal levels of serotonin and aggressive behaviors [33, 34]. While this is a possible mechanism it is important to remember that suicide is a complex behavior that is influenced by myriad factors. While low serum testosterone certainly cannot be said to be directly causal in this patient with multiple risk factors, it is important to consider that this hormone has very strong influence on male behaviors on the whole and might be an important consideration when studying suicide in men.

References

[1] Henriksson, MM, Aro HM, Marttunen MJ, Heikkinen ME, Isometsa ET, Kuoppasalmi KI, Lonnqvist JK, Mental disorders and comorbidity in suicide. *Am. J. Psychiatry* 1993, 150, (6), 935-40.

[2] Rutz W, von Knorring L, Pihlgren H, Rihmer Z, Walinder J, Prevention of male suicides: lessons from Gotland study. *Lancet* 1995, 345, (8948), 524.

[3] Murphy GE, Why women are less likely than men to commit suicide. *Compr. Psychiatry* 1998, 39, (4), 165-75.

[4] Piccinelli M, Wilkinson, G., Gender differences in depression. Critical review. *Br. J.Psychiatry* 2000, 177, 486-92.

[5] Mann JJ, Apter A, Bertolote J, Beautrais A, Currier D, Haas A, Hegerl U, Lonnqvist J, Malone K, Marusic A, Mehlum L, Patton G, Phillips M, Rutz W, Rihmer Z, Schmidtke A, Shaffer D, Silverman M, Takahashi Y, Varnik A, Wasserman D, Yip P, Hendin H, Suicide prevention strategies: a systematic review. *JAMA* 2005, 294, (16), 2064-74.

[6] Kiraly DD, Sher L, Low Testosterone in a Young Combat Veteran with Dual Diagnosis and Suicidal Behavior: A case study. *Int. J. Adolesc. Med. Health* 2015, 27, (2), In Press.

[7] Kiraly DD, Sher L, Suicidal Behavior in a Medical Professional with Comorbid Depression and Substance Use Disorder: An Educational Case Report. *Int. J. Adolesc. Med. Health* 2015, 27, (2), In Press.

[8] Sher L, Stanley BH, Harkavy-Friedman JM, Carballo JJ, Arendt M, Brent DA, Sperling D, Lizardi D, Mann JJ, Oquendo MA, Depressed patients with co-occurring alcohol use disorders: a unique patient population. *J. Clin. Psychiatry* 2008, 69, (6), 907-15.

[9] Sullivan LE, Fiellin DA, O'Connor PG, The prevalence and impact of alcohol problems in major depression: a systematic review. *Am. J. Med.* 2005, 118, (4), 330-41.

[10] Sher L, Oquendo MA, Richardson-Vejlgaard R, Makhija NM, Posner K, Mann JJ, Stanley BH, Effect of acute alcohol use on the lethality of suicide attempts in patients with mood disorders. *J. Psychiatr. Res.* 2009, 43, (10), 901-5.

[11] Conner KR, Houston RJ, Swogger MT, Conwell Y, You S, He H, Gamble SA, Watts A, Duberstein PR, Stressful life events and suicidal behavior in adults with alcohol use disorders: role of event severity, timing, and type. *Drug Alcohol Depend.* 2012, 120, (1-3), 155-61.

[12] Kaplan MS, McFarland BH, Huguet N, Conner K, Caetano R, Giesbrecht N, Nolte KB, Acute alcohol intoxication and suicide: a gender-stratified analysis of the National Violent Death Reporting System. *Inj. Prev.* 2013, 19, (1), 38-43.

[13] Nemtsov A, Suicides and alcohol consumption in Russia, 1965-1999. Drug Alcohol Depend 2003, 71, (2), 161-8.

[14] Razvodovsky YE, Alcohol and suicide in Belarus. *Psychiatr. Danub.* 2009, 21, (3), 290-6.

[15] Kolves K, Varnik A, Tooding LM, Wasserman D, The role of alcohol in suicide: a case-control psychological autopsy study. *Psychol. Med.* 2006, 36, (7), 923-30.

[16] Kang HK, Bullman TA, Risk of suicide among US veterans after returning from the Iraq or Afghanistan war zones. *JAMA* 2008, 300, (6), 652-3.

[17] Lineberry TW, O'Connor SS, Suicide in the US Army. *Mayo Clin. Proc.* 2012, 87, (9), 871-8.

[18] Rice TR, Sher L, Suicidal behavior in war veterans. *Expert. Rev. Neurother.* 2012, 12, (5), 611-24.

[19] Sher L, Braquehais MD, Casas M, Posttraumatic stress disorder, depression, and suicide in veterans. *Cleve Clin. J. Med.* 2012, 79, (2), 92-7.

[20] Sher L, Alcohol, testosterone and suicide. *Aust. N. Z. J. Psychiatry* 2014, 48, (7), 688-689.

[21] Badr FM, Bartke A, Dalterio S, Bulger W, Suppression of testosterone production by ethyl alcohol. Possible mode of action. *Steroids* 1977, 30, (5), 647-55.

[22] Rivier C, Alcohol rapidly lowers plasma testosterone levels in the rat: evidence that a neural brain-gonadal pathway may be important for decreased testicular responsiveness to gonadotropin. *Alcohol. Clin. Exp. Res.* 1999, 23, (1), 38-45.

[23] Barnett G, Chiang CW, Licko V, Effects of marijuana on testosterone in male subjects. *J. Theor. Biol.* 1983, 104, (4), 685-92.

[24] Dalterio S, Bartke A, Burstein S, Cannabinoids inhibit testosterone secretion by mouse testes in vitro. *Science* 1977, 196, (4297), 1472-3.

[25] Pope HG, Cohane GH, Kanayama G, Siegel AJ, Hudson JI, Testosterone gel supplementation for men with refractory depression: a randomized, placebo-controlled trial. *Am. J. Psychiatry* 2003, 160, (1), 105-11.

[26] Zarrouf FA, Artz S, Griffith J, Sirbu C, Kommor M, Testosterone and depression: systematic review and meta-analysis. *J. Psychiatr. Pract.* 2009, 15, (4), 289-305.

[27] Markianos M, Tripodianakis J, Istikoglou C, Rouvali O, Christopoulos M, Papageorgopoulos P, Seretis A, Suicide attempt by jumping: a study of gonadal axis hormones in male suicide attempters versus men who fell by accident. *Psychiatry Res.* 2009, 170, (1), 82-5.

[28] Tripodianakis J, Markianos M, Rouvali O, Istikoglou C, Gonadal axis hormones in psychiatric male patients after a suicide attempt. *Eur. Arch. Psychiatry Clin. Neurosci.* 2007, 257, (3), 135-9.

[29] Sher L, Grunebaum MF, Sullivan GM, Burke AK, Cooper TB, Mann JJ, Oquendo MA, Association of testosterone levels and future suicide attempts in females with bipolar disorder. *J. Affect. Disord.* 2014, 166, 98-102.

[30] Sher L, Grunebaum MF, Sullivan GM, Burke AK, Cooper TB, Mann JJ, Oquendo MA, Testosterone levels in suicide attempters with bipolar disorder. *J. Psychiatr. Res.* 2012, 46, (10), 1267-71.

[31] Butterfield MI, Stechuchak KM, Connor KM, Davidson JR, Wang C, MacKuen CL, Pearlstein AM, Marx CE, Neuroactive steroids and suicidality in posttraumatic stress disorder. *Am. J. Psychiatry* 2005, 162, (2), 380-2.

[32] Perez-Rodriguez MM, Lopez-Castroman J, Martinez-Vigo M, Diaz-Sastre C, Ceverino A, Nunez-Beltran A, Saiz-Ruiz J, de Leon J, Baca-Garcia E, Lack of association between testosterone and suicide attempts. *Neuropsychobiology* 2011, 63, (2), 125-30.

[33] Birger M, Swartz M, Cohen D, Alesh Y, Grishpan C, Kotelr M, Aggression: the testosterone-serotonin link. *Isr. Med. Assoc J.* 2003, 5, (9), 653-8.

[34] Rupprecht R, Neuroactive steroids: mechanisms of action and neuro psychopharmacological properties. *Psychoneuroendocrinology* 2003, 28, (2), 139-68.

Section Four: Men's Health: Medical Issues and Future Directions

In: Neurobiology of Men's Mental Health
Editors: Leo Sher and Timothy R. Rice

ISBN: 978-1-63483-191-8
© 2015 Nova Science Publishers, Inc.

Chapter 22

Testosterone and Male Reproductive Health

Omer A. Raheem, MD, MSc, MCh Urol, MRCS, and Tung-Chin Hsieh*, MD

Department of Urology, University of California San Diego School of Medicine, La Jolla, CA, US

Abstract

Testosterone is one of the most important hormones that plays an essential role in male fertility and the development of secondary sexual characteristics in males. Its production is a hierarchy-regulated process involving the hypothalamic-pituitary-gonadal axis via the secretion of the gonadotropin releasing hormone (GnRH), which ultimately controls the excretion of Testosterone from the Leydig cells of the testicular interstitium in a feedback mechanism. At puberty, testosterone takes a central role in initiation of spermatogenesis and development of male gametes which is paramount for male fertility. Hypogonadism can lead to impaired spermatogensis and is found in up to 30% of male infertility patients. It can be due to deficiencies of gonadotropins and steroidogenic enzymes, congenital abnormalities, chronic illness, history of exogenous testosterone administration, or insult or neoplasm of the brain. Effective treatment of hypogonadism in male infertility involves accurate diagnosis and careful hormone restoration based on our understanding of the hypothalamc-pituitary-gonadal physiology. Advancement in genetic research has identified multiple genetic defects resulting in hypogonadism. Greater understanding of the endocrinologic mechanism of spermatogenesis will further help infertile patients. This chapter aims to highlight and streamline some of the complex conditions that result in inadequate androgen activity as well as end organ effects that result in subsequent male infertility.

Keywords: testosterone, male, hypogonadism, infertility

---

* Corresponding author address: T. Mike Hsieh, MD, UC San Diego Health System, 200, West Arbor Drive #8897, San Diego, CA 92103-8897, USA. Email: t7hsieh@ucsd.edu.

List of Abbreviations

AAS	anabolic-androgen steroid
ACTH	adenocorticotropic hormone
AIDS	acquired immune deficiency syndrome
AMH	anti-mullerian hormone
AR	androgen receptor
CAG	cytosine adenine guanine
CAH	congenital adrenal hyperplasia
c-AMP	cyclic adenosine monophosphate
CNS	central nervous system
DHT	dihydrotestosterone
DNA	deoxyribonucleic acid
E2	Estradiol
FSH	follicle stimulating hormone
GnRH	gonadotropin releasing hormone
hCG	human chorionic gonadotropin
HH	hypogonadotropic hypogonadism
hMG	human menopausal gonadotropin
HPG	hypothalamic-pituitary-gonadal
ICSI	intracytoplasmic sperm injection
KS	Kleinfelter's syndrome
LDL	low density lipoprotein
LH	leuteinizing hormone
MRI	magnetic resonance imaging
PCT	post cycle therapy
SHBG	sex hormone binding globulin
TRT	testosterone replacement therapy

1. Introduction

Testosterone is an essential male hormone. In utero, its role in genital tract development is paramount for proper male genital formation. When puberty begins, testosterone takes a central role in initiation of spermatogenesis and development of secondary sexual characteristics. Testosterone is required to maintain maleness in the adult. The male hormonal reproductive axis is a finely tuned apparatus with exquisite control over its biological effects. This chapter explores testosterone in the natural physiologic state and the pathology of inadequate testosterone or hypogonadism, with particular emphasis on its effects related to spermatogenesis and fertility.

2. Synthesis and Kinetics of Testosterone

Testosterone is a steroid hormone. In men, it is almost exclusively produced by the Leydig cells of the testis. In adult men, the average testosterone production is 7000 μg/day [1]. As steroid hormones, testosterone is lipid soluble and able to traverse the cell membrane where it then binds to its androgen receptor (AR). Once bound, this AR-testosterone complex is then able to enter the nucleus of the cell where it interacts with deoxyribonucleic acid (DNA) at androgen response elements. As the bound AR associates with DNA, the recruitment of co-activators helps to form the transcriptional apparatus. This completed apparatus performs androgen-regulated gene transcription which leads to downstream protein synthesis [2].

Once testosterone is excreted into the circulation, the majority of testosterone is bound to plasma proteins. The primary proteins that bind testosterone are sex hormone binding globulin (SHBG) and albumin. The majority of testosterone is bound to albumin (54-68%); slightly less is bound to SHBG (30-44%), and only 0.5-3% remains unbound or as free testosterone [3, 4]. SHBG is produced by the liver and avidly binds to testosterone, rendering it biologically unavailable [5]. Albumin's association with testosterone is much weaker. Albumin-bound testosterone and the small percentages of unbound testosterone compose what is termed "bioavailable testosterone." These bioavailable testosterone molecules have the ability to enter target tissues and end-organs, bind to the AR, and initiate protein synthesis.

3. Hypothalamic-Pituatary-Gonadal Axis

Testosterone's synthesis is hierarchy-regulated process involving the hypothalamic-pituitary-gonadal axis (HPG) via the secretion of the gonadotropin releasing hormone (GnRH), which ultimately controls the excretion of Testosterone from the Leydig cells of the testicular interstitium in a feedback mechanism. Testosterone production begins with the release of the GnRH from neurons within the medial basal hypothalamus. The axon terminals release this hormone into the hypothalamic pituitary portal system. This network of capillaries delivers GnRH to the anterior pituitary and stimulates the production of the gonadotropins, luteinizing hormone (LH) and follicle stimulating hormone (FSH). The levels of GnRH released are in such low concentrations that it cannot be measured directly from peripheral blood [6]. GnRH is released in a pulsatile manner, though the mechanisms controlling this release are not well understood. The pulsatile release of GnRH causes a pulsatile release of the gonadotropins. Due to the half-life of FSH (2-3 hours), only the LH pulses can be measured in serial peripheral blood samples. Normal men will demonstrate 12-16 LH pulses over a twenty four hour period i.e., roughly every 20 minutes [7]. The pulsatile nature of GnRH release is significant for its effect on the anterior pituitary. The effect of exogenous, continuous administration of low dose GnRH results in suppression of gonadotropin release [8]. The administration of a high dose GnRH agonist also inhibits gonadotropin release, and subsequent testosterone synthesis. This understanding of Testosterone's kinetics has been used for decades in the hormonal manipulation of prostate cancer pioneered by Huggins and Hodges in 1941.

The gonadotropins LH and FSH belong to the glycoprotein hormone family along with thyroid stimulating hormone (TSH) and human chorionic gonadotropin (hCG). This family shares a common alpha subunit, while each individual peptide in the hormone family has a unique beta subunit that confers its specificity for a particular receptor in the target end-organ [6].

LH binds to a G coupled protein receptor on the cell surface of the Leydig cell. This causes an increase in cyclic adenosine monophosphate (c-AMP). The increase of c-AMP has several effects with most directed at the rate limiting steps of steroidogenesis. In humans, cholesterol is synthesized within Leydig cells and circulating LDL cholesterol is used to generate testosterone through an enzymatic process. The pulsatile release of LH from the anterior pituitary results in a pulsatile response of testosterone production by the testis [9]. Testosterone production also exhibits a circadian rhythm, particularly in younger men. This rhythm can result in a difference from peak to nadir of 140ng/dl. Peak testosterone levels occur at approximately eight am and nadir levels at approximately eight pm [10]. The circadian changes in testosterone levels are present in older men as well, but with significantly smaller variation from peak to nadir (i.e., approximately 60ng/dl).

Indifferent to other naturally secreted hormones, Testosterone production is regulated by a negative feedback loop. Testosterone and its aromatized metabolite Estradiol (E_2) act at both the level of the hypothalamus and the anterior pituitary to decrease secretion of GnRH and the gonadotropins. GnRH producing neurons have been shown to lack the androgen or estrogen receptors. The exact mechanism of this negative feedback is not well established but animal model studies suggest an indirect feedback from surrounding neurons [11]. Although, the anterior pituitary is not directly responsive to testosterone, Testosterone must be converted to E_2, which in turn provides its negative feedback signal [12]. Figure 1 illustrates the complete pathway for the control of LH-dependent testosterone production.

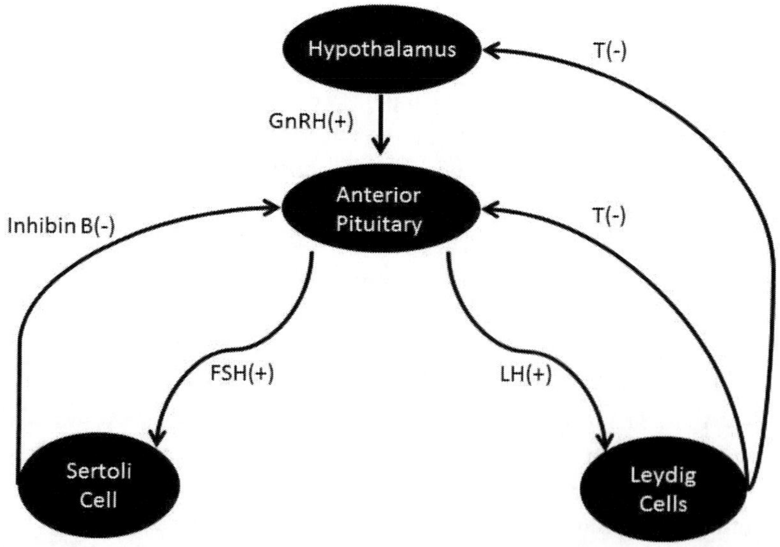

Keys: T=testosterone, LH=Luteinizing hormone, FSH=Follicle Stimulating hormone, GnRH=gonadotropin releasing hormone.

Figure 1. Diagram of positive and negative feedback loop of Hypothalamus-Pituitary-Gonadal axis.

The control of FSH and spermatogenesis varies slightly. Testosterone and FSH exert their effects on Sertoli cells of the seminiferous tubules. The Sertoli cells in turn act as supportive cells to spermatogonia and aid in spermatogenesis. In response to FSH stimulation and spermatogenesis, Sertoli cells produce the glycoprotein Inhibin B. Inihibin B has a negative feedback effect on the anterior pituitary decreasing FSH production [13].

4. Testosterone and Spermatogenesis

Male sexual differentiation begins between the seventh and twelfth week of gestation. The undifferentiated gonads develop into fetal testes under the influence of a gene complex located on the short arm of the Y chromosome (SRY). The fetal testes produce two hormones: testosterone and anti-mullerian hormone (AMH). Testosterone is important for the development of the Wolffian duct structures, which ultimately form the epididymis, vas deferens, and seminal vesicles. Under the influence of intratesticular testosterone, the number of gonadocytes per seminiferous tubule will increase three-fold during the first month of pregnancy. Testosterone is further converted into a potent form, dihydrotestosterone (DHT), which is important for the formation of the external genitalia. AMH is responsible for the regression of the Mullerian duct system.

5. Hypogonadism

Hypogonadism is most commonly defined biochemically as a serum testosterone less than 300ng/dl. This is not a universal diagnosis. However, the current American Endocrine Society guidelines state that "clinicians should use the lower limit of a normal range for healthy young men established in their own laboratory" [14]. Physical signs of postpubertal hypogonadism include increased body fat, reduced muscle bulk and strength, low bone mineral density, loss of body hair and a decreased need for shaving. Patients may present with symptoms of decreased energy or motivation, depressed mood, decreased libido, diminished work performance, poor concentration, and sleep disturbances [14]. The prevalence of low testosterone, furthermore, increases with age: one population based study found that 39% of men over the age of 45 had a testosterone of 300ng/dl or less [15]. The amount of free testosterone decreases with aging to a greater extent than does total testosterone as a consequence of increased SHBG [16].

Hypogonadism has also been associated with other disease states including chronic opioid use (72%), obesity (52%), diabetes (50%), AIDS (50%), hypertension (42%), hyperlipidemia (40%), erectile dysfunction (19%). Having low serum testosterone has been shown to be a risk factor for metabolic syndrome and type 2 diabetes [21]. A recent study was designed to determine the impact of testosterone levels on survival rates of males over the age 40. After adjusting for age, medical comorbidity, and other clinical covariates, low testosterone levels continued to be associated with increased mortality. The hazard ratio was 1.88 (95% CI 1.34-2.63, P <.001) [22].

5.1. Hypogonadotropic Hypogonadism

Hypogonadotropic hypogonadism (HH) is characterized by failure of gonadal function secondary to deficient gonadotropin secretion. It is often the result of either a pituitary or a hypothalamic defect and is commonly associated with a structural lesion or functional defect affecting this region. Biochemically, low serum levels of sex steroids occur in the presence of inappropriately low or normal levels of LH and FSH. Gonadotropin deficiency commonly is due to a deficiency of hypothalamic GnRH.

During embryonic development, GnRH neurons originate outside the central nervous system (CNS), and they migrate along the olfactory nerve complex before reaching the arcuate nucleus in the hypothalamus [23]. Disruption of this pathway can lead to Kallmann's syndrome in which HH is associated with anosmia secondary to agenesis of the olfactory bulbs. Under normal physiology, GnRH is synthesized by hypothalamic neurons and secreted in a pulsatile fashion into the hypophyseal portal circulation. In the anterior pituitary, the binding of GnRH to its receptor stimulates the synthesis and pulsatile secretion of FSH and LH. In the male, pulsatile secretion of gonadotropins increases a few days after birth, generating testosterone levels near the adult range for the first few months of life [24]. During childhood, the hypothalamic-pituitary-gonadal system enters a quiescent period with FSH and LH pulses detected at low amplitude [25]. During puberty, the pulsatile secretion of gonadotropins resumes with a gradual increase in both amplitude and frequency. In the male, LH pulses can be detected approximately every 2 hours; whereas in the female the frequency varies between 60 minutes and 6 hours, depending on the ovulation cycle [26]. The change in LH release pattern is due to the increased pulsatile release of GnRH at the onset of puberty, which ultimately leads to increasing steroidogenesis, gametogenesis, and development of secondary sexual characteristics. The precise mechanism controlling the onset of puberty is still not fully known.

The clinical presentation of HH depends on the time of onset, the severity of the defect, etiology (congenital vs. acquired), and the presence of associated conditions. Most cases of congenital HH are diagnosed during the second or third decade of life when patients present with pubertal failure, primary amenorrhea, or infertility. In some cases, the diagnosis can be made before puberty on the basis of the presence of microphallus, cryptorchidism, and a positive family history [27]. Except in the neonatal period, when gonadotropins and sex steroids levels are expected to be elevated, HH generally cannot be confirmed until puberty.

HH with associated anosmia is suggestive of Kallman's syndrome. Anomalies frequently associated with X-linked Kallman's syndrome include cryptorchidism, bimanual synkinesis, and unilateral renal agenesis. When a child is too young to undergo olfactory testing, the absence of olfactory bulbs on magnetic resonance imaging (MRI) can lead to early diagnosis. However, a normal MRI does not rule out the disease, since normal olfactory bulbs can be seen in up to 25% of Kallman's syndrome patients [28]. Both autosomal and X-linked patterns of inheritance have been recorded, but only the X-linked form has been characterized at the genetic level. The deficit observed in Kallman's syndrome appears to be caused by a functional deficit in GnRH secretion. Studies have shown a spectrum of gonadotropin release patterns among these patients ranging from apulsatile LH to a low-amplitude LH pulse [29].

Low levels of sex steroids in the presence of low or normal LH and FSH levels confirm the diagnosis of HH. Acute stimulation with GnRH (GnRH test) is of little value in differentiating between a hypothalamic and a pituitary defect. Most patients with HH show little or no response to an initial dose of GnRH, but may respond to repeated injections in the presence of a hypothalamic defect. The poor response to the initial stimulation is likely due to down regulation of GnRH receptors after prolonged GnRH deficiency [30]. The complete failure of a response is suggestive of the absence of LH/FSH secretory neurons or severe GnRH receptor defects.

Acquired forms of HH can be caused by structural or anatomical lesions involving the hypothalamic pituitary region; i.e., CNS tumors, infiltrative processes, pituitary microadenomas, pituitary infarction, head trauma or radiation therapy of the head and neck. Even though gonadotropin failure commonly manifests itself as secondary hypogonadism, multiple pituitary hormone deficiencies are often present.

Functional hypogonadism is characterized by GnRH deficiency in the absence of an anatomical cause and is typically reversible. This hormone deficiency is generally manifested as hypothalamic amenorrhea with associated weight loss secondary to anorexia nervosa, excessive exercise programs or psychological stress. Variable patterns of abnormal LH pulsatility and low leptin levels have been reported [31]. The role of leptin in the pathophysiology of the disease remains unclear [32]. HH is less commonly diagnosed in adult men because the hypogonadal symptoms are less evident than in women.

The first insight into the genetics of HH was the identification of the KAL1 gene in patients with X-linked Kallman's syndrome [33]. Recent advances in genetic research have identified multiple mutations involved in the development and function of the hypothalamic pituitary axis. Genetic defects have been shown to contribute to abnormalities in GnRH receptors, gonadotropin subunits, and steroidogenic factors [34]. Table 1 lists the identified genetic causes of HH. Despite improved understanding, the pathogenic genetic defect can be found in only approximately 30% of current HH cases, suggesting that many mutations remain to be discovered.

Table 1. Identified genetic causes of Hypogonadotropic Hypogonadism [13]

Gene	Locus	Phenotype
KAL 1	Xp22.3	X-linked Kallmann's syndrome
GnRH	8p11	GnRH deficiency (mice)
GnRHR	4q21	GnRH resistance
LHβ	19q13	Biologically inactive LH
FSHβ	11p13	Low serum FSH
HESX-1	3p21	Septo-optic dysplasia
LHX3	9q34	Pituitary hypoplasia
PROP-1	5q35	Pituitary hypoplasia (ACTH spared)
SF-1	9p33	Primary adrenal failure and gonadotropin deficiency
DAX-1	7q31	Primary adrenal failure with hypothalamic/pituiary defect
Leptin	7q31	Obesity, IHH, hypoleptinemia and hyperinsulinemia
Leptin-R	1p31	Obesity, IHH, hyperleptinemia and hyperinsulinemia
PC-1	5q15	Obesity, IHH, hypocortisolemia, hypoinsulinemia and high proinsulin.

Treatment of HH involves sex steroid therapy to induce pubertal development before maintenance therapy is begun to prevent the sequelae of hypogonadism. Testosterone replacement usually starts with a low dose of long-acting intramuscularly injected testosterone esters (i.e., testosterone cypionate or enanthate). However, this approach is plagued with side effects associated with large fluctuations in testosterone levels and supra-physiologic levels in the first few days after injection. Topical testosterone preparations are often used for maintenance replacement treatment with the advantage of being a noninvasive option while also mimicking the normal circadian rhythm of testosterone secretion. Use of a long-acting subcutaneous testosterone implant has been shown to be an effective alternative treatment option for induction of puberty and maintenance therapy and has a high rate of patient satisfaction and compliance [35].

When fertility is desired, spermatogenesis can be induced using gonadotropins or GnRH therapy. Even though fertility can be achieved with hCG therapy alone in a patient with partial HH, in most instances the addition of FSH is required as well [36]. Successful restoration of fertility has been reported with human menopausal gonadotropin (hMG) and hCG therapy; as an alternative to hMG, recombinant FSH has been shown to be equally effective, although more expensive [37]. In patients with hypothalamic HH, pulsatile GnRH therapy can be used as an alternative to gonadotropin treatment with equal or superior efficacy. GnRH is administered subcutaneously or intravenously using a portable infusion pump at a dose frequency of 120 minutes, but is a relatively complex treatment option [38]. During all treatment regimens, testicular volume, testosterone, and gonadotropin levels are monitored, and the dosage is adjusted accordingly. Testicular volume increases faster in response to pulsatile GnRH therapy than to gonadotropins, while semen parameters are comparable after 12 to 18 months of treatment. Time to initiation of spermatogenesis is variable among all modalities of therapy; fertility is generally achieved after four to 20 months of therapy in the majority of men. A history of bilateral cryptorchidism is a poor prognostic indicator of fertility potential with medical therapy [39].

HH is a disease of secondary hypogonadism as a result of deficiencies in the hypothalamic pituitary system. The clinical presentation is variable with infertility occurring in both congenital and acquired forms. Treatment involves induction of puberty and restoration of fertility using gonadotropins and sex steroids. Continued genetic research has the potential to further advance our understanding of the disease and to help optimize treatment of these patients [40].

5.2. Congenital Adrenal Hyperplasia

The adrenal cortex is capable of producing androgens. In pre-menopausal women it is responsible for greater than 50% of circulating androgens [41]. In men its contribution is minimal compared to testicular testosterone production. Suppression of the HPG axis with adrenal androgens is possible in the pathologic state; however, it is most notable in congenital adrenal hyperplasia (CAH).

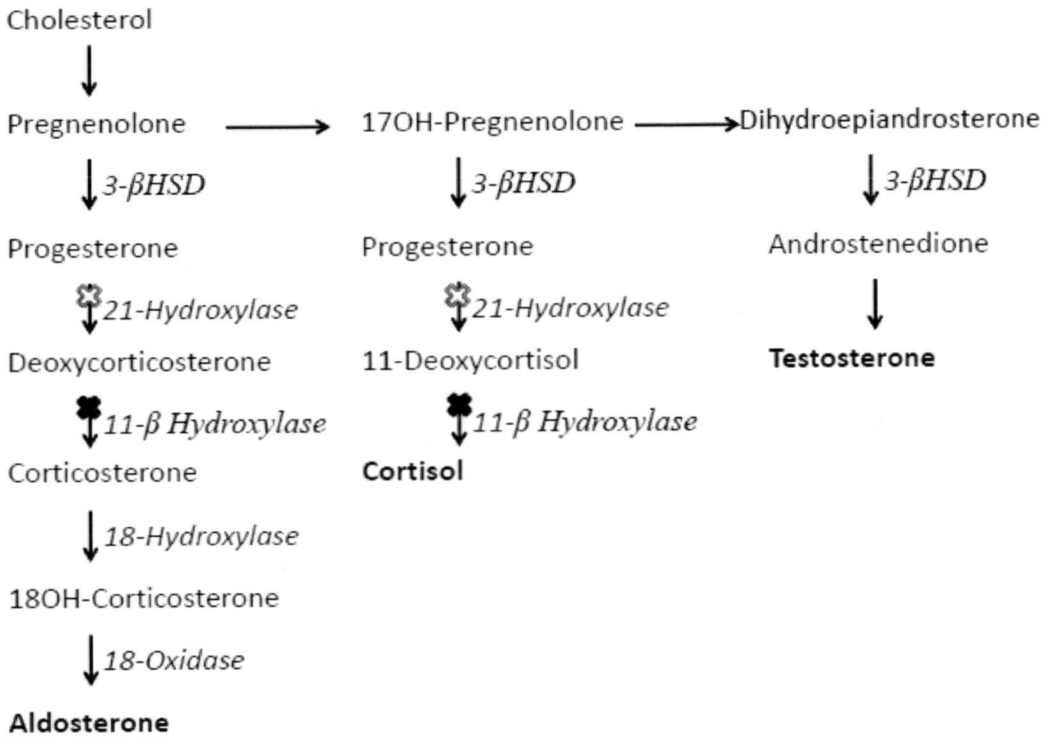

Figure 2. Enzymatic pathway for steroid hormone production. Open X is site of enzyme loss in 21-hydroxylase deficiency CAH. Solid X is site of enzyme loss with 11-β-Hydroxylase deficiency CAH. 3-βHSD= 3-βHydroxysteroid Dehydrogenase.

CAH is a group of autosomal recessive disorders of adrenal corticosteroid biosynthesis. CAH is the result of defects in any one of a number steroidogenic enzymes [42]. The two most common enzyme defects are 21-hydroxylase deficiency and 11-β-hydroxylase deficiency (Figure 2). Deficiency of 21-hydroxylase accounts for approximately 90-95% of cases of CAH, while 7% of cases are due to 11-β-hydroxylase deficiency [43, 44]. The steroid pathway is outlined in Figure 2 with the 21-hydroxylase and 11-β-hydroxylase enzyme defects highlighted. Other enzymatic defects have been described that result in CAH but are exceedingly rare. The diagnosis of CAH is usually made shortly after birth, but can also be made antenatally. In the neonate it is critical to diagnose these children as early as possible, as up to 75% of neonates with 21-hydroxylase deficiency may have the salt wasting form which can lead to electrolyte imbalance and death [45].

Without a full complement of the enzymes necessary for their synthesis, there is a glucocorticoid and mineralocorticoid deficit. The body tries to overcome this by increasing adrenal stimulation with adrenocorticotropic hormone (ACTH). The initial adrenal hormone products (please name them), unable to continue down either of these pathways for glucocorticoid or mineralocorticoid, are shunted toward the production of androgens.

Twenty one-hydroxylase deficiency presents with three different phenotypes: salt wasting, simple virilizing, and non-classic. The salt wasting and simple virilizers are diagnosed in childhood, and children with these defects are the more severely affected. The non-classical patients may make it into adulthood prior to diagnosis. In affected females the affect can range from mild virilization to more severe ambiguous genitalia. Affected males will have phenotypically normal genitalia. Short stature is characteristic due to the early closure of the epiphyseal plates. Another manifestation of adrenal deficiency is the presence of testicular adrenal rest tumors that may resemble Leydig cell tumors [46]. These tumors may be large and easily palpable. The origin of these tumors is thought to be from aberrant adrenal tissue. This tissue is sensitive to ACTH stimulation. Proper treatment with glucocorticoids causes these lesions to regress. These tumors can become very large and be locally destructive to normal testicular tissue, resulting in infertility and hypogonadism [47].

Eleven β-hydroxylase deficiency has a similar phenotype to the simple virilizers of 21-hydroxylase deficiency. Salt wasting has rarely been reported in those with 11-β-hydroxylase deficiency. Patients with 11-β-hydroxylase deficiency may be hypertensive, which is not seen in those with 21-hydroxylase deficiency. Treatment for these patients with 21-hydroxylase deficiency is with glucocorticoid replacement and, when necessary, mineralocorticoids as well [44]. With treatment, the over production of ACTH is stopped. Without this overstimulation, the production of adrenal androgens returns to normal levels as the byproducts of incomplete glucocorticoid synthesis are no longer available for the androgen pathway. The goal of treatment is to find the right balance to suppress the adrenal axis but not over-treat and cause the patient to suffer the effects of glucocorticoid excess. Patients should be counseled that they will require life-long treatment.

5.3. Hypogonadism Secondary to Congenital Abnormalities

5.3.1. Prader-Willi Syndrome

Prader-Willi syndrome is a result of maternal uni-parental disomy of chromosome 15. Uni-parental disomy results when both chromosomes of a pair or areas of a chromosome have been inherited from a single parent. This is the result of a trisomy that is reduced to disomy leaving both copies from the same parent. This has been termed trisomy rescue. This disease is the result of phenotypic differences based on the uni-parental origin of the gene expressed, a concept known as imprinting [48].

The clinical features of Prader-Willi syndrome are low birth weight and severe hypotonia and difficulty with feeding in early infancy. At an early age these patients develop hyperphagia and obesity. Other characteristic features include short stature, small hands, almond shaped eyes, and a triangular mouth. Most of these patients have mental retardation. Both sexes demonstrate hypogonadism with incomplete pubertal development and genital hypoplasia [49].

5.3.2. Klinefelter's Syndrome

The endocrine disorder that is manifested by tall stature and eunuchoid proportions, gynecomastia, azoospermia, and increased urinary secretion of FSH was first described by Harry Klinefelter and colleagues in 1942 [50]. In 1956, Jacob and Strong found a chromatin-positive patient with Klinefelter's syndrome (KS) with a 47,XXY karyotype. With a reported prevalence of 0.1 to 0.2% in the general population and 10% of infertile men with non-obstructive azoospermia, KS is the most common type of human chromosomal aneuploidy [51]. Large population-based studies indicate that only 25% of KS men are ever diagnosed during their lifetimes and fewer than 10% are diagnosed before puberty [52, 53].

The clinical presentation of KS patients varies according to age. Before puberty, only discrete physical changes might be appreciated: e.g., slightly smaller than normal testicular volume or long-leggedness. Sexual development of a hypogonadal child may not differ from that of a normal prepubertal boy. In adolescence and after puberty, the characteristics of small firm testes and skeletal disproportion begin to develop. Patients with KS are generally of average height or taller because of increased leg length. This stature before puberty suggests that it is not due to androgen deficiency, but is related to the underlying chromosomal abnormality. In contrast to the typical tall stature, the arm span rarely exceeds the patient's height. About half of the patients develop painless gynecomastia of various degrees. However, the risk of breast cancer is not higher than in the general population. After reaching the age of 25, 70% of the patients develop symptoms related to hypogonadism, (e.g., low libido, decreased potency, osteoporosis, and decreased muscle mass) [54]. The cognitive phenotype reflects not a general reduction of intelligence but deficits in very specific cognitive domains, mainly language and executive function. Increased psychiatric disturbances, criminal behavior, and mental retardation described in early reports have not been observed in prospective studies based on chromosomal analysis [55].

Serum testosterone concentrations are below normal in 80% of adult patients with a 47,XXY karyotype [56]. The exact mechanism of hypogonadism is unknown and demonstrates a variable degree of Leydig-cell dysfunction. On average, the levels of E_2 and SHBG are elevated, further decreasing the bioavailable free testosterone. Concentrations of LH and FSH are high in most cases. FSH shows the most discrimination, with little overlap when compared to levels in normal men, a consequence of damaged seminiferous tubules. Inhibin-B levels are generally within normal range in prepubertal boys with KS but decrease significantly after puberty due to the disappearance of germ cells and Sertoli cells [57].

When testosterone levels are low in patients with KS, life-long testosterone replacement therapy (TRT) is indicated and should be initiated as early as possible to avoid symptoms and sequelae of hypogonadism. Early TRT results in increased masculinity, strength, libido, and bone mineral density. It also has a positive effect on mood and cognitive function while reducing fatigue and irritability. Long-acting preparations of TRT, such as subcutaneous implantable testosterone pellets, have been shown to be an effective treatment option for non-compliant adolescent patients with KS [58].

Despite the emphasis on TRT to correct KS-associated hypogonadism, exogenous T also has a negative effect on fertility. TRT inhibits spermatogenesis at the level of spermatogonial differentiation. In KS patients who have undergone a successful sperm retrieval procedure, administration of hCG can increase testosterone levels, indicating the presence of functional testicular tissues. Various medications, such as aromatase inhibitors and gonadotropins, have been shown to raise testosterone levels in KS patients but they are not associated with improved sperm retrieval outcomes [59]. With the advent of intracytoplasmic sperm injection (ICSI), fertility is now possible for azoospermic men with successful testicular sperm retrieval. Pregnancy from ICSI can be successful in patients with KS, and a live birth rate of 20% has been reported [60].

There still is no consensus regarding preservation of fertility in patients with KS. Germ cell deterioration is a progressive process that starts in infancy. Cryopreservation of semen samples or testicular tissue from early pubertal boys who have low numbers of spermatozoa is possible and should be offered to appropriate KS patients. Cryopreservation of sperm or testicular tissue should also be offered before initiation of TRT.

5.4. Exogenous Testosterone-associated Infertility

Testosterone is the most important androgen in the male body. It is an anabolic-androgen steroid (AAS) with a one-to-one ratio of androgenic and anabolic activity. Anabolism is defined as any state in which nitrogen is retained in the lean body mass through stimulation of protein synthesis and/or reduction of protein breakdown. Testosterone promotes growth, increased muscle size, and bone metabolism. Characteristically, chemically engineered AAS have weaker AR receptor binding than do pure androgenic steroids. Structural modifications in synthetic AAS have been made to maximize the anabolic effects while minimizing the androgenic activity. However, all AAS' are virilizing when administered at high doses for long periods of time. AAS' are available in both oral and injectable formulations. Medical indications for AAS therapy include AIDS-associated cachexia, catabolic state of severe burn, androgen deficiency, and chronic obstructive pulmonary disease [61].

The true prevalence of non-medical use of AAS is unknown. It is estimated that 3-12% of male athletes in the U.S. have used steroids, while the incidence is 14% among college athletes and 30-75% among males involved in professional sports and recreational bodybuilding [62]. For the purpose of optimizing enhancement, various regimens combining oral and injectable formulations are used during cycles of 4-12 weeks. The goal is to maintain supraphysiologic levels for extended periods while minimizing the side effects of individual agents. High testosterone levels can also occupy cortisol receptors, creating an anti-catabolic effect. Anti-estrogens and hCG are sometimes used in an attempt to restart endogenous testosterone production (post cycle therapy, PCT) [63].

Side effects of AAS use are well documented. Acne, alopecia, and lower urinary tract symptoms are related to the strong androgenic effects of DHT. Gynecomastia is a result of aromatization of high levels of testosterone. Hepatic dysfunction is often observed in users of oral alkylating agents. Cardiovascular side effects of steroid abuse include hypertension, polycythemia, and ventricular dysfunction. Renal failure from rhabdomyolysis and membrano-progressive glomerulonephritis has been reported in heavy users. Aggressive behavior, depression, mood swings, and even psychosis are some of the psychiatric patterns reported to be related to AAS. Mortality related to use of exogenous AAS is estimated to be 4.6 times higher in users than non-users [64].

Exogenous administration of testosterone results in a negative feedback on the HPG axis inducing a hypogonadotropic state. Despite the high serum levels of androgens associated with AAS use, there is insufficient intra-testicular testosterone concentration to maintain spermatogenesis. Patients with a history of AAS abuse often develop hypogonadal symptoms and small testes with oligospermia or azoospermia and abnormalities in sperm motility and morphology.

Histologically, administration of exogenous testosterone is associated with alterations of Leydig cells, along with abnormalities of cellular morphology. Decreased steroidogenesis of the testicular tissue was found in immunohistochemical studies. Specific maturation impairment has been described (i.e., lack of advancement beyond spermatids) [65]. The proliferation of Leydig cells never recovers to baseline level, even long after discontinuation of AAS use. Even with hCG post cycle therapy, qualitative impairment of the sperm may persist, suggesting an irreversible insult to the testicular tissues.

The management of male infertility after administration of exogenous testosterone has been extensively reported [66]. Spontaneous recovery of spermatogensis has been reported in certain populations 4-18 months after discontinuation. However, 4-10% remained azoospermic, and only 46% returned to baseline sperm concentration [67]. There is no consensus regarding the optimal management of persistent infertility after cessation of exogenous testosterone use. Most regimens parallel the treatment of HH by induction of spermatogenesis with gonadotropins or gonadotropin analogs, e.g., hCG, hMG, or recombinant FSH. Additionally, clomiphene citrate and anti-estrogens have been utilized to stimulate hypothalamic-pituitary function. Successful fertility restoration has been reported in azoospermic patients as long as five years after discontinuation of testosterone therapy. The favorable outcome in reversing exogenous testosterone-induced infertility is likely related to the endocrinological nature of the process.

For patients with an indication for TRT, fertility preservation with the goal of preserving intra-testicular testosterone and spermatogenesis has been reported. Concomitant weekly administration of intramuscular testosterone with hCG can maintain intra-testicular testosterone levels similar to pretreatment concentrations. Additionally, concomitant hCG can also preserve spermatogenesis in patients undergoing treatment with transdermal or intramuscular TRT [68].

The varied and widespread abuse of AAS has made the study of optimal management of exogenous testosterone-induced male infertility difficult. In the absence of accurate identification of the affected patients, proper comparison of treatment regimens is not possible. However, our understanding of the treatment for HH has led to success in reversing the underlying endocrine dysfunction. In patients undergoing TRT, preservation of spermatogenesis is now possible with concomitant administration of hCG. Despite success in reversing exogenous testosterone-associated infertility, persistent gonadal dysfunction and increased morbidity and mortality have been identified. Emphasis on preventing AAS abuse is important, and close follow-up of affected individuals is crucial to their health.

Conclusion

Testosterone is an essential male hormone. Its presence is required for proper male development and spermatogenesis. This chapter highlights some of the conditions that result in inadequate androgen activity as well as end organ effects that result in subsequent infertility. Our understanding of the complex role that androgens play in men's health is still being expanded, especially in regards to the complex process of spermatogenesis. A greater understanding of the regulation of spermatogenesis would further our ability to counsel and treat patients presenting with infertility.

References

[1] Felig, P. and L.A. Frohman, *Endocrinology & metabolism.* 4th ed2001, New York: McGraw-Hill, Health Professions Division. xxxiii, 1562 p.

[2] Li, J. and F. Al-Azzawi, Mechanism of androgen receptor action. *Maturitas,* 2009. 63(2): p. 142-8.

[3] Dunn, J.F., B.C. Nisula, and D. Rodbard, Transport of steroid hormones: binding of 21 endogenous steroids to both testosterone-binding globulin and corticosteroid-binding globulin in human plasma. *The Journal of clinical endocrinology and metabolism,* 1981. 53(1): p. 58-68.

[4] Vermeulen, A., Physiology of the testosterone-binding globulin in man. *Annals of the New York Academy of Sciences,* 1988. 538: p. 103-11.

[5] Avvakumov, G.V., et al., Structural analyses of sex hormone-binding globulin reveal novel ligands and function. *Molecular and cellular endocrinology,* 2010. 316(1): p. 13-23.

[6] Melmed, S. and R.H. Williams, *Williams textbook of endocrinology.* 12th ed2011, Philadelphia: Elsevier/Saunders. xviii, 1897 p.

[7] Santoro, N., M. Filicori, and W.F. Crowley, Jr., Hypogonadotropic disorders in men and women: diagnosis and therapy with pulsatile gonadotropin-releasing hormone. *Endocrine reviews,* 1986. 7(1): p. 11-23.

[8] Matsumoto, A.M., K.M. Gross, and W.J. Bremner, The physiological significance of pulsatile LHRH secretion in man: gonadotrophin responses to physiological doses of pulsatile versus continuous LHRH administration. *International journal of andrology,* 1991. 14(1): p. 23-32.

[9] Veldhuis, J.D., et al., Operating characteristics of the male hypothalamo-pituitary-gonadal axis: pulsatile release of testosterone and follicle-stimulating hormone and their temporal coupling with luteinizing hormone. *The Journal of clinical endocrinology and metabolism,* 1987. 65(5): p. 929-41.

[10] Bremner, W.J., M.V. Vitiello, and P.N. Prinz, Loss of circadian rhythmicity in blood testosterone levels with aging in normal men. *The Journal of clinical endocrinology and metabolism,* 1983. 56(6): p. 1278-81.

[11] Oakley, A.E., D.K. Clifton, and R.A. Steiner, Kisspeptin signaling in the brain. *Endocrine reviews,* 2009. 30(6): p. 713-43.

[12] Pitteloud, N., et al., Inhibition of luteinizing hormone secretion by testosterone in men requires aromatization for its pituitary but not its hypothalamic effects: evidence from the tandem study of normal and gonadotropin-releasing hormone-deficient men. *The Journal of clinical endocrinology and metabolism,* 2008. 93(3): p. 784-91.

[13] de Kretser, D.M., et al., The role of activin, follistatin and inhibin in testicular physiology. *Molecular and cellular endocrinology,* 2004. 225(1-2): p. 57-64.

[14] Bhasin, S., et al., Testosterone therapy in men with androgen deficiency syndromes: an Endocrine Society clinical practice guideline. *The Journal of clinical endocrinology and metabolism, 2010.* 95(6): p. 2536-59.

[15] Mulligan, T., et al., Prevalence of hypogonadism in males aged at least 45 years: the HIM study. *International journal of clinical practice,* 2006. 60(7): p. 762-9.

[16] Gray, A., et al., Age, disease, and changing sex hormone levels in middle-aged men: results of the Massachusetts Male Aging Study. *The Journal of clinical endocrinology and metabolism,* 1991. 73(5): p. 1016-25.

[17] Daniell, H.W., Hypogonadism in men consuming sustained-action oral opioids. The journal of pain: *Official journal of the American Pain Society,* 2002. 3(5): p. 377-84.

[18] Dobs, A.S., et al., Pharmacokinetic characteristics, efficacy, and safety of buccal testosterone in hypogonadal males: a pilot study. *The Journal of clinical endocrinology and metabolism,* 1998. 83(1): p. 33-9.

[19] Bodie, J., et al., Laboratory evaluations of erectile dysfunction: an evidence based approach. *The Journal of urology,* 2003. 169(6): p. 2262-4.

[20] Grinspoon, S., et al., Effects of androgen administration in men with the AIDS wasting syndrome. A randomized, double-blind, placebo-controlled trial. *Annals of internal medicine,* 1998. 129(1): p. 18-26.

[21] Laaksonen, D.E., et al., Testosterone and sex hormone-binding globulin predict the metabolic syndrome and diabetes in middle-aged men. *Diabetes care,* 2004. 27(5): p. 1036-41.

[22] Shores, M.M., et al., Low serum testosterone and mortality in male veterans. *Archives of internal medicine,* 2006. 166(15): p. 1660-5.

[23] Seminara, S.B., F.J. Hayes, and W.F. Crowley, Jr., Gonadotropin-releasing hormone deficiency in the human (idiopathic hypogonadotropic hypogonadism and Kallmann's syndrome): pathophysiological and genetic considerations. *Endocrine reviews,* 1998. 19(5): p. 521-39.

[24] Winter, J.S., et al., Pituitary-gonadal relations in infancy. I. Patterns of serum gonadotropin concentrations from birth to four years of age in man and chimpanzee. *The Journal of clinical endocrinology and metabolism,* 1975. 40(4): p. 545-51.

[25] Mitamura, R., et al., Diurnal rhythms of luteinizing hormone, follicle-stimulating hormone, and testosterone secretion before the onset of male puberty. *The Journal of clinical endocrinology and metabolism,* 1999. 84(1): p. 29-37.

[26] Grumbach M, S.D., Puberty: ontogeny, neuroendocrinology, physiology and disorders. *Textbook of Endocrinology1998,* Philadelphia: Saunders.

[27] Quinton, R., et al., Idiopathic gonadotrophin deficiency: genetic questions addressed through phenotypic characterization. *Clinical endocrinology,* 2001. 55(2): p. 163-74.

[28] Quinton, R., et al., The neuroradiology of Kallmann's syndrome: a genotypic and phenotypic analysis. *The Journal of clinical endocrinology and metabolism,* 1996. 81(8): p. 3010-7.

[29] Oliveira, L.M., et al., The importance of autosomal genes in Kallmann syndrome: genotype-phenotype correlations and neuroendocrine characteristics. *The Journal of clinical endocrinology and metabolism,* 2001. 86(4): p. 1532-8.

[30] Silverira L, M.G., Stewart P, Thomas M, Bouloux P. Novel homozygous splice site GnRH receptor mutations associated with severe familial hypogonadotropic hypogonadism. in *83rd annual meeting of The Endocrine Society.* 2001.

[31] Perkins, R.B., J.E. Hall, and K.A. Martin, Aetiology, previous menstrual function and patterns of neuro-endocrine disturbance as prognostic indicators in hypothalamic amenorrhoea. *Human reproduction,* 2001. 16(10): p. 2198-205.

[32] Laughlin, G.A. and S.S. Yen, Hypoleptinemia in women athletes: absence of a diurnal rhythm with amenorrhea. *The Journal of clinical endocrinology and metabolism,* 1997. 82(1): p. 318-21.

[33] Bianco, S.D. and U.B. Kaiser, The genetic and molecular basis of idiopathic hypogonadotropic hypogonadism. Nature reviews. *Endocrinology,* 2009. 5(10): p. 569-76.

[34] Semple, R.K. and A.K. Topaloglu, The recent genetics of hypogonadotrophic hypogonadism - novel insights and new questions. *Clinical endocrinology,* 2010. 72(4): p. 427-35.

[35] Silveira, L.F., G.S. MacColl, and P.M. Bouloux, Hypogonadotropic hypogonadism. *Seminars in reproductive medicine,* 2002. 20(4): p. 327-38.
[36] Buchter, D., et al., Pulsatile GnRH or human chorionic gonadotropin/human menopausal gonadotropin as effective treatment for men with hypogonadotropic hypogonadism: a review of 42 cases. European journal of endocrinology / *European Federation of Endocrine Societies,* 1998. 139(3): p. 298-303.
[37] Liu, P.Y., et al., Efficacy and safety of recombinant human follicle stimulating hormone (Gonal-F) with urinary human chorionic gonadotrophin for induction of spermatogenesis and fertility in gonadotrophin-deficient men. *Human reproduction,* 1999. 14(6): p. 1540-5.
[38] Jones, T.H. and J.F. Darne, Self-administered subcutaneous human menopausal gonadotrophin for the stimulation of testicular growth and the initiation of spermatogenesis in hypogonadotrophic hypogonadism. *Clinical endocrinology,* 1993. 38(2): p. 203-8.
[39] Kirk, J.M., et al., Gonadal function and response to human chorionic and menopausal gonadotrophin therapy in male patients with idiopathic hypogonadotrophic hypogonadism. *Clinical endocrinology,* 1994. 41(1): p. 57-63.
[40] TC Hsieh, MG McIntyre, LI Lipshultz Testosterone and Male Infertility. Clinical Urologic Endocrinology, *Principles for Men's Health.* ISBN: 9781447144052.
[41] Longcope, C., Adrenal and gonadal androgen secretion in normal females. *Clinics in endocrinology and metabolism,* 1986. 15(2): p. 213-28.
[42] Merke, D.P. and S.R. Bornstein, Congenital adrenal hyperplasia. *Lancet,* 2005. 365(9477): p. 2125-36.
[43] White, P.C., K.M. Curnow, and L. Pascoe, Disorders of steroid 11 beta-hydroxylase isozymes. *Endocrine reviews,* 1994. 15(4): p. 421-38.
[44] White, P.C. and P.W. Speiser, Congenital adrenal hyperplasia due to 21-hydroxylase deficiency. *Endocrine reviews,* 2000. 21(3): p. 245-91.
[45] Kohn, B., et al., Splicing mutation in CYP21 associated with delayed presentation of salt-wasting congenital adrenal hyperplasia. *American journal of medical genetics,* 1995. 57(3): p. 450-4.
[46] Claahsen-van der Grinten, H.L., et al., Testicular adrenal rest tumours in congenital adrenal hyperplasia. Best practice & research. *Clinical endocrinology & metabolism,* 2009. 23(2): p. 209-20.
[47] Stikkelbroeck, N.M., et al., High prevalence of testicular adrenal rest tumors, impaired spermatogenesis, and Leydig cell failure in adolescent and adult males with congenital adrenal hyperplasia. *The Journal of clinical endocrinology and metabolism,* 2001. 86(12): p. 5721-8.
[48] Kliegman, R. and W.E. Nelson, *Nelson textbook of pediatrics.* 19th ed2011, Philadelphia, PA: Elsevier/Saunders. p.
[49] Buiting, K., Prader-Willi syndrome and Angelman syndrome. *American journal of medical genetics.* Part C, Seminars in medical genetics, 2010. 154C(3): p. 365-76.
[50] Klinefelter H, R.E., Albright F, Syndrome characterized by gyencomastia, aspermatogeneis without leydigism, increased excretion of follicle stimulating hormone. *J. Clin Endocrinol* 1942(2): p. 615-27.
[51] Lanfranco, F., et al., Klinefelter's syndrome. *Lancet,* 2004. 364(9430): p. 273-83.

[52] Abramsky, L. and J. Chapple, 47,XXY (Klinefelter syndrome) and 47,XYY: estimated rates of and indication for postnatal diagnosis with implications for prenatal counselling. *Prenatal diagnosis,* 1997. 17(4): p. 363-8.

[53] Bojesen, A., S. Juul, and C.H. Gravholt, Prenatal and postnatal prevalence of Klinefelter syndrome: a national registry study. *The Journal of clinical endocrinology and metabolism,* 2003. 88(2): p. 622-6.

[54] Vignozzi, L., et al., Clinical and therapeutic aspects of Klinefelter's syndrome: sexual function. *Molecular human reproduction,* 2010. 16(6): p. 418-24.

[55] Geschwind, D.H., et al., Neurobehavioral phenotype of Klinefelter syndrome. *Mental retardation and developmental disabilities research reviews,* 2000. 6(2): p. 107-16.

[56] Salbenblatt, J.A., et al., Pituitary-gonadal function in Klinefelter syndrome before and during puberty. *Pediatric research,* 1985. 19(1): p. 82-6.

[57] Christiansen P, A.A., Skakkebaek N, Longitudinal studies of inhibin B levels in boys and young adults with Klinefelter syndrome. *J Clin Endocrinol Metab* 2003(88): p. 888-91.

[58] Moskovic, D.J., et al., Subcutaneous Implantable Testosterone Pellets Overcomes Noncompliance in Adolescent Men with Klinefelter's Syndrome: Case Report and Review of the Literature. *Journal of andrology,* 2011.

[59] Madgar, I., et al., Prognostic value of the clinical and laboratory evaluation in patients with nonmosaic Klinefelter syndrome who are receiving assisted reproductive therapy. *Fertility and sterility*, 2002. 77(6): p. 1167-9.

[60] Staessen, C., et al., PGD in 47,XXY Klinefelter's syndrome patients. *Human reproduction update,* 2003. 9(4): p. 319-30.

[61] Shahidi, N.T., A review of the chemistry, biological action, and clinical applications of anabolic-androgenic steroids. *Clinical therapeutics,* 2001. 23(9): p. 1355-90.

[62] Parr, M.K., U. Flenker, and W. Schanzer, Sports-related issues and biochemistry of natural and synthetic anabolic substances. *Endocrinology and metabolism clinics of North America,* 2010. 39(1): p. 45-57, viii.

[63] de Souza, G.L. and J. Hallak, Anabolic steroids and male infertility: a comprehensive review. *BJU international,* 2011. 108(11): p. 1860-5.

[64] Feinberg, M.J., A.R. Lumia, and M.Y. McGinnis, The effect of anabolic-androgenic steroids on sexual behavior and reproductive tissues in male rats. *Physiology & behavior,* 1997. 62(1): p. 23-30.

[65] Johnsen, S.G., Testicular biopsy score count--a method for registration of spermatogenesis in human testes: normal values and results in 335 hypogonadal males. *Hormones,* 1970. 1(1): p. 2-25.

[66] Turek, P.J., et al., The reversibility of anabolic steroid-induced azoospermia. *The Journal of urology,* 1995. 153(5): p. 1628-30.

[67] Contraceptive efficacy of testosterone-induced azoospermia in normal men. World Health Organization Task Force on methods for the regulation of male fertility. *Lancet,* 1990. 336(8721): p. 955-9.

[68] Mills, J., Recovery of testosterone-induced azoospermia with hCG, in American Urology Association annual meeting 2008.

In: Neurobiology of Men's Mental Health
Editors: Leo Sher and Timothy R. Rice

ISBN: 978-1-63483-191-8
© 2015 Nova Science Publishers, Inc.

Chapter 23

TESTOSTERONE AND MALE CARDIOVASCULAR HEALTH

Omer A. Raheem, MD, MSc, MCh Urol, MRCS, and Tung-Chin Hsieh*, MD

Department of Urology, University of California San Diego School of Medicine, La Jolla, CA, US

ABSTRACT

Cardiovascular disease is the leading cause of death in the United States. Many of the risk factors for development of cardiovascular disease are preventable, including smoking, dyslipidemia, hypertension, diabetes, abdominal obesity, psychosocial factors, inadequate daily consumption of fruits and vegetables, regular alcohol consumption, and lack of regular physical activity. Others risk factors are non-modifiable and include age, family history and gender. Historically, testosterone has been associated with increased cardiovascular risk. However, contemporary published literature shows conflicting data on testosterone and cardiovascular disease. Deficient biological levels of testosterone can ultimately leads to hypogonadism. Furthermore, published data from modern research indicate that hypogonadism is closely related to the development of various cardiovascular risk factors, including hyperlipidemia and insulin resistance. Several studies have demonstrated beneficial effects of testosterone supplementation therapy on reversing symptoms of hypogonadism and improving cardiovascular disease risk profiles. In this chapter, we perform a critical analysis on the association between testosterone and cardiovascular disease.

Keywords: testosterone, cardiovascular disease, hypogonadism

---

* Corresponding author address: T. Mike Hsieh, MD, UC San Diego Health System, 200 West Arbor Drive #8897, San Diego, CA 92103-8897, USA. Email: t7hsieh@ucsd.edu.

1. INTRODUCTION

Cardiovascular disease (CVD) is the leading cause of death in most developed countries, with an estimated 17.3 million deaths worldwide per year [1]. Although mortality from CVD has decreased considerably in recent years, CVD and its complications remain highly prevalent and are significant burdens to the health system [2, 3]. The American Heart Association (AHA) projected that CVD care costs would triple from $272.5 billion in 2010 to an estimated $818.1 billion in 2030 [1].

Numerous risk factors for CVD are preventable, including smoking, dyslipidemia, hypertension, diabetes, abdominal obesity, psychosocial factors, inadequate daily consumption of fruits and vegetables, regular alcohol consumption, and lack of regular physical activity [4]. Others risk factors are non-modifiable and include age [5], family history of CVD [6-8], and gender [9]. The lifetime risk of coronary heart disease (CHD) at age 40 is one in two for men and one in three for women [7, 10]. Men are at greater risk for CVD than premenopausal women [11]. Higher CVD mortality may be partially attributed to behavioral and physical characteristics of males, including increased risk-taking behavior (smoking, drinking), endocrine and metabolic factors like fat distribution, and lower male engagement in preventive care [12]. However, postmenopause, a woman's risk is similar to a man's [13]. This finding may suggest a potential association of CVD to sex hormones.

Historically, testosterone (T) has been associated with increased CVD risk. However, modern literature shows conflicting data on T and CVD. T may have a protective effect on the cardiovascular (CV) system via its effects on vascular reactivity [14], immune modulation [15], arterial wall stiffness [16], and the endothelium [17]. Contemporary studies report that hypogonadism is associated with higher CVD incidence as well as insulin resistance, diabetes mellitus, and metabolic syndrome (MetS) [18]. In fact, low serum T has been associated with increased mortality in male veterans [19]. Contrarily, other studies demonstrated no association between T and incident CVD [20, 21]. In this chapter, we aim to providing a critical analysis on the association between T and CVD.

2. TESTOSTERONE AND CARDIOVASCULAR DISEASE

2.1. Testosterone and Coronary Heart Disease

Several studies have demonstrated that patients with congestive heart disease (CHD) and coronary artery disease (CAD) as diagnosed by coronary angiography have lower levels of T compared to control subjects [22-24], and that hypogonadism is an independent risk factor for developing CHD [22]. In addition, T has been reported to have an inverse relationship with CAD severity [25-28] with epidemiologic evidence demonstrating a fivefold decrease in risk of severe atherosclerotic CAD between the lowest and highest quartiles of total T (TT) [23]. This inverse correlation between T and CAD severity is present in both men and postmenopausal women with CAD [29].

Men with myocardial infarction (MI) have been reported to have lower T and an increased estrogen (E) to T ratio when compared to controls and T may have a regulatory effect on lipid metabolism and thus on ischemic heart disease, with a possible interdependent effect with insulin [30, 31]. TT and bioavailable T have been reported to decrease following acute MIs in men, in association with adverse changes in the fibrinolytic profile including an elevation in plasminogen activator inhibitor-1 (PAI-I) activity and reduction of tissue plasminogen activator activity, independent of insulin changes [32].

T deficiency is prevalent in patients with chronic congestive heart failure (CHF), with a prevalence of 39% [33, 34]. Chronic CHF has been associated with significantly lower free testosterone (FT) [35]. New York Heart Association Functional Classification is the most commonly used classification system that divides patients into four categories by limitations in physical exercise; higher New York Heart Association classes and impaired left ventricular function have been associated with lower levels of TT and FT even after adjustment for associated comorbidities [33, 34, 36]. Endogenous T has also been linked to increased exercise tolerance in patients with CHF. Bocchi et al., investigated the correlation between T and hemodynamics in CHF patients with erectile dysfunction (ED) and reported a direct correlation between T levels and exercise tolerance, as reflected by an increased distance at 6-minute cardiopulmonary walk test. Interestingly, T was inversely correlated with diastolic blood pressure and right and left ventricle ejection fraction [37].

Several studies have identified an inverse relationship between T levels and mortality in men with CHD including CAD and CHF [38, 39]. In a study of 208 men with CHF, Jankowska et al., reported that both TT and estimated free T (eFT) were independent prognosticators of mortality [34]. Wehr et al., reported that patients with FT in the lowest quartile also had increased CHF mortality compared to those in the highest quartile; however, no significant association between TT and CV events or cardiac disease was noted [36].

In contrast, multiple published studies have demonstrated a lack of association or an adverse association between T and CAD or identified relationships between other sex hormones. A longitudinal case control study reported no significant difference in T between low-risk men who developed CHD and those who did not [40]. Similarly, Contoreggi et al., reported that plasma levels of T, FT, and dehydroepiandrosterone were not predictive of the risk for developing CAD in men and that cholesterol and systolic blood pressure are more important CAD risk factors [41]. Luria et al., reported that T levels were similar between patients with acute MIs and proven CAD and healthy individuals; however, although T levels were similar in all groups, estradiol (E2) levels were substantially elevated in male patients with MI and chronic angina pectoris, leading to the hypothesis that increased peripheral conversion of T to E may contribute to MI and CAD [42]. Contrarily, Kajinami et al., reported that E2 was lower and FT was elevated in patients with proven CAD and controls without but not to a significant degree; however, the ratio of FT to E2 was significantly higher than in the controls, leading to the hypothesis that T may inhibit early atherogenesis by conversion to E2 [43].

In a prospective population-based study of 1,009 older White males, sex hormones measured (T, androstenedione, estrone, E2) did not have a significant association with known CVD at baseline or with subsequent CV or ischemic heart disease mortality [44]. In a longitudinal study of CVD in male twins followed for 20 years, no significant difference in T levels was present in discordant monozygotic twins and similar T, E, and sex hormone–binding globulin were present in both positive and negative concordant twins [45].

2.2. Testosterone and Cerebrovascular Disease: Stroke and Transient Ischemic Attack

Low TT and FT have been reported to be predictive of increased incidence of stroke or TIA, even after adjusting for conventional risk factors for CVD [46]. Multiple studies have demonstrated that low T and FT are associated with increased carotid intima-media thickness (IMT), which serves as a measure of cerebrovascular atherosclerosis [47, 48]. A recent study of middle-aged Japanese men reported that low serum FT but not TT was associated with carotid artery IMT after multivariate analysis with age, BMI, and other clinically relevant factors [49]. T's beneficial effect on carotid IMT is further demonstrated by Aversa et al., who demonstrated a significant decrease in carotid IMT after a 12-month course of transdermal T [50].

In contrast, a Honolulu–Asia Aging Study, although E2 increased the risk of stroke, T had no association to stroke risk [51]. In a retrospective longitudinal study, no association was observed for FT or TT with carotid IMT in young to middle-aged women [52]. Similarly, T was not associated with the progression or level of carotid IMT or adventitial diameter in perimenopausal women [53].

2.3. Testosterone, Aortic Atherosclerosis/Aneurysm, and Peripheral Vascular Disease

In a large population-based cohort study, Yeap et al., reported that higher FT reduces the odds ratio of abdominal aortic aneurysms in elderly men, suggesting that impaired gonadal function may be involved in abnormal arterial dilation as well as vascular disease in older men [54]. Interestingly, Vaidya et al., reported that T was negatively associated with baseline waist-to-hip ratio, which has been strongly associated with prevalent atherosclerosis, in men but not in postmenopausal women [55]. Using ankle brachial index (ABI) as a measure of PAD (ABI < 0.9), Tivesten et al., demonstrated that elderly men with lower T and higher E had significantly lower ABI and thus lower extremity PAD [56]. Using intermittent claudication as a measure of PAD, Yeap et al., reported that lower T had an inverse association with symptoms and thus lower-extremity PAD even after multivariate adjustment for age, smoking BMI, and other relevant clinical factors [57].

Interestingly, Maggio et al., reported that high T levels are independently associated with the presence of PAD as defined by ABI > 0.90 in older women but not in men [58]. Although Haring et al., identified an inverse relationship between ABI and TT, the relationship was insignificant in multivariable logistic models [59]. Finally, a Cochrane review of three studies with a total of 109 subjects with intermittent claudication or critical leg ischemia reported that T replacement therapy did not improve walking distance and other measures of PAD (venous filling time, muscle blood flow, plethysmography); however, authors indicated that this might reflect limited data available rather than the lack of a real effect [60].

3. Testosterone and Metabolic Syndrome

Metabolic syndrome (MetS) encompasses a combination of medical disorders, which increase the risk of developing CVD and diabetes. A meta-analysis of 20 studies reported that even after adjusting for age and BMI, MetS is independently associated with hypogonadism [61]. Contrarily, a more recent epidemiologic study of elderly men and women did not demonstrate an inverse relationship between T and MetS after adjusting for lipids, BMI, inflammation, and insulin resistance [62]. T has been inversely associated with all the components of MetS: central obesity [55], low high-density lipoprotein [63] and impaired fasting glucose [64]. In contrast, other studies did not demonstrate an association between lipid profile (high-density lipoprotein, low-density lipoprotein, TG, and total cholesterol) and T in both healthy men and men with CAD [65]. In addition to the independent components of MetS, low levels of TT and sex hormone–binding globulin have also been associated with an increased likelihood of having MetS, independent of other known CV risk factors and insulin resistance [66].

4. Testosterone Supplementation Therapy

4.1. Effect of Testosterone Supplemetation on ST Depression/Ischemia

T supplementation therapy has been demonstrated to have a beneficial effect on exercise induced myocardial ischemia in men with CAD [67]. In a double-blinded, randomized study, men who received 4 to 8 weeks of testosterone cypionate treatment had significant decreases in post-exercise ST segment depression on EKG compared to placebo, indicating a decrease in myocardial ischemia [67]. Cornoldi et al., reported that a 12-week course of low-dose T in diabetic males with CAD decreased the number of anginal attacks, silent ischemic episodes, and total ischemic burden as evidenced in ambulatory EKG compared to placebo [68]. In contrast, Thompson et al., reported that acute administration of T in 32 men with provocable myocardial ischemia did not improve cardiac function, demonstrating no difference on exercise/pharmacologic stress test in peak heart rate, systolic blood pressure, maximal rate pressure product, perfusion imaging scores, or the onset of ST segment depression; they concluded that acute T supplementation does not have a beneficial or adverse effect on the onset and magnitude of stress-induced myocardial ischemia in men with stable CAD [69].

4.2. Effect of Testosterone Supplementation on CHF Exercise Tolerance

As low T independently predicts reduced exercise capacity and poor clinical outcomes in CHF patients, several studies have investigated the effect of T supplementation on exercise tolerance in patients with CHF. T supplementation in elderly men with CHF has been reported to increase exercise capacity as demonstrated by incremental shuffle walk test as well as improve symptoms of CHF by at least one functional class [70]. Toma et al., concluded that T supplementation produced significant improvements in exercise capacity compared to placebo with no adverse effects as demonstrated by an increase in 6-minute walk test, incremental shuffle walk test, and peak oxygen consumption [71]. In fact, the differences were greater than that seen with other therapies including angiotensin-converting–enzyme inhibitors, beta-blockers, and cardiac resynchronization therapy [71]. Whereas Malkin et al., did not find an increase in skeletal muscle bulk or handgrip strength [70], Iellamo et al., observed an improvement [72]. Studies looking at T therapy in men with CVD have been listed in Table 1.

5. Risks of Testosterone Therapy

T supplementation is not without adverse effects. Supratherapeutic doses of anabolic steroids can reduce cardiac function and induce hypertension, ventricular remodeling, MI, and sudden cardiac death [73, 74]. Additionally, T causes salt and water retention, particularly in older men, which may contribute to edema, hypertension, and CHF [75].

An elevated hematocrit is the most common side effect of T therapy [76]. Hematocrit and hemoglobin have been reported to increase in a linear, dose-dependent fashion with T [77]. In a meta-analysis of 19 randomized controlled trials investigating the adverse effects of T therapy, elderly T treated men were 4 times more likely to have increased hematocrit of >50% than placebo-treated men; despite this increase, no significant difference in CV events was noted [78]. Several studies have not identified any changes in erythropoietin (EPO) levels [79]. However, a more recent study demonstrated that T-induced erythropoiesis is associated with both EPO stimulation and reduced ferritin and hepicidin, proposing a possible recalibration of the EPO set point and increased utilization of iron for erythropoiesis in T-treated men [78].

The Testosterone in Older Men With Mobility Limitations trial (TOM) was prematurely stopped due to a higher incidence of cardiac, respiratory, and dermatological events in patients receiving T supplementation versus placebo [79]. In the TOM trial, elderly community-dwelling men with low serum T and limited mobility were randomly assigned to receive 6 months of daily T 100 mg or placebo transdermal gel to assess the effect of T therapy on exercise tolerance. The trial was stopped after 209 of the planned 252 men were enrolled due to the imbalance in cardiac events between the two groups (23 vs. 5). Despite the alarming increase in cardiac events in men receiving T therapy versus placebo, diversity of the cardiac events that occurred and the lack of structured ascertainment of CV events limit the clinical value of the study; moreover, the population studied and the utilization of higher T dosages may limit the study's generalizability [79].

Two retrospective cohort Veterans Affairs studies reported differing conclusions regarding the association of T therapy and mortality [80, 81]. Shores et al., investigated the association between T therapy and total mortality in men with serum TT levels less than 250 ng/dl and older than 40 years. A total of 398 T-treated men, mean age 60.9 years, were compared with 633 untreated men, mean age 62.8 years. T users were younger, had higher BMIs, and had lower T levels. After adjusting for these factors and other relevant covariates, T-treated men had a 39% reduction in mortality risk (hazard ratio [HR] = 0.61; 95% confidence interval [CI] = 0.42, 0.88; p = .008) versus untreated men [82]. Contrarily Vigen et al., conducted a retrospective cohort study of men who underwent coronary angiography and had T levels less than 300 ng/dl and demonstrated an increase in adverse effects (HR = 1.29; 95% CI = 1.04, 1.58) in men treated with

T. Of the 1,223 T-treated men, mean age 60.6 years, 67 died, 420 had MIs, and 486 had strokes compared to 681 deaths, 420 MIs, and 519 strokes among the 7,486 untreated men, mean age 63.8. The hazard risk remained unchanged even after adjustment for CAD [83]. Finkle et al., published recent secondary data analysis of a large Claims and Encounters Database of 55,593 men who filled a prescription for T supplementation, compared with 167,279 men who filled a prescription for a PDE5-I as control group, aimed at determining whether T supplementation was associated with increased cardiovascular events. Overall, Finkle's study findings are similar to those from retrospective cohort analyses [81, 83]. In Finckle's study the rates of MI increased in the 3 months after filling a T supplementation prescription (from 3.48 per 1000 person years before to 4.75 per 1,000 person-years after; rate ratio: 1.36; 95% confidence interval [CI], 1.03–1.81). The increase was more pronounced in the subset of men ≥65 yr of age (5.27 per 1000 person-years before to 11.52 person-years after; rate ratio: 2.19; 95% CI, 1.27–3.77) and this study concluded that in men with pre-existing diagnosed heart disease, the risk of MI following initiation of T supplementation prescription is substantially increased. Despite inherited bias in this study such as utilization of PDE5-I as control group, it raises concerns that physicians should consider adding to their patient counseling a discussion about putative cardiovascular risks associated with T supplementation with appropriate monitoring during treatment [84]. A recent critical analysis by Su et al., highlighted the current association between testosterone and cardiovascular disease in men [85].

6. Mechanistic Effect of Testosterone on Cardiovascular System

6.1. Endothelial Dysfunction

Endothelial dysfunction is considered to be the first step in the formation of atherosclerotic lesions. T has been demonstrated to have a protective effect on endothelial function [86]. T has been inversely correlated with vascular cell adhesion molecule-1, which is produced by endothelial cells (ECs) and up-regulated when ECs undergo inflammatory and malignant stimulation [86]. Foresta et al., reported that hypogonadal men had lower levels of endothelial progenitor cells (EPCs), which are important in endothelial regeneration, and higher levels of an osteocalcin-positive subpopulation of EPCs, which are highly correlated with atherosclerosis progression, compared to eugonadal men [87]. In contrast, Florvaag et al., reported that despite the frequency of T deficiency in men with CHF, it did not have an effect on EPCs [33]. Incubating ECs and smooth muscle cells (SMCs) with androgens in vitro, Nheu et al., reported that T can have potentially beneficial or harmful effects, noting that although T induces DNA synthesis and EC growth via activation of MAP (mitogen-activated protein) kinase ERK1/2 activity, T also enhances SMC proliferation, which can contribute to atherosclerosis [88]. Finally, T has been shown to significantly reduce endoplasmic reticulum stress and superoxide generation in human umbilical vein ECs, both of which have been implicated in atherosclerosis; however, when combined with aromatase inhibitors, the protective effect of T was lost, suggesting an E2-mediated mechanism [89].

Table 1. Summary of randomized clinical trials of Testosterone Replacement Therapy in patients with cardiovascular disease

Study	Type of study	Age, yrs	Sex	No. of intervention-treated subjects	No. of placebo-treated subjects	Inclusion criteria	Treatment	Dose	Treatment duration	Findings	Values
Jaffe (1977)	R, DB	35-71	M	25	25	ST segment depression following two-step exercise test	IM testosterone cypionate	200 mg weekly	4 weeks	↓ ST depression	3.9 ± 0.9^a mm ($31.7\% \pm 6.5\%$)
Jaffe (1977)	R, DB	35-71	M	25	25	ST segment depression following two-step exercise test	IM testosterone cypionate	200 mg weekly	8 weeks	↓ ST depression	6.0 ± 1.0^a mm ($51.2\% \pm 7.0\%$)
Thompson et al., (2002)	R, C	69 ± 6	M	34	34	Inducible myocardial ischemia on recent stress perfusion imaging study	IV testosterone	Individualized: to 2× baseline or 6× baseline	Acute	↔ time to ST depression	Placebo: 4.8 ± 2.2 minutes
											2× T: 4.9 ± 2.2 minutes
											6× T: 4.8 ± 2.3 minutes
Thompson et al., (2002)	R, C	69 ± 6	M	34	34	Inducible myocardial ischemia on recent stress perfusion imaging study	IV testosterone	Individualized: to 2× baseline or 6× baseline	Acute	↔ exercise time	Placebo: 9.3 ± 2 minutes
											2× T: 9.2 ± 2 min
											6× T: 9.3 ± 2 minutes
Malkin et al., (2006)	R, DB, PC	64 ± 9.9	M	37	39	CHF	Testosterone patch (Androderm)	5 mg qhs	12 months	↑ exercise time	25 ± 15 minutes; $p = .006$
Cornoldi et al., (2010)	R, DB, PC	74 ± 7	M	43	44	Diabetic with CAD	IM testosterone undecanoate	40 mg tid	12 weeks	↓ episodes per patient, silent episodes, total ischemic burden (seconds/24 hours)	

Study	Type of study	Age, yrs	Sex	No. of intervention-treated subjects	No. of placebo-treated subjects	Inclusion criteria	Treatment	Dose	Treatment duration	Findings	Values
Iellamo et al., (2010)	R, DB, PC	68.2 ± 6.85	F	20	12	CHF	Testosterone patch (Intrinsa)	300 μg biweekly	24 weeks	↑ peak oxygen consumption, distance in 6-minute walk test	$p < .05$

Note. R = randomized; DB = double-blind; M = male; F = female; IM = intramuscular; C = crossover; IV = intravenous; CAD = coronary artery disease; PC = placebo controlled; qhs = at bedtime; q = every; T = testosterone; SB = single-blind; CHF = congestive heart failure. Ages are represented in ranges or *M* ± *SD*, unless otherwise denoted. Continuous values are *M* ± *SD* unless otherwise denoted.

a. Mean ± standard error of mean.

T's proposed anti-anginal and anti-ischemial effects have been partially attributed to its effects on the vascular system. Although T's vasodilatory effects are well recognized, the exact mechanism of action is yet to be elucidated. This direct relaxation response to T has been attributed to potassium conductance in the form of non–ATP (adenosine triphosphate)-sensitive potassium channel [90], ATP-sensitive potassium channel [89], and large-conductance calcium-activated potassium channel opening action [91]. T has also been reported to induce vasodilation by reducing calcium influx into vascular smooth muscle by acting as a selective and potent inhibitor of L-type calcium channels at physiologic levels and as an inhibitor of T-type channels at supraphysiologic levels [92]. Supporting T's direct vasodilatory effect, despite inducing coronary artery dilation and increasing coronary blood flow in men with CAD, short-term (3-minute) intracoronary administration of T had no effect on acetylcholine-mediated increase in coronary blood flow or vasoreactivity, suggesting a lack of effect on stimulated endothelial nitric oxide [14].

Contrarily, other studies have suggested an endothelium-dependent mediated mechanism behind T's vasodilatory effect [93]. Both acute and long-term administration of T in men with CAD resulted in increases in brachial artery flow-mediated reactivity, which induces shear stress release of nitric oxide and subsequently leads to vasodilation [94]. Arterial wall stiffness is an independent predictor of CVD risk. Low T levels have been associated with endothelial dysfunction [95]. This inverse relationship has been demonstrated using both pulse pressure [94] and pulse wave velocity [96] as reflections of arterial wall stiffness. In fact, the association between T and CVD mortality was lost in male hemodialysis patients after adjusting for pulse wave velocity, suggesting that endothelial dysfunction may be a possible explanation of T's inverse association with CVD [97]. In contrast, long-term (8-week) administration of T increased myocardial perfusion in unobstructed coronary arteries and decreased radial and aortic augmentation indexes, indicating decreased arterial wall stiffness; however, no effect was observed on global perfusion or on endothelial function [98].

6.2. Inflammation

Atherosclerosis is mediated by an ongoing inflammatory response, which is induced by cytokines and other inflammatory markers. Cytokines cause cellular and local arterial wall inflammation and may lead to vascular smooth muscle apoptosis, degradation of the fibrin cap, and plaque rupture, thereby leading to platelet adhesion, thrombus formation, and ultimately angina or MI [15]. An elevation in inflammatory markers or cytokines has been identified to be predictive of outcomes in patients with CVD [99]. TT has been negatively associated with inflammatory markers such as macrophage inflammatory protein 1-alpha, 1-beta, and tumor necrosis factor alpha in young men, suggesting a low-grade inflammatory state [100]. Accordingly, T has been reported to have immune-modulating characteristics, which, due to inflammation's role in atherosclerosis, may play into the development of CVD.

T may suppress the expression of inflammatory cytokines. T supplementation has been reported to suppress the expression of high-sensitivity CRP and interleukin-6 in patients who underwent coronary artery stent implantation, leading to the hypothesis that T's anti-inflammatory property could potentially attenuate major CV events after stent placement [101]. In a randomized, placebo-controlled, crossover study of T supplementation in hypogonadal men, T supplementation reduced levels of proinflammatory cytokines TNF-α

(tumor necrosis factor alpha) and IL-1β (interleukin-1 beta) while suppressing levels of cytokine interleukin-10 [102,103]. Similarly, Corcoran et al.,demonstrated that T suppresses macrophage TNF-α and IL-1β expression [104]. Hernandez-Mijares et al., did not find a significant difference in IL-6 or TNF-α between diabetic males with hypogonadism compared to eugonadal men [105].

6.3. Clotting

Fibrinogen is a known risk factor for CVD as well as an inflammatory biomarker [106]; it increases CVD risk through its effects on atherogenesis, thrombogenesis, and ischemia by increasing plasma and blood viscosity [107]. T is negatively correlated with fibrinogen [27]. A recent study found that patients with prostate cancer on androgen deprivation therapy had elevated levels of fibrinogen compared to healthy controls [108]. In addition to fibrinogen, PAI-1, another cloting mediator and risk factor for ischemic heart disease, has been negatively correlated with T [27]. Contrarily, a double-blinded, randomized placebo-controlled trial of T supplementation in men with chronic stable angina demonstrated no changes in fibrinogen or PAI-1, suggesting that T supplementation does not affect blood coagulation status [2]. Moreover, a study comparing chemically or surgically castrated males to eugonadal controls showed that castrated men had less platelet thromboxane A2 receptors, suggesting that the inhibition of T production may attenuate platelet aggregation responses [109].

7. Summary of the Effect of Testosterone on Cardiovascular Disease

Hypogonadism is a growing concern of our aging population; it affects 20% of men older than 60 years, 30% of men older than 70 years, and 50% of men older than 80 years [110]. Adding to this, low T has been linked to increased risk for major adverse cardiac events in hypertensive patients [110]. With increased awareness among health providers, there has been great emphasis on studying the association of T to CVD. The chapter primiraly highlights the important role that T plays in CVD, pointing toward a potential beneficial effect of T on CVD with conflicting studies directed primarily at CAD.

Most of the studies examining T and CVD were observational, cross-sectional, or retrospective studies that cannot demonstrate cause and effect. Whether this inverse association between T and CVD is a result of T or a result of the disease process remains unclear. There also is no consensus on the cutoff T value of hypogonadism in the published literature. T supplementation results in positive short- and long-term physiological and biochemical changes in patients with CVD. T supplementation may reduce the CV risk factors clustered in MetS without any significant hematological or prostate adverse effect [50]. Other than the TOM trial [81], published clinical trials and meta-analyses investigating the benefits and risks of T therapy have not demonstrated significant adverse CV events in the acute setting.

Although the results in Shores et al., [82] are reassuring, the more recent study by Vigen et al., [83] raises real concerns regarding the long-term risks of T supplementation therapy. Both studies incorporated multivariate analyses to guard against potential confounders with the latter study utilizing a sophisticated weighted analysis to further improve their analyses. However, as with any retrospective study, unmeasured bias and error may be invariably inherited. Future randomized control studies are needed to better delineate the risks and benefits of T supplementation in CVD and establish the optimal protocol for T therapy while comparing acute versus chronic adverse effects.

Conclusion

Cardiovascular disease is the leading cause of mortality in the United States. Growing evidence of contemporary literature suggests a cardiovascular-protective effect of testosterone in men, and current studies have begun to shed light on the mechanism of action of androgens on the CV system. T therapy can reverse symptoms of hypogonadism while potentially reducing CVD risk factors. However, due to conflicting data, future studies should further explore the long-term safety of T therapy. Only by correcting modifiable risk factors and restoring androgen imbalance can optimal management of CVD be achieved.

References

[1] Laslett, L. J., Alagona, P., Jr., Clark, B. A., III, Drozda, J. P., Jr., Saldivar, F., Wilson, S. R., ... Hart, M. (2012). The worldwide environment of cardiovascular disease: Prevalence, diagnosis, therapy, and policy issues: A report from the American College of Cardiology. *Journal of the American College of Cardiology, 60*(25 Suppl.), S1-49.

[2] Smith, A. M., English, K. M., Malkin, C. J., Jones, R. D., Jones, T. H., & Channer, K. S. (2005). Testosterone does not adversely affect fibrinogen or tissue plasminogen activator (tPA) and plasminogen activator inhibitor-1 (PAI-1) levels in 46 men with chronic stable angina. *European Journal of Endocrinology, 152*, 285-291.

[3] Smolina, K., Wright, F. L., Rayner, M., & Goldacre, M. J. (2012). Determinants of the decline in mortality from acute myocardial infarction in England between 2002 and 2010: Linked national database study. *British Medical Journal, 344*, d8059.

[4] Yusuf, S., Hawken, S., Ounpuu, S., Dans, T., Avezum, A., Lanas, F., ... Lisheng, L. (2004). Effect of potentially modifiable risk factors associated with myocardial infarction in 52 countries (the INTERHEART study): Case-control study. *Lancet, 364*, 937-952.

[5] Savji, N., Rockman, C. B., Skolnick, A. H., Guo, Y., Adelman, M. A., Riles, T., & Berger, J. S. (2013). Association between advanced age and vascular disease in different arterial territories: A population database of over 3.6 million subjects. *Journal of the American College of Cardiology, 61*, 1736-1743.

[6] Andresdottir, M. B., Sigurdsson, G., Sigvaldason, H., & Gudnason, V. (2002). Fifteen percent of myocardial infarctions and coronary revascularizations explained by family

history unrelated to conventional risk factors. The Reykjavik Cohort Study. *European Heart Journal, 23*, 1655-1663.

[7] Lloyd-Jones, D. M., Nam, B. H., D'Agostino, R. B., Sr., Levy, D., Murabito, J. M., Wang, T. J., ... O'Donnell, C. J. (2004). Parental cardiovascular disease as a risk factor for cardiovascular disease in middle-aged adults: A prospective study of parents and offspring. *Journal of the American Medical Association, 291*, 2204-2211.

[8] Sesso, H. D., Lee, I. M., Gaziano, J. M., Rexrode, K. M., Glynn, R. J., & Buring, J. E. (2001). Maternal and paternal history of myocardial infarction and risk of cardiovascular disease in men and women. *Circulation, 104*, 393-398.

[9] Kappert, K., Bohm, M., Schmieder, R., Schumacher, H., Teo, K., Yusuf, S., ... Unger, T. (2012). Impact of sex on cardiovascular outcome in patients at high cardiovascular risk: Analysis of the Telmisartan Randomized Assessment Study in ACE-Intolerant Subjects With Cardiovascular Disease (TRANSCEND) and the Ongoing Telmisartan Alone and in Combination With Ramipril Global End Point Trial (ONTARGET). *Circulation, 126*, 934-941.

[10] Lloyd-Jones, D. M., Larson, M. G., Beiser, A., & Levy, D. (1999). Lifetime risk of developing coronary heart disease. *Lancet, 353*, 89-92.

[11] Yang, X. P., & Reckelhoff, J. F. (2011). Estrogen, hormonal replacement therapy and cardiovascular disease. *Current Opinion in Nephrology and Hypertension, 20*, 133-138.

[12] Tudiver, F., & Talbot, Y. (1999). Why don't men seek help? Family physicians' perspectives on help-seeking behavior in men. *Journal of Family Practice, 48*, 47-52.

[13] Yang, X. C., Jing, T. Y., Resnick, L. M., & Phillips, G. B. (1993). Relation of hemostatic risk factors to other risk factors for coronary heart disease and to sex hormones in men. *Arteriosclerosis and Thrombosis, 13*, 467-471.

[14] Webb, C. M., McNeill, J. G., Hayward, C. S., de Zeigler, D., & Collins, P. (1999). Effects of testosterone on coronary vasomotor regulation in men with coronary heart disease. *Circulation, 100*, 1690-1696.

[15] Malkin, C. J., Pugh, P. J., Jones, R. D., Jones, T. H., & Channer, K. S. (2003). Testosterone as a protective factor against atherosclerosis--immunomodulation and influence upon plaque development and stability. *The Journal of Endocrinology, 178*(3), 373-380.

[16] Dockery, F., Bulpitt, C. J., Donaldson, M., Fernandez, S., & Rajkumar, C. (2003). The relationship between androgens and arterial stiffness in older men. *Journal of the American Geriatrics Society, 51*, 1627-1632.

[17] Pearson, L. J., Yandle, T. G., Nicholls, M. G., & Evans, J. J. (2008). Regulation of endothelin-1 release from human endothelial cells by sex steroids and angiotensin-II. *Peptides, 29*, 1057-1061.

[18] Corona, G., Rastrelli, G., Vignozzi, L., Mannucci, E., & Maggi, M. (2011). Testosterone, cardiovascular disease and the metabolic syndrome. *Best Practice & Research. Clinical Endocrinology & Metabolism, 25*, 337-353.

[19] Shores, M. M., Matsumoto, A. M., Sloan, K. L., & Kivlahan, D. R. (2006). Low serum testosterone and mortality in male veterans. *Archives of Internal Medicine, 166*, 1660-1665.

[20] Arnlov, J., Pencina, M. J., Amin, S., Nam, B. H., Benjamin, E. J., Murabito, J. M., ... Vasan, R. S. (2006). Endogenous sex hormones and cardiovascular disease incidence in men. *Annals of Internal Medicine, 145*, 176-184.

[21] Haring, R., Teng, Z., Xanthakis, V., Coviello, A., Sullivan, L., Bhasin, S., ... Vasan, R. S. (2013). Association of sex steroids, gonadotrophins, and their trajectories with clinical cardiovascular disease and all-cause mortality in elderly men from the Framingham Heart Study. *Clinical Endocrinology, 78*, 629-634.

[22] Cao, J., Zou, H., Zhu, B. P., Wang, H., Li, J., Ding, Y., & Li, X. Y. (2010). Sex hormones and androgen receptor: Risk factors of coronary heart disease in elderly men. *Chinese Medical Sciences Journal, 25*, 44-49.

[23] Chute, C. G., Baron, J. A., Plymate, S. R., Kiel, D. P., Pavia, A. T., Lozner, E. C., ... MacDonald, G. J. (1987). Sex hormones and coronary artery disease. *American Journal of Medicine, 83*, 853-859.

[24] Sieminska, L., Wojciechowska, C., Swietochowska, E., Marek, B., Kos-Kudla, B., Kajdaniuk, D., & Nowalany-Kozielska, E. (2003). Serum free testosterone in men with coronary artery atherosclerosis. *Medical Science Monitor, 9*(5), CR162-CR166.

[25] Hu, X., Rui, L., Zhu, T., Xia, H., Yang, X., Wang, X., ... Jiang, H. (2011). Low testosterone level in middle-aged male patients with coronary artery disease. *European Journal of Internal Medicine, 22*, e133-136.

[26] Li, L., Guo, C. Y., Jia, E. Z., Zhu, T. B., Wang, L. S., Cao, K. J., ... Yang, Z. J. (2012). Testosterone is negatively associated with the severity of coronary atherosclerosis in men. *Asian Journal of Andrology, 14*, 875-878.

[27] Phillips, G. B., Pinkernell, B. H., & Jing, T. Y. (1994). The association of hypotestosteronemia with coronary artery disease in men. *Arteriosclerosis and Thrombosis, 14*, 701-706.

[28] Rosano, G. M., Sheiban, I., Massaro, R., Pagnotta, P., Marazzi, G., Vitale, C., ... Fini, M. (2007). Low testosterone levels are associated with coronary artery disease in male patients with angina. *International Journal of Impotence Research, 19*, 176-182.

[29] Kaczmarek, A., Reczuch, K., Majda, J., Banasiak, W., & Ponikowski, P. (2003). The association of lower testosterone level with coronary artery disease in postmenopausal women. *International Journal of Cardiology, 87*, 53-57.

[30] Lichtenstein, M. J., Yarnell, J. W., Elwood, P. C., Beswick, A. D., Sweetnam, P. M., Marks, V., ... Riad-Fahmy, D. (1987). Sex hormones, insulin, lipids, and prevalent ischemic heart disease. *American Journal of Epidemiology, 126*, 647-657.

[31] Sewdarsen, M., Jialal, I., Vythilingum, S., & Desai, R. (1986). Sex hormone levels in young Indian patients with myocardial infarction. *Arteriosclerosis, 6*, 418-421.

[32] Pugh, P. J., Channer, K. S., Parry, H., Downes, T., & Jone, T. H. (2002). Bio-available testosterone levels fall acutely following myocardial infarction in men: Association with fibrinolytic factors. *Endocrine Research, 28*, 161-173.

[33] Florvaag, A., Oberle, V., Fritzenwanger, M., Kretschmar, D., Betge, S., Goebel, B., ... Jung, C. (2012). Testosterone deficiency in male heart failure patients and its effect on endothelial progenitor cells. *The Aging Male, 15*, 180-186.

[34] Jankowska, E. A., Biel, B., Majda, J., Szklarska, A., Lopuszanska, M., Medras, M., ... Ponikowski, P. (2006). Anabolic deficiency in men with chronic heart failure: Prevalence and detrimental impact on survival. *Circulation, 114*, 1829-1837.

[35] Kontoleon, P. E., Anastasiou-Nana, M. I., Papapetrou, P. D., Alexopoulos, G., Ktenas, V., Rapti, A. C., ... Nanas, J. N. (2003). Hormonal profile in patients with congestive heart failure. *International Journal of Cardiology, 87*, 179-183.

[36] Wehr, E., Pilz, S., Boehm, B. O., Marz, W., Grammer, T., & Obermayer-Pietsch, B. (2011). Low free testosterone is associated with heart failure mortality in older men referred for coronary angiography. *European Journal of Heart Failure, 13*, 482-488.

[37] Bocchi, E. A., Carvalho, V. O., & Guimaraes, G. V. (2008). Inverse correlation between testosterone and ventricle ejection fraction, hemodynamics and exercise capacity in heart failure patients with erectile dysfunction. *International Brazilian Journal of Urology, 34*, 302-310.

[38] Malkin, C. J., Pugh, P. J., Morris, P. D., Asif, S., Jones, T. H., & Channer, K. S. (2010). Low serum testosterone and increased mortality in men with coronary heart disease. *Heart, 96*, 1821-1825.

[39] Ponikowska, B., Jankowska, E. A., Maj, J., Wegrzynowska-Teodorczyk, K., Biel, B., Reczuch, K., ... Ponikowski, P. (2010). Gonadal and adrenal androgen deficiencies as independent predictors of increased cardiovascular mortality in men with type II diabetes mellitus and stable coronary artery disease. *International Journal of Cardiology, 143*, 343-348.

[40] Heller, R. F., Miller, N. E., Wheeler, M. J., & Kind, P. R. (1983). Coronary heart disease in "low risk" men. *Atherosclerosis, 49*, 187-193.

[41] Contoreggi, C. S., Blackman, M. R., Andres, R., Muller, D. C., Lakatta, E. G., Fleg, J. L., & Harman, S. M. (1990). Plasma levels of estradiol, testosterone, and DHEAS do not predict risk of coronary artery disease in men. *Journal of Andrology, 11*, 460-470.

[42] Luria, M. H., Johnson, M. W., Pego, R., Seuc, C. A., Manubens, S. J., Wieland, M. R., & Wieland, R. G. (1982). Relationship between sex hormones, myocardial infarction, and occlusive coronary disease. *Archives of Internal Medicine, 142*, 42-44.

[43] Kajinami, K., Takeda, K., Takekoshi, N., Matsui, S., Tsugawa, H., Kanemitsu, S., ... Fukuda, A. (2004). Imbalance of sex hormone levels in men with coronary artery disease. *Coronary Artery Disease, 15*, 199-203.

[44] Barrett-Connor, E., & Khaw, K. T. (1988). Endogenous sex hormones and cardiovascular disease in men. A prospective population-based study. *Circulation, 78*, 539-545.

[45] Mikulec, K. H., Holloway, L., Krasnow, R. E., Javitz, H., Swan, G. E., Reed, T., ... Carmelli, D. (2004). Relationship of endogenous sex hormones to coronary heart disease: A twin study. *Journal of Clinical Endocrinology & Metabolism, 89*, 1240-1245.

[46] Yeap, B. B., Hyde, Z., Almeida, O. P., Norman, P. E., Chubb, S. A., Jamrozik, K., ... Hankey, G. J. (2009). Lower testosterone levels predict incident stroke and transient ischemic attack in older men. *Journal of Clinical Endocrinology & Metabolism, 94*, 2353-2359.

[47] De Pergola, G., Pannacciulli, N., Ciccone, M., Tartagni, M., Rizzon, P., & Giorgino, R. (2003). Free testosterone plasma levels are negatively associated with the intima-media thickness of the common carotid artery in overweight and obese glucose-tolerant young adult men. *International Journal of Obesity and Related Metabolic Disorders, 27*, 803-807.

[48] Fukui, M., Kitagawa, Y., Nakamura, N., Kadono, M., Mogami, S., Hirata, C., ... Yoshikawa, T. (2003). Association between serum testosterone concentration and carotid atherosclerosis in men with type 2 diabetes. *Diabetes Care, 26*, 1869-1873.

[49] Tsujimura, A., Yamamoto, R., Okuda, H., Yamamoto, K., Fukuhara, S., Yoshioka, I., ... Nonomura, N. (2012). Low serum free testosterone level is associated with carotid intima-media thickness in middle-aged Japanese men. *Endocrine Journal, 59*, 809-815.

[50] Aversa, A., Bruzziches, R., Francomano, D., Rosano, G., Isidori, A. M., Lenzi, A., & Spera, G. (2010). Effects of testosterone undecanoate on cardiovascular risk factors and atherosclerosis in middle-aged men with late-onset hypogonadism and metabolic syndrome: Results from a 24-month, randomized, double-blind, placebo-controlled study. *Journal of Sexual Medicine, 7*, 3495-3503.

[51] Abbott, R. D., Launer, L. J., Rodriguez, B. L., Ross, G. W., Wilson, P. W., Masaki, K. H., ... Petrovitch, H. (2007). Serum estradiol and risk of stroke in elderly men. *Neurology, 68*, 563-568.

[52] Calderon-Margalit, R., Schwartz, S. M., Wellons, M. F., Lewis, C. E., Daviglus, M. L., Schreiner, P. J., ... Siscovick, D. S. (2010). Prospective association of serum androgens and sex hormone-binding globulin with subclinical cardiovascular disease in young adult women: The "Coronary Artery Risk Development in Young Adults" women's study. *Journal of Clinical Endocrinology & Metabolism, 95*, 4424-4431.

[53] El Khoudary, S. R., Wildman, R. P., Matthews, K., Thurston, R. C., Bromberger, J. T., & Sutton-Tyrrell, K. (2012). Endogenous sex hormones impact the progression of subclinical atherosclerosis in women during the menopausal transition. *Atherosclerosis, 225*, 180-186.

[54] Yeap, B. B., Hyde, Z., Norman, P. E., Chubb, S. A., & Golledge, J. (2010). Associations of total testosterone, sex hormone-binding globulin, calculated free testosterone, and luteinizing hormone with prevalence of abdominal aortic aneurysm in older men. *Journal of Clinical Endocrinology & Metabolism, 95*, 1123-1130.

[55] Vaidya, D., Dobs, A., Gapstur, S. M., Golden, S. H., Cushman, M., Liu, K., & Ouyang, P. (2012). Association of baseline sex hormone levels with baseline and longitudinal changes in waist-to-hip ratio: Multi-ethnic study of atherosclerosis. *International Journal of Obesity, 36*, 1578-1584.

[56] Tivesten, A., Mellstrom, D., Jutberger, H., Fagerberg, B., Lernfelt, B., Orwoll, E., ... Ohlsson, C. (2007). Low serum testosterone and high serum estradiol associate with lower extremity peripheral arterial disease in elderly men. The MrOS Study in Sweden. *Journal of the American College of Cardiology, 50*, 1070-1076.

[57] Yeap, B. B., Alfonso, H., Chubb, S. A., Handelsman, D. J., Hankey, G. J., Golledge, J., ... Norman, P. E. (2013). Lower plasma testosterone or dihydrotestosterone, but not estradiol, is associated with symptoms of intermittent claudication in older men. *Clinical Endocrinology, 79*(5), 725-732.

[58] Maggio, M., Cattabiani, C., Lauretani, F., Artoni, A., Bandinelli, S., Schiavi, G., ... Ceda, G. P. (2012). The relationship between sex hormones, sex hormone binding globulin and peripheral artery disease in older persons. *Atherosclerosis, 225*, 469-474.

[59] Haring, R., Travison, T. G., Bhasin, S., Vasan, R. S., Wallaschofski, H., Davda, M. N., ... Murabito, J. M. (2011). Relation between sex hormone concentrations, peripheral arterial disease, and change in ankle-brachial index: findings from the Framingham Heart Study. *Journal of Clinical Endocrinology & Metabolism, 96*, 3724-3732.

[60] Price, J., & Leng, G. C. (2012). Steroid sex hormones for lower limb atherosclerosis. *Cochrane Database of Systematic Reviews,* (10), CD000188.

[61] Corona, G., Monami, M., Rastrelli, G., Aversa, A., Tishova, Y., Saad, F., ... Maggi, M. (2011). Testosterone and metabolic syndrome: A meta-analysis study. *Journal of Sexual Medicine, 8*, 272-283.

[62] Chrysohoou, C., Panagiotakos, D., Pitsavos, C., Siasos, G., Oikonomou, E., Varlas, J., ... Stefanadis, C. (2013). Low total testosterone levels are associated with the metabolic syndrome in elderly men: The role of body weight, lipids, insulin resistance, and inflammation; the Ikaria study. *Review of Diabetic Studies, 10*, 27-38.

[63] Zmuda, J. M., Cauley, J. A., Kriska, A., Glynn, N. W., Gutai, J. P., & Kuller, L. H. (1997). Longitudinal relation between endogenous testosterone and cardiovascular disease risk factors in middle-aged men. A 13-year follow-up of former Multiple Risk Factor Intervention Trial participants. *American Journal of Epidemiology, 146*, 609-617.

[64] Colangelo, L. A., Ouyang, P., Liu, K., Kopp, P., Golden, S. H., Dobs, A. S., ... Gapstur, S. M. (2009). Association of endogenous sex hormones with diabetes and impaired fasting glucose in men: Multi-ethnic study of atherosclerosis. *Diabetes Care, 32*, 1049-1051.

[65] Davoodi, G., Amirezadegan, A., Borumand, M. A., Dehkori, M. R., Kazemisaeid, A., & Yaminisharif, A. (2007). The relationship between level of androgenic hormones and coronary artery disease in men. *Cardiovascular Journal of Africa, 18*, 362-366.

[66] Kupelian, V., Hayes, F. J., Link, C. L., Rosen, R., & McKinlay, J. B. (2008). Inverse association of testosterone and the metabolic syndrome in men is consistent across race and ethnic groups. *Journal of Clininical Endocrinology & Metabolism, 93*, 3403-3410.

[67] Jaffe, M. D. (1977). Effect of testosterone cypionate on postexercise ST segment depression. *British Heart Journal, 39*, 1217-1222.

[68] Cornoldi, A., Caminiti, G., Marazzi, G., Vitale, C., Patrizi, R., Volterrani, M., ... Rosano, G. (2010). Effects of chronic testosterone administration on myocardial ischemia, lipid metabolism and insulin resistance in elderly male diabetic patients with coronary artery disease. *International Journal of Cardiology, 142*, 50-55.

[69] Thompson, P. D., Ahlberg, A. W., Moyna, N. M., Duncan, B., Ferraro-Borgida, M., White, C. M., ... Heller, G. V. (2002). Effect of intravenous testosterone on myocardial ischemia in men with coronary artery disease. *American Heart Journal, 143*, 249-256.

[70] Malkin, C. J., Pugh, P. J., West, J. N., van Beek, E. J., Jones, T. H., & Channer, K. S. (2006). Testosterone therapy in men with moderate severity heart failure: A double-blind randomized placebo controlled trial. *European Heart Journal, 27*, 57-64.

[71] Toma, M., McAlister, F. A., Coglianese, E. E., Vidi, V., Vasaiwala, S., Bakal, J. A., ... Ezekowitz, J. A. (2012). Testosterone supplementation in heart failure: A meta-analysis. *Circulation. Heart Failure, 5*, 315-321.

[72] Iellamo, F., Volterrani, M., Caminiti, G., Karam, R., Massaro, R., Fini, M., ... Rosano, G. M. (2010). Testosterone therapy in women with chronic heart failure: A pilot double-blind, randomized, placebo-controlled study. *Journal of the American College of Cardiology, 56*, 1310-1316.

[73] D'Andrea, A., Caso, P., Salerno, G., Scarafile, R., De Corato, G., Mita, C., ... Calabro, R. (2007). Left ventricular early myocardial dysfunction after chronic misuse of anabolic androgenic steroids: A Doppler myocardial and strain imaging analysis. *British Journal of Sports Medicine, 41*, 149-155.

[74] Karila, T. A., Karjalainen, J. E., Mantysaari, M. J., Viitasalo, M. T., & Seppala, T. A. (2003). Anabolic androgenic steroids produce dose-dependant increase in left ventricular mass in power atheletes, and this effect is potentiated by concomitant use of growth hormone. *International Journal of Sports Medicine, 24*, 337-343.

[75] Johannsson, G., Gibney, J., Wolthers, T., Leung, K. C., & Ho, K. K. (2005). Independent and combined effects of testosterone and growth hormone on extracellular water in hypopituitary men. *Journal of Clinical Endocrinology & Metabolism, 90*, 3989-3994.

[76] Shahani, S., Braga-Basaria, M., Maggio, M., & Basaria, S. (2009). Androgens and erythropoiesis: Past and present. *Journal of Endocrinological Investigation, 32*, 704-716.

[77] Coviello, A. D., Kaplan, B., Lakshman, K. M., Chen, T., Singh, A. B., & Bhasin, S. (2008). Effects of graded doses of testosterone on erythropoiesis in healthy young and older men. *Journal of Clinical Endocrinology & Metabolism, 93*, 914-919.

[78] Calof, O. M., Singh, A. B., Lee, M. L., Kenny, A. M., Urban, R. J., Tenover, J. L., & Bhasin, S. (2005). Adverse events associated with testosterone replacement in middle-aged and older men: A meta-analysis of randomized, placebo-controlled trials. *Journals of Gerontology. Series A, Biological Sciences and Medical Sciences, 60*, 1451-1457.

[79] Maggio, M., Snyder, P. J., Ceda, G. P., Milaneschi, Y., Luci, M., Cattabiani, C., ... Ferrucci, L. (2013). Is the haematopoietic effect of testosterone mediated by erythropoietin? The results of a clinical trial in older men. *Andrology, 1*, 24-28.

[80] Bachman, E., Travison, T. G., Basaria, S., Davda, M. N., Guo, W., Li, M., ... Bhasin, S. (2013). Testosterone induces erythrocytosis via increased erythropoietin and suppressed hepcidin: Evidence for a new erythropoietin/hemoglobin set point. *Journals of Gerontology. Series A, Biological Sciences and Medical Sciences*.

[81] Basaria, S., Coviello, A. D., Travison, T. G., Storer, T. W., Farwell, W. R., Jette, A. M., ... Bhasin, S. (2010). Adverse events associated with testosterone administration. *New England Journal of Medicine, 363*, 109-122.

[82] Shores, M. M., Smith, N. L., Forsberg, C. W., Anawalt, B. D., & Matsumoto, A. M. (2012). Testosterone treatment and mortality in men with low testosterone levels. Journal of Clinical Endocrinology & Metabolism, 97, 2050-2058.

[83] Vigen, R., O'Donnell, C. I., Baron, A. E., Grunwald, G. K., Maddox, T. M., Bradley, S. M., ... Ho, P. M. (2013). Association of testosterone therapy with mortality, myocardial infarction, and stroke in men with low testosterone levels. Journal of the American Medical Association, 310, 1829-1836.

[84] Finkle WD, Greenland S, Ridgeway GK, Adams JL, Frasco MA, Cook MB, Fraumeni JF Jr, Hoover RN. Increased risk of non-fatal myocardial infarction following testosterone therapy prescriptionin men. PLoS One. 2014; 9(1):e85805.

[85] Su JJ, Park SK, Hsieh TM. The effect of testosterone on cardiovascular disease: a critical review of the literature. Am J Mens Health. 2014; 8(6):470-91.

[86] Fu, L., Gao, Q. P., & Shen, J. X. (2008). Relationship between testosterone and indexes indicating endothelial function in male coronary heart disease patients. *Asian Journal of Andrology, 10*, 214-218.

[87] Foresta, C., De Toni, L., Selice, R., Garolla, A., & Di Mambro, A. (2010). Increased osteocalcin-positive endothelial progenitor cells in hypogonadal male patients. *Journal of Endocrinological Investigation, 33*, 439-442.

[88] Nheu, L., Nazareth, L., Xu, G. Y., Xiao, F. Y., Luo, R. Z., Komesaroff, P., & Ling, S. (2011). Physiological effects of androgens on human vascular endothelial and smooth muscle cells in culture. *Steroids, 76*, 1590-1596.

[89] Haas, M. J., Raheja, P., Jaimungal, S., Sheikh-Ali, M., & Mooradian, A. D. (2012). Estrogen-dependent inhibition of dextrose-induced endoplasmic reticulum stress and superoxide generation in endothelial cells. *Free Radical Biology & Medicine, 52*, 2161-2167.

[90] Yue, P., Chatterjee, K., Beale, C., Poole-Wilson, P. A., & Collins, P. (1995). Testosterone relaxes rabbit coronary arteries and aorta. *Circulation, 91*, 1154-1160.

[91] Seyrek, M., Yildiz, O., Ulusoy, H. B., & Yildirim, V. (2007). Testosterone relaxes isolated human radial artery by potassium channel opening action. *Journal of Pharmacological Sciences, 103*, 309-316.

[92] Yildiz, O., Seyrek, M., Gul, H., Un, I., Yildirim, V., Ozal, E., ... Bolu, E. (2005). Testosterone relaxes human internal mammary artery in vitro. *Journal of Cardiovascular Pharmacology, 45*, 580-585.

[93] Scragg, J. L., Jones, R. D., Channer, K. S., Jones, T. H., & Peers, C. (2004). Testosterone is a potent inhibitor of L-type Ca(2+) channels. *Biochemical and Biophysical Research Communications, 318*, 503-506.

[94] Kang, S. M., Jang, Y., Kim, J., Chung, N., Cho, S. Y., Chae, J. S., & Lee, J. H. (2002). Effect of oral administration of testosterone on brachial arterial vasoreactivity in men with coronary artery disease. *American Journal of Cardiology, 89*, 862-864.

[95] Laurent, S., Tropeano, A. I., & Boutouyrie, P. (2006). Pulse pressure reduction and cardiovascular protection. *Journal of Hypertension, 24*(3 Suppl.), S13-18

[96] Corona, G., Mannucci, E., Lotti, F., Fisher, A. D., Bandini, E., Balercia, G., ... Maggi, M. (2009). Pulse pressure, an index of arterial stiffness, is associated with androgen deficiency and impaired penile blood flow in men with ED. *Journal of Sexual Medicine, 6*, 285-293.

[97] Fukui, M., Ose, H., Kitagawa, Y., Yamazaki, M., Hasegawa, G., Yoshikawa, T., & Nakamura, N. (2007). Relationship between low serum endogenous androgen concentrations and arterial stiffness in men with type 2 diabetes mellitus. *Metabolism, 56*, 1167-1173.

[98] Kyriazis, J., Tzanakis, I., Stylianou, K., Katsipi, I., Moisiadis, D., Papadaki, A., ... Daphnis, E. (2011). Low serum testosterone, arterial stiffness and mortality in male haemodialysis patients. *Nephrology, Dialysis, Transplantation, 26*, 2971-2977.

[99] Webb, C. M., Elkington, A. G., Kraidly, M. M., Keenan, N., Pennell, D. J., & Collins, P. (2008). Effects of oral testosterone treatment on myocardial perfusion and vascular function in men with low plasma testosterone and coronary heart disease. *American Journal of Cardiology, 101*, 618-624.

[100] Libby, P., Ridker, P. M., & Maseri, A. (2002). Inflammation and atherosclerosis. *Circulation, 105*, 1135-1143.

[101] Bobjer, J., Katrinaki, M., Tsatsanis, C., Giwercman, Y. L., & Giwercman, A. (2013). Negative association between testosterone concentration and inflammatory markers in young men: A nested cross-sectional study. *PLoS One, 8*(4), e61466.

[102] Guler, N., Batyraliev, T., Dulger, H., Ozkara, C., Tuncer, M., Aslan, S., ... Agirbasli, M. (2006). The effects of short term (3 weeks) testosterone treatment on serum inflammatory markers in men undergoing coronary artery stenting. *International Journal of Cardiology, 109*, 339-343.

[103] Malkin, C. J., Pugh, P. J., Morris, P. D., Kerry, K. E., Jones, R. D., Jones, T. H., & Channer, K. S. (2004). Testosterone replacement in hypogonadal men with angina improves ischaemic threshold and quality of life. *Heart, 90*, 871-876.

[104] Corcoran, M. P., Meydani, M., Lichtenstein, A. H., Schaefer, E. J., Dillard, A., & Lamon-Fava, S. (2010). Sex hormone modulation of proinflammatory cytokine and C-reactive protein expression in macrophages from older men and postmenopausal women. *Journal of Endocrinology, 206*, 217-224.

[105] Hernandez-Mijares, A., Garcia-Malpartida, K., Sola-Izquierdo, E., Banuls, C., Rocha, M., Gomez-Martinez, M. J., ... Victor, V. M. (2010). Testosterone levels in males with type 2 diabetes and their relationship with cardiovascular risk factors and cardiovascular disease. *Journal of Sexual Medicine, 7*, 1954-1964.

[106] Danesh, J., Lewington, S., Thompson, S. G., Lowe, G. D., Collins, R., Kostis, J. B., ... Wood, A. (2005). Plasma fibrinogen level and the risk of major cardiovascular diseases and nonvascular mortality: An individual participant meta-analysis. *Journal of the American Medical Association, 294*, 1799-1809.

[107] Kaptoge, S., White, I. R., Thompson, S. G., Wood, A. M., Lewington, S., Lowe, G. D., & Danesh, J. (2007). Associations of plasma fibrinogen levels with established cardiovascular disease risk factors, inflammatory markers, and other characteristics: Individual participant meta-analysis of 154,211 adults in 31 prospective studies: The fibrinogen studies collaboration. *American Journal of Epidemiology, 166*, 867-879.

[108] Ziaran, S., Goncalves, F. M., & Breza, J., Sr. (2013). Patients with prostate cancer treated by ADT have significantly higher fibrinogenemia than healthy control. *World Journal of Urology, 31*, 289-292.

[109] Ajayi, A. A., & Halushka, P. V. (2005). Castration reduces platelet thromboxane A2 receptor density and aggregability. *QJM: Monthly Journal of the Association of Physicians, 98*, 349-356.

[110] Harman, S. M., Metter, E. J., Tobin, J. D., Pearson, J., & Blackman, M. R. (2001). Longitudinal effects of aging on serum total and free testosterone levels in healthy men. Baltimore Longitudinal Study of Aging. *Journal of Clinical Endocrinology & Metabolism, 86*, 724-731.

In: Neurobiology of Men's Mental Health
Editors: Leo Sher and Timothy R. Rice
ISBN: 978-1-63483-191-8
© 2015 Nova Science Publishers, Inc.

Chapter 24

STEROID SECRETION AND PSYCHOLOGICAL WELL-BEING IN MEN 40+

Andreas Walther*, MSc,
and Ulrike Ehlert, Prof. Dr. rer. nat.
Clinical Psychology and Psychotherapy, University of Zurich,
Zurich, Switzerland
University Research Priority Program, Dynamics of Healthy Aging

ABSTRACT

Current demographic statistics show that modern societies are rapidly aging. Aging in general is accompanied by an increased susceptibility to physical and mental disease as well as an increased risk of mortality. Health care costs are expected to rise enormously. Therefore, successful aging with less disease burden and dependency is emerging as one of the most crucial health care goals of the upcoming decades. Over the last 20 years, research on successful aging in men has increasingly focused on age-related hormonal changes. A gradual reduction in testosterone and dehydroepiandrosterone beginning around the age of 40 has been well documented. At the same age, other steroid hormones including estradiol and cortisol also show an age-dependent progressive change. These changing patterns of steroid hormone secretion play an essential role for general health and psychological well-being in males over 40.

In the following chapter, we will discuss the underlying biological mechanisms of age-related hormonal alterations and elucidate their influence in the context of psychological, sexual, cognitive and physical areas of life in aging men. Considering these widespread consequences due to age-related hormonal changes, we will also discuss the clinical implications for men facing these conditions. Apparent good health or specific lifestyle factors were shown to slow down age-related hormonal changes. Since clinical conditions cannot be prevented for every male by maintaining good health or establishing a healthier lifestyle, hormone replacement represents a reasonable opportunity to overcome certain psychological, sexual, cognitive and physical

* Corresponding author: M.Sc. Andreas Walther, University of Zurich, Department of Psychology, Clinical Psychology and Psychotherapy, Binzmühlestrasse 14 / Box 26, CH-8050 Zurich, Switzerland, Telephone: +41 (0) 44 635 73 53, a.walther@psychologie.uzh.ch

impairments caused by age-related steroid alterations. Besides individual psychological support for the successful management of severe symptoms of age-related hormonal changes, targeted psychoeducation at a population-based level would, in the long term, be one possibility to inform men about their options and to support them in countering these largely neglected disorders.

INTRODUCTION

Current demographic statistics show that modern societies are rapidly aging (Revision of the official United Nations population estimates and projections from 2012). Aging in general is accompanied by an increased susceptibility to physical and mental disease, as well as an increased risk of mortality [1-4]. As a result, health care costs are expected to rise enormously.

An age-related increase in mild to moderate depressive symptoms has been documented for males up to the age of 80 [4-7]. For example, from the 50-64 age group to the over-65 age group, there is an 11.1% rise in mild depressive symptoms [4]. In parallel to this trend, the epidemiology of suicide rates also shows a steady increase from young men to men up to the age of over 75 [8, 9]. Moreover, there is a strong correlation between aging and increased sexual dysfunction. Men aged 70-80 years are nearly seven times more likely to experience erection difficulties than men aged 50-59 years [10, 11]. In addition, the diminishment of cognitive abilities increases during the aging process [12]. Cognitive skills such as speed of processing, working memory, and long-term memory are associated with a continuous decline with age [12]. Furthermore, age-related bodily changes such as an increase in fat mass, a decrease in muscle mass and a loss of bone mineral density typically start to occur around the age of 40 [13-15]. These changes lead to a higher general frailty and increased risk of bone breakage after falls [16]. As a consequence of this combination of psychological, sexual, cognitive and physical changes, one of the greatest fears of aging people is the loss of the ability to live independently [17].

Therefore, successful aging with less disease burden and dependency is emerging as one of the most crucial health care goals of the upcoming decades [18]. Over the last 20 years, research on successful aging in men has increasingly focused on age-related hormonal changes and their consequences for physical and mental health. A gradual reduction in testosterone (T) starting around the age of 40 years, often referred to "andropause" or "late onset hypogonadism," has been well documented [19, 20]. *Andropause* defines a syndrome that requires the presence of both clinical symptoms and biochemical conditions. A decrease in sexual satisfaction or a decline in the general feeling of well-being, along with diminished serum levels of total (tT) and free T (fT), constitutes the andropause syndrome in older men [21]. *Late onset hypogonadism* is more strongly related to sexual functioning. It is defined similarly, in terms of tT serum levels below 11nmol/l, a fT level below 220pmol/l and the following three sexual symptoms: decreased frequency of morning erection, decreased frequency of sexual thoughts, and erectile dysfunction [22]. In the following, the term andropause will be used for the age-related continuous drop in T and the associated sexual symptomatology.

Although research has shown that andropause and late onset hypogonadism are strongly related to the decline in the marker T, it is important not to neglect other relevant steroid changes and their comparable impact on the aforementioned areas such as psychological well-being, sexual health, cognition and physical health. In the following, we will therefore provide an overview of how these different areas of life are affected by changes in steroid hormones over the course of the lifetime, and will derive the clinical implications for men facing such conditions. Furthermore, an overview of the efficiency and safety of hormone replacement protocols for specific psychological, sexual, cognitive and physical conditions will be given. Finally, we will discuss the need for systematic psychological support for the successful management of severe symptoms of andropause and the consequences of the accompanying changes in steroid secretion.

Hormonal Changes in Men Aged over 40 Years

Testosterone (T) is the major male sex hormone and affects various biological systems and psychological dimensions. T is mainly produced in the Leydig cells of the testes and most of the circulating T is bound to sex hormone-binding globulin (SHBG) (44%), albumin (50%) or cortisol-binding globulin (4%) [23]. Only 2% of the circulating T remains free. Active or bioavailable T (bT) consists of free and albumin-bound T (52%) [24]. In 1991, Gray and colleagues conducted a meta-analysis indicating that healthy men have higher overall T levels than ill men [25]. Furthermore, for healthy men, a continuous T decline related to age has been reported, with an annual decline of 0.4% for tT and 1.2% for fT. Data from 250 healthy non-obese and 50 obese men confirmed the age-related T decline, even after controlling for body mass index (BMI), smoking or interactions with other hormones [26]. While these studies were based on cross-sectional data, the Baltimore Longitudinal Study of Aging and the Massachusetts Male Aging Study (MMAS) were the first studies to report an age-related T decline for longitudinal data [19, 20]. Besides baseline, follow-up was assessed in the MMAS seven to ten years later for 1156 healthy men aged 40 to 70. Feldman and colleagues [20] reported an annual decline of 1.6% for tT and of 2-3% for fT, resulting in a 30% loss of tT and 40-60% loss of fT for men over the age of 60 years. This more pronounced decline in fT is due to a parallel increase in sex hormone-binding globulin (SHBG) levels, lying at 1.6% per year [19, 20].

Dehydroepiandrosterone (DHEA) is synthesized by the adrenal glands, and works as a precursor hormone of different androgens and estrogens [24]. DHEA affects different body systems (e.g., immunity, musculoskeletal integrity, and cardiovascular health) and is assumed to be anti-ageing [27]. A gradual decline in DHEA with age has been well documented [20, 28-31]. Only 10-20% of the amount of DHEA found in young individuals remains in elderly people aged 70-80 [32-34]. This decline seems to be associated with an increased risk of osteoporosis, atherosclerosis and decreased immune function [32].

Estradiol (E2) is the main biologically active estrogen in the human hormone system. While in men, only about 20% of E2 is produced by the Leydig cells, most of the circulating E2 is metabolized via the aromatase enzyme in peripheral tissues from androgens and mainly from T [35]. There is an age-related increase in aromatase activity [36] and an age-related increase in subcutaneous fat [37], which is a main site of aromatization of T to E2 [38]. Thus, when assessing age effects on E2 in men, fat mass or body mass index (BMI) should be controlled for. While in one study in 810 men aged 24-90 years, bioavailable E2 (bE2) was shown to decrease independently, total E2 (tE2) levels decreased with age only when controlling for confounders (weight and BMI) [39]. Additionally, an age-related decline in bE2 levels was reported in 419 men aged 24-79 years, but no such decline was found for tE2 levels (even after adjusting for fat mass) [35, 40]. In 403 healthy men aged between 73 and 94, Van den Beld and colleagues [41] reported that both tE2 and free E2 (fE2) levels decreased with increasing age. These findings were confirmed by Orwoll and colleagues [42]. Recent studies using modern hormone analysis methods have shown an age-related decline for tE2 [43, 44] and fE2 levels [45]. However, Kozloski and colleagues [46] found no age effect for salivary fE2, and an age-related increase in tE2 has even been reported [45]. To summarize, so far, the literature shows inconsistent findings for E2 level changes in aging men. Two of the major reasons for these inconsistencies are the low levels of circulating E2 in males and the differing sensitivity of the assays used to analyze different body fluids for the hormonal level.

Table 1. Age-related changes in steroid and SHBG concentrations in men

Hormone	Change across the lifespan	Change in %	References
Testosterone (T)	↓	Decline in tT: 1.6% per year Decline in fT: 2-3% per year	Feldman et al., 2002; Gray et al., 1991; Harman et al., 2001
Estradiol (E2)	(↓)	Moderate age-related decline	Ferrini & Barrett-Connor, 1998 Vermeulen et al., 2002
Dehydroepi-androsterone (DHEA)	↓	Decline of 80-90% in DHEA in 70-80-year old-men compared to young individuals	Genazzani et al., 2007; Parker, 1999
Cortisol (C)	↑	Increase in mean C levels by 20-50% between 20-80 years of age	Karlamangla et al., 2013; Kern et al., 1996; Van Cauter et al., 1996; Van Cauter et al., 2000
Progesterone (P)	≈	Remains unchanged	Kozloski et al., 2014; Oettel & Mukhopadhyay, 2004
Sex hormone-binding globulin (SHBG)	↑	Increase of 1.6% per year after the age of 40	Feldman et al., 2002; Harman et al., 2001

Declining concentrations are indicated by ↓; increasing concentrations are indicated by ↑; no change is indicated by ≈; moderate changes are represented in brackets (e.g., (↓)).

Cortisol (C) is a primary agent of the neuroendocrine response to stress as well as an important factor in regulatory processes of the immune system, metabolic activity or cognition [47]. In the MMAS, no age-related change in C levels was observed [20], probably due to varying time points of blood sampling within four hours after awakening. Van Cauter and colleagues [48] reported an increase in mean plasma C levels of 20-50% between the ages of 20 and 80 years, and replicated these findings for 24-hour plasma cortisol profiles [49] and samples drawn during nocturnal sleep [50]. In a recent study with controlled sampling time points, an increase in the salivary C levels with age was reported for 1693 subjects, while older age and male gender were independently associated with higher cortisol peak, nadir, and area under the curve (AUC) [51]. Similar findings for hair cortisol measurements were reported in a study of 654 participants aged 47-82, where hair cortisol concentrations (HCC) increased with age and were higher for males than for females [52].

Progesterone (P), an important precursor of different steroid hormones, is mainly produced by men in the Leydig cells and to a smaller extent in the adrenals. P also exerts effects on several organ systems [53]. In the Luric-Jenapharm study, serum P levels of 1015 men aged 20-80 were measured cross-sectionally, and no age-related changes were found [53]. The National Social Life, Health, and Aging Project, a US-longitudinal study, reported no age-related P decline for 1220 men in wave one and for1325 men in wave two [46]. The age-related changes in steroid concentrations in men are summarized in table 1.

Biological Mechanisms Underlying Age-Related Hormonal Alterations

Recent literature favors two explanations for the biological causes of the reported T and E2 decline in aging males, which will be summarized in the following. First, an attrition of the T and E2-producing testicular Leydig cells and a diminished secretory capacity occurs as a function of age [54]. While a healthy man at the age of 20 is provided with more than 700 million Leydig cells, a reduction rate of 80 million per decade of life leads to an amount of 300 million Leydig cells at the age of 70 [55, 56]. Along with this degeneration and dissolution of Leydig cells, a decreased secretory capacity of T and E2 for elderly compared to young men has been reported from various stimulation tests [57-60].

Second, at the hypothalamic-pituitary level, age-related changes in the amplitude and frequency of gonadotropin-releasing hormone (GnRH) pulses [61] and subsequent luteinizing hormone (LH) and follicle stimulating hormone (FSH) secretion occur in men [62, 63]. Leydig cell function, including T secretion, is regulated through LH [64]. While an age-related decrease in LH pulse amplitudes is observed, the pulse frequency seems to increase [65-68]. One explanation seems to be the diminished GnRH release with age provoked by the hypothalamic GnRH pulse generator [62, 63]. This assumption is supported by the fact that stimulation with exogenous GnRH shows no difference in the subsequent LH secretion in young compared to old men, indicating equal pituitary responsiveness [64]. The paradoxical findings of an age-related decrease in LH pulse amplitudes and consistently elevated mean LH levels in the elderly [20] can be explained by the observation of increased plasma half-life of LH and increased LH pulse frequency in elderly men [63, 64, 68, 69]. Taken together, these age-related biological mechanisms lead to markedly lower T and slightly lower E2 levels in older compared to younger men.

Although T is the major source (about 80%) of plasma E2, due to its conversion via the aromatase enzyme, the age-related T decline is only slightly depicted in plasma E2 levels. The aromatase activity and the body fat mass, which is a main site of aromatization of T to E2, increase with age [36, 37], and both physiological conditions increase T conversion to E2 in older men. Therefore, a smaller decrease in E2 compared to T in elderly men is observed [35, 70].

The underlying biological mechanism which leads to a continuous DHEA decline is not fully understood [32]. Since steroids produced in the adrenals such as C do not decline [49, 51], the DHEA decline seems to be related to changes in the zona reticularis, which is the inner layer of the adrenals and the production site of DHEA and its sulfate (DHEAS) [32]. There are several potential causes which may explain this age-related reduction, such as a reduced number of LDL receptors, which are responsible for the import of plasma cholesterol, reduced hydroxymethylglutaryl coenzyme A reductase, which regulates cholesterol biosynthesis, reduced levels of ACTH receptors, or impaired signal transduction [32].

In line with these tissue-specific age-related alterations, a morphological reduction of the entire zona reticularis in relation to the other layers (zona fasciculata and glomerulosa) in aging men has been documented in some studies [71, 72], while others did not find such a decrease [73].

Van Cauter and colleagues [74] termed the elevation of basal C levels a hallmark of aging. An age-related C increase is due to different alterations of the hypothalamic-pituitary-adrenal (HPA) axis [75]. This has been demonstrated in several animal studies: In male rat brains, a reduction of corticosteroid receptors of type I (mineralocorticoid receptor) and of type II (glucocorticoid receptor) has been found as a function of age [76]. In the male rat, mRNA and protein levels of both receptors decrease with age. Furthermore, an increased secretion of corticotropin-releasing hormone (CRH) from the hypothalamus, along with a functional decrement in corticosteroid negative feedback, has been reported [76]. In humans, both an age-related elevation of the number of CRH neurons and subsequently in the CRH secretion has been observed at the hypothalamic level [77]. Additionally, an age-related increase in adrenocorticotropin-releasing hormone (ACTH) and C secretion has been documented [52, 78-80], but has not been confirmed by all authors [e.g., 81]. Overall, the findings indicate an age-related hyperactive basal HPA axis regulation [80], while at the hypothalamic level, a decreased sensitivity for the negative feedback loop seems to maintain an elevated output of the initial releasing hormone CRH. In addition to the aforementioned central nervous system (CNS) modifications, another possible cause of higher C levels in the elderly might be age-related slower metabolic clearance rates of CRH, ACTH or C [79], or lower ACTH efficacy in older men [82].

To date, no age-related P decrease in males has been observed, although the main production site of P, the Leydig cells in the testes, experiences a continuous decline with age [55]. This indicates that P production in men can be maintained into old age, even though the number of Leydig cells strongly decreases. Whether this is due to compensatory processes of other body systems or a specific protection mechanism for P secretion remains unclear.

CONSEQUENCES OF AGE-RELATED HORMONAL CHANGES

Psychological Consequences: Well-being and Depression

Age-related changes in steroid hormones have a strong relationship with psychological well-being and depression.

As T levels decline in aging men, and mild to moderate depressive symptoms increase as a function of age, a variety of studies investigated the relationship between T and depression. In the Rancho Bernado study, which comprised 856 men aged between 50 and 89 years, higher depression scores with lower bT levels were found [83]. Seidman and colleagues [84] reported a relationship between low T levels and elderly dysthymic men. A cross-sectional study in 3987 men aged between 70 and 89 years showed that after adjusting for age and comorbidities, men in the lowest quintile of fT had a 2.7-fold increased risk of developing depression compared to men in the highest quintile [85]. In a longitudinal observation with a 2-year follow-up, the incidence of developing depression in men aged 45 or older with repeated low T levels was three times higher (21.7%) than in men with normal T levels (7.1%) [86]. A longitudinal study with a three-year follow-up replicated these findings in a sample of 608 men aged 65 years or older, since very low fT levels (<170 pmol/l) were associated with depressive symptoms, and low fT levels (<220 pmol/l) were predictive of the onset of depressive symptoms [87]. Furthermore, a simultaneous assessment of the CAG repeat length polymorphism in the androgen receptor encoding gene site, tT levels and self-reported depression showed that low T levels were associated with depression in men with the shorter allele only [88]. Although there is a large body of literature demonstrating an influence of T on well-being and depression in men, Amiaz and Seidman [89] conclude after reviewing the literature that there is no consistent relationship between T and mood. To summarize, recent findings indicate that in subgroups of men, an association between low T levels and depression is more pronounced. Treatment-resistant depressive men, men with major depression and HIV infection, hypogonadal men, dysthymic men, and elderly men (> 60 years) are at risk of lower T levels [7].

Due to the direct conversion of T to E2 via the enzymatic activity of aromatase, it is more difficult to study E2-induced effects exclusively. Suppression studies using aromatase-blockade for diagnostic purposes are extensive, and to a certain extent critical for human subjects. Although clinical evidence for the anti-depressant effect of estrogens in women is well documented [90, 91], comparable studies in men are rare. Almeida and colleagues [92] reported a significant decline in self-reported mood in men with prostate cancer following androgen blockade therapy (depletion of T and E2 levels). In 60 hypogonadal men (serum T levels $\leq$ 300 ng/dL) aged 60 or older, higher endogenous serum E2 levels were independently associated with a greater sense of well-being [93]. A study of 120 older community dwellers failed to report significant findings for the influence of E2 on mood or well-being, although there was a positive trend for self-reported quality of life [94]. Taken together, only a small number of results are available to interpret the effect of E2 on mood in men, but there is a tendency that higher E2 levels in elderly men seem to be associated with better well-being and quality of life. Nevertheless, there is substantial literature which stands in conflict with this finding [e.g., 83, 95].

Due to the widespread effects of DHEA, various studies have been conducted to elucidate the influence of DHEA on well-being and depression. A review by Davis and colleagues [96] suggested a positive relationship between DHEA and psychological well-being especially in elderly people. In depressed subjects, significantly lower levels of salivary DHEA levels have been found compared to non-depressed subjects [97]. A recent study in 170 healthy adults divided into the three age groups young (18–30), middle-aged (31–45) and old (>46) reported parallel findings of an age-related decline in DHEAS and an increase in depression, daily hassles and stress scores [98]. A study examining 41 healthy men who were enrolled in a military survival school showed a positive relationship between DHEA and DHEAS levels and stress tolerance and superior military performance, and a negative relationship with dissociative symptoms [99]. In addition, Izawa and colleagues [100] found that lower DHEA levels during a psychosocial stress test were correlated with increased negative mood during and after the stress test. The substantial DHEA decline with age seems to entail fundamental disadvantages, which the aging male needs to somehow compensate in order to preserve an overall state of psychological functionality and positive mood.

Table 2. The influence of steroid concentrations on psychological well-being and depression in men

Hormone	Psychological well-being	Depression	References
Testosterone (T)	↑	↓	Amiaz & Seidman, 2008; Joshi et al., 2010; Yeap, 2014
Estradiol (E2)	(↑)	(↓)	Almeida et al., 2004; Barrett-Connor, 1999; Borst et al., 2014
Dehydroepi-androsterone (DHEA)	↑	↓	Abraham et al., 2013; Bloch et al., 1999; Davis et al., 2011; Michael et al., 2000; Morales et al., 1994
Cortisol (C)	↓	↑	Hsiao et al., 2011; Lok et al., 2012; McKay & Zakzanis, 2010; Pariante, 2009; Stetler & Miller, 2011

Decreased well-being or depression is indicated by ↓; increased well-being or depression is indicated by ↑; moderate associations are represented in brackets (e.g., (↓)).

C, as the secretory end product of the HPA axis, is strongly related to depression. Elevated plasma or saliva cortisol levels are considered as main indicators of HPA axis hyperactivity and have been well documented in depressed patients [101], and seem to disappear after full remission of depressive symptoms [102-105]. Based on meta-analytic findings, Stetler and Miller [101] concluded that small-to-moderate elevations in C and ACTH and a reduction in CRH levels are associated with depression. Additionally, they found that elevated C levels are more likely to occur in specific subgroups of depressed subjects: older, hospitalized, melancholic depression, endogenous depression, depression with psychotic features and taking antidepressant medication. As a result of the age-related C increase and the higher prevalence of mild-to-moderate depressive symptoms, it seems important to acknowledge the substantial impact of elevated C levels in elderly men on well-being and depression. The influence of steroid concentrations on psychological well-being and depression in aging men is summarized in table 2.

Sexual Consequences: Erectile Function and Sexual Desire

Age-related biological alterations seem to increase the risk of erectile dysfunction (ED). In men below the age of 40 years, the prevalence of ED ranges from 1-9%, increasing in the decade from 40 to 49 years to 2-30%, for older men (60 to 69 years) to 20-40%, and for men over 70 years to 50-75% [106]. As age is strongly associated with ED, and in parallel, hormonal changes in T, E2, DHEA and C also start at around 40 years, several studies have attempted to disentangle the specific relationship of these steroid hormones and ED or decreased sexual desire (SD).

Martin and colleagues [107] reported an association of low plasma T levels with moderate to severe ED and low SD in a sample of 1195 men aged 35-80. In a population-based study with 1744 men aged 75-95, low serum T levels were also reported to be associated with a lack of interest in sex, but not with other sexual problems [108]. Shi and colleagues [109] identified low tT and serum fT levels as independent risk factors for ED and low fT levels as predictor of ED in 476 men aged between 40 and 70. Similar findings were reported in a study of 136 men aged 20-49, where erectile function was associated with fT, but not with other studied hormones such as DHEAS, E2, LH or the binding protein SHBG [110]. Higher endogenous T levels might therefore be considered protective against ED and hypoactive SD.

A recent study demonstrated low T levels as the primary reason for decreased erectile function, but concomitantly elevated E2 levels seemed to show an additive impairment effect [111]. These results are supported by a study in rabbits, which reported that ED induced by a high-fat diet is rather more associated with high E2 levels than with low T levels. Furthermore, ED induced by a high-fat diet in rabbits could be restored with both T supplementation and estrogen receptor (ER) blockade reducing E2 action [112]. However, Finkelstein and colleagues [113] described contradictory evidence in terms of a suppression of endogenous E2 results in decreased erectile function and SD, raising the prospect of an optimal E2 level. A recent study examining T supplementation in 423 men reported that elevated levels of serum E2 were associated with increased SD [114]. To summarize, in males, the role of E2 in the development and maintenance of ED is not yet fully understood and requires further investigation, since E2 and ED as well as SD seem to be more strongly correlated than was previously assumed [115].

Basar and colleagues [95] reported an association between sexual dysfunction and low DHEAS levels for 348 men aged 21-76. Although in the MMAS [116], DHEAS showed a strong negative correlation with the prevalence of ED in men aged 40-70, the results on the impact of DHEA and DHEAS on male sexual function are still the subject of controversial debate [31].

There are only a small number of studies on the specific relationship between C and sexual function, although evidence of a negative impact of stress, resulting in an over-reactive

HPA axis and its suppressing influence on the hypothalamic-pituitary-gondal axis, is fairly consistent [117, 118]. This effect has been shown in patients with Cushing syndrome (chronic hypercortisolism), where patients show a decreased SD [119]. On the other hand, serum C levels were shown to decrease with increased sexual arousal, suggesting an inhibitory role for C in the mechanism of male sexual response [120]. Although in some studies no difference was observed between healthy men and patients with ED with respect to ACTH and C levels [121, 122], other studies did demonstrate such an association [123]. Kobori and colleagues [124] reported for 105 depressed men aged 30-72 that in patients without antidepressant intake, salivary C levels showed significant negative associations with different dimensions of self-reported sexual functioning such as erectile function, SD or intercourse satisfaction. Furthermore, a study with 31 men suffering from ED reported that higher levels of perceived stress were related to decreased orgasmic and erectile function, while elevated salivary C levels 45 min after awakening showed a negative association with SD [123]. In a review, Chrousos [125] discussed the effects of poor quality of life, chronic stress and depression in terms of causing chronic activation of the stress system, leading to aggravation of ED. Taken together, elevated C levels impair sexual function in men, and the fact of rising C levels with age, independently of stress, should be taken into account in the treatment of elderly men with ED.

The described results indicate the importance of assessing a broad spectrum of hormones in patients with ED and hypoactive SD. These data may help to identify suboptimal ranges of hormone concentrations and to differentiate between specific subgroups of men with ED (older, hypogonadal, obese, diabetic, with metabolic syndrome) for the implementation of specific treatment. The influence of steroid concentrations on erectile function and sexual desire in aging men is summarized in table 3.

Table 3. The influence of steroid concentrations on erectile function and sexual desire in men

Hormone	Erectile Function	Sexual Desire	References
Testosterone (T)	↑	↑	Hyde et al., 2012; Jastrzebska et al., 2014; Martin et al., 2012; Shi et al., 2014
Estradiol (E2)	mixed	mixed	El-Sakkar, 2013; Finkelstein et al., 2013; Ramasamy et al., 2014; Vignozzi et al., 2014
Dehydroepi-androsterone (DHEA)	(↑)	(↑)	Basar et al., 2005; Baulieu et al., 2000; Levy, 1994
Cortisol (C)	↓	↓	Derouet et al., 2002; Gray et al., 1998; Kalaitzidou et al., 2013; Kobori et al., 2009; Starkman et al., 1981; Ückert et al., 2003

Decreased erectile function or sexual desire is indicated by ↓; increased erectile function or sexual desire is indicated by ↑; moderate associations are represented in brackets (e.g., (↓)).

Cognitive Consequences: Cognition and Dementia

Cognition is at increased risk with advancing age. Recent research has referred to specific hormone concentrations in the aging male with respect to preserving a functional level of cognition and disease absence or with regard to possible therapeutic targets against dementia.

A positive association of bT with memory and concentration was shown in a sample of 547 men (55-89 years) [83]. In a sample of 371 men (> 50 years), better scores on cognitive functioning in three different cognitive tests were associated with higher bT levels [126]. In 400 men aged 40-80, associations were revealed for tT and bT with respect to cognitive processing capacity and speed. Interestingly, a curvilinear relationship with optimal T levels in the 3rd and 4th quintiles of values was shown [127]. In 1107 men aged 35-90, a positive association between higher fT levels and both memory and visuospatial ability was found [128]. In contrast, in 450 men aged 35-80, low fT levels were associated with better performance on spatial visualization tasks compared to high fT levels [129]. In a study with 2932 men (70-89 years), serum fT levels of 210 pmol/l or more were associated with better cognitive performance [130]. However, there is substantial literature which conflicts with these findings, showing no relationship between endogenous T concentrations and cognition for adjusted models (controlling for educational attainment and physical health measures) [131].

In terms of dementia, the predicted prevalence is sexually dimorphic, and 19-29% lower for men [132]. When focusing on men with Alzheimer's dementia (AD), lower ratios of tT:SHBG were found than in the healthy control group, while low fT levels were an independent predictor of AD [133, 134]. Rosario and colleagues [135] reported that in male brains analyzed post mortem, a significant decrease in brain levels of T was observed in cases with mild neuropathological changes as well as those with advanced AD neuropathology. By contrast, a study measuring fT and fE2 levels revealed no significant effects of these hormones on executive function or global cognitive function despite a large sample size and adequate hormone measurement [136]. Taken together, higher T levels seem to be protective against age-related cognitive decline in men and should be considered as a possible therapeutic treatment for males suffering from dementia.

There are few available studies on the association between E2 and cognition in males. In 547 men aged 59-89, higher levels of tE2 and bE2 were related to decreased global cognitive function after a 4-year follow-up [83]. In a cross-sectional study of 310 men aged 50 or older, Yaffe and colleagues [126] reported that higher tE2 but not bE2 levels were associated with decreased cognitive function. Another longitudinal study supported these findings by demonstrating in 242 elderly men (73-91 years) that higher levels of tE2 and bE2 are associated with increased risk of cognitive decline [137]. A longitudinal study conducted over 6 years examined 2974 men aged 71-93 and revealed that higher levels of bE2 were associated with increased risk of AD and with lower cognitive abilities, indicating a more rapid cognitive decline for higher than for lower E2 levels [138]. By contrast, Zimmermann and associates [139] reported a positive association between tE2 levels and a verbal memory test in 185 elderly men. Furthermore, a prospective cohort study observed no relationship between cognition and tE2 or fE2 levels [136]. In conclusion, these studies support the perspective that in middle-aged and older men, lower E2 levels are associated with better cognitive performance [140]. Therefore, a moderate age-related decline in endogenous E2 levels in older men does not seem to be one of the main biological mechanisms supporting a more rapid decline in cognitive function.

Table 4. The influence of steroid concentrations on cognitive function and dementia in men

Hormone	Cognitive Function	Dementia	References
Testosterone (T)	↑	↓	Barrett-Connor et al., 1999; Fonda et al., 2005; Hogervorst et al., 2004; LeBlanc et al., 2010; Moffat et al., 2004; Muller et al., 2005; Thilers et al., 2006; Rosario et al., 2011; Yaffe et al., 2002; Yeap et al., 2008; Yonker et al., 2006
Estradiol (E2)	↓	↑	Barrett-Connor et al., Geerlings et al., 2006; LeBlanc et al., 2010; 1999; Muller et al., 2009; Yaffe et al., 2002; Yeap, 2014; Zimmermann et al., 2011
Dehydroepian drosterone (DHEA)	↑	↓	Fonda et al., 2005; Goldman & Glei, 2007; Haren et al., 2008; Hildreth et al., 2013; Moffat et al., 2000; Sanders et al., 2010; Sorwell & Urbanski, 2010; Valenti et al., 2009
Cortisol (C)	(↓)	(↑)	Csernansky et al., 2006; Elgh et al., 2006; Fonda et al., 2005; Franz et al., 2011; Hartman et al., 1997; Karlamangla et al., 2005; Lupien et al., 1994; Lupien et al., 1998; Moriarty et al., 2014; Pulopulos et al., 2014; Rasmouson et al., 2001; Schrijvers et al., 2011; Singh-Manoux et al., 2014;

Decreased cognitive function or dementia is indicated by ↓; increased cognitive function or dementia is indicated by ↑; moderate associations are represented in brackets (e.g., (↓)).

The strong age-related decline in DHEA and DHEAS in older people seems to be associated with a parallel decline in cognitive function [141]. In an analysis of 410 men (>65 years), low DHEAS levels at baseline were strongly associated with larger cognitive decline at a 3-year follow-up [142]. These findings were replicated by the documentation of a positive association between DHEAS and cognitive function in 124 men (50-65 years) [143]. A sex-specific relationship for men only was reported for low DHEAS levels and worse working memory and executive function in a sample of 49 men and 54 women aged 60- 88 [144]. Another study supported the beneficial sex-specific effects of DHEA on cognition in men, with longitudinal data showing an association between baseline DHEAS concentrations and the prediction of the subsequent 3-year cognitive decline in men only (n = 472, mean age 66) [145]. However, data from the large sample of 981 older men of the MMAS were unable to replicate any association between DHEA or DHEAS and cognitive function [131]. Moffat and colleagues [146] also reported no significant effect of DHEAS levels on several dimensions of cognition in 883 men aged 22-91. Furthermore, a 10-year longitudinal study found no relationship between DHEAS levels and cognitive decline in men aged 65 years or older [147].

As the underlying mechanisms in cognitive aging remain poorly understood, HPA axis dysregulation is one possible explanatory pathway, with C as peripheral marker. Derived from animal models, Sapolsky and colleagues [148, 149] hypothesized that HPA dysregulation impairs cognitive functions such as memory and learning processes. Increasing cortisol levels with age and elevated basal cortisol levels in 19 healthy adults (11 men and 8 women) aged 60-80 years were related to inferior performance in memory and attention tests [150]. These authors also reported prolonged cortisol elevations associated with memory impairment and decreased hippocampal volume in 11 subjects (mean age 76.5) [151]. Several groups found that dementia patients showed elevated serum cortisol levels compared to healthy controls [152-154]. In 33 patients (19 men, mean age 75 years) with very mild and mild AD, higher plasma C levels were associated with more rapid disease progression [155]. In a sample of 538 men and women (70-79 years), a significant negative effect of urinary cortisol on cognitive function was detected independently of gender over a 7-year follow-up [156]. Supporting these results, data from the MMAS showed that fully adjusted models relate higher serum C levels to lower cognitive functioning [131]. By contrast, in 3341 subjects (1420 men; >50 years), Schrijvers and colleagues [157] reported no association between serum morning cortisol levels and cognitive function at baseline or annual follow-up. There was no association between serum morning cortisol and dementia at an average of 7.1-year follow-up. Several studies investigated the relationship between the circadian rhythm of C secretion and cognitive function or dementia. The findings show an association between a flatter slope and worse verbal fluency and a greater decline in memory and executive function [158-160]. In 778 middle-aged twin men, Franz and colleagues [161] reported that an overall higher C output per day was associated with poorer performance on several cognitive domains (executive functioning, processing speed and visual-spatial memory). In addition, a 35-year longitudinal examination of this study revealed that general cognitive ability at age 20 was a significant predictor of midlife cortisol levels. Furthermore, a specific genetic polymorphism in the apolipoprotein E gene (APOE-ε4 allele) has been related to modulating the relationship between C and cognition [162]. While some studies reported an unhealthy diurnal C secretion profile (low morning C, high evening C and flatter slope) to be associated with faster decline in memory or verbal fluency in the risk-allele carriers [163, 164], others failed to replicate these findings for the association with dementia [157]. Taken together, C does not seem to be a strong contributor to cognitive decline in older men [164], or there appears to be a U-shaped relationship between C levels and cognitive function [165]. Low long-term cortisol levels (measured in scalp hair) were associated with worse performance in working memory, learning, and short- and long-term verbal memory in 14 healthy men (56-77 years). However, higher daily salivary cortisol output was related to worse cognitive performance [166]. Taken together, although there is considerable conflicting evidence, the findings indicate an essential role of C in the age-related cognitive decline, suggesting higher serum, salivary and urinary cortisol levels to be related with overall worse cognitive performance and increased risk of dementia as well as disease progression in older men. The influence of steroid concentrations on cognition and dementia in aging men is summarized in table 4.

PHYSICAL CONSEQUENCES: BODY COMPOSITION AND FRAILTY

Lifetime risk of a fracture after the age of 50 years has been reported to lie at 20% in men [167], while other studies report a lifetime fracture risk of 24% for men at the age of 45 [168]. Age-related sarcopenia, a clinical condition defined as muscle mass two standard deviations below the sex-specific level of young-normal mean, is prevalent in up to 15% of over-65-year-olds [169]. The age-related decrease in muscle mass and bone mineral density (BMD) as well as the age-related increase in fat mass plays a key role in frailty, suggesting hormonal changes as a therapeutic target to prevent general frailty in older men.

A cross-sectional study examined 121 males and 180 females aged 65-97 years and revealed that muscle mass and strength decreased significantly with increasing age in both sexes, whereas only in men was muscle mass significantly positively associated with serum fT levels [170]. Roy and colleagues [171] reported similar findings for 262 men aged 24-90 years from the Baltimore Longitudinal Study of Aging. A prospective cohort study demonstrated in 3616 community-dwelling men (70-88 years) that lower fT was independently associated with frailty at baseline and at 6-year follow-up [108]. In 3690 men (>70 years), low baseline serum T levels were related to higher all-cause mortality assessed 10 years later [172].

Along with the strong evidence that T is related to body composition and frailty, an essential influence of E2 on these parameters has been described [173]. Finkelstein and colleagues [113] reported an independent effect of low E2 levels on increased fat mass. An examination of serum sex steroid levels in young and older men in relation to longitudinal changes in BMD revealed that bE2 levels play a key role in both the acquisition of peak bone mass by promoting bone anabolism in young men and in bone loss by inhibiting bone resorption in elderly men [174, 175]. The authors also conclude that an age-related decrease in bE2 levels under a certain concentration (40 pmol/l) may be a major cause of bone loss in elderly men. This perspective is further supported by other studies, which consistently replicated the influence of E2 levels on BMD [176-178]. Another large cross-sectional study of 2908 men with a mean age of 75.4 years reported serum E2 to be an independent predictor of BMD in all bone sites studied [179]. Overall, cross-sectional observational as well as prospective studies in men have documented that serum E2 is positively associated with BMD, and higher endogenous E2 levels in older men are associated with a decreased risk of fractures [173, 180]. Taken together, low levels of E2 in older men seem to be associated with increased fat mass and decreased BMD, and therefore low E2 levels are a considerable risk factor for bone fractures after falls.

Since DHEA is considered as an anabolic hormone, the relationship between DHEA and muscle mass and general frailty has been addressed. Voznesensky and colleagues [181] reported a significant negative association between serum DHEA levels and frailty (assessed as weight loss, grip strength, sense of exhaustion, walking speed and physical activity) in 898 adults (591 men; mean age 74.6 years). Low DHEAS levels were significantly related to an increase in fat mass in 170 subjects aged 18-60 or over [98]. In a longitudinal study, lower serum DHEAS levels were associated with increased odds of frailty at a 10-year follow-up in 254 subjects aged 65-70 [182].

Table 5. The influence of steroid concentrations on body composition and frailty in men

Hormone	Body Composition	Frailty	References
Testosterone (T)	↑	↓	Baumgartner et al., 1999; Hyde et al., 2012; Roy et al., 2002; Yeap et al., 2014
Estradiol (E2)	↑	↓	Amin et al., 2000; Finkelstein et al., 2013; Hoppe et al., 2011; Khosla et al., 2001; Mellström et al., 2006; Mellström et al., 2008; Nguyen et al., 2014; Szulc et al., 2001; van Pottelbergh et al., 2003
Dehydroepiandro sterone (DHEA)	↑	↓	Abraham et al., 2013; Baylis et al., 2012; Voznesensky et al., 2009
Cortisol (C)	↓	↑	Baylis et al., 2012; Feller et al., 2014; Holanda et al., 2012; Warriner & Saag, 2013

Worse body composition or decreased frailty is indicated by ↓; better body composition or increased frailty is indicated by ↑.

Due to the fact that C is a catabolic hormone, and owing to the negative impact of glucocorticoids through its direct action on bone cells and indirect effects on calcium absorption, several studies investigated the relationship between C and frailty [183]. Holanda and colleagues [184] reported for 69 residents of long-stay institutions (37.7% men; mean age 77.5 years) a positive association between salivary C levels and frailty assessed as weight loss, fatigue, slowness, weakness and low physical activity. In a longitudinal study, a higher C:DHEAS ratio was associated with increased odds of frailty at a 10-year follow-up in 254 subjects aged 65-70 years [182]. As chronically elevated C levels are associated with reduced bone formation, persistent bone destruction and increased osteocyte and osteoblast apoptosis, glucocorticoid-induced osteoporosis seems to be particularly responsible for fractures in elderly people [183]. Elevated C levels are also associated to worse body composition [52]. Taking the age-related increase in C levels into account C plays a key role in frailty, risk of fractures and worse body composition in elderly men. The influence of steroid concentrations on body composition and frailty in aging men is summarized in table 5.

Clinical Implications

Hormone Replacement

In men aged over 40 years, age-related hormonal changes are common and derive from altered secretion from the testes, adrenals and pituitary gland [185]. One of the cornerstones of the treatment of age-related disorders is hormone replacement. In the attempt to counter the widespread effects of hormone decline, supplementation of T and DHEA in men has become quite common. Therefore, it is surprising that studies on specific indications, long-term effects, and risks of such replacement therapies are broadly lacking [24].

T replacement therapy (TRT) is the most frequently used hormone replacement therapy in men and especially in older men with symptoms of andropause. The primary aim of TRT is to restore serum T levels to the normal range, relieving men from hypogonadal symptoms, recovering quality of life, reducing disability, compressing major illnesses into a narrow age range, and prolonging longevity [186].

With regard to affective disorders, several intervention studies in eugonadal and hypogonadal men provide evidence that T substitution in men with low or low-normal T levels improves mood and reduces depressive symptoms [140]. A recent meta-analysis of T supplementation in depressed men concluded that T supplementation has a positive effect on mood, especially in dysthymic men and men with minor depression [187]. As recent findings showed that in specific subgroups of men, associations between low T levels and depression are quite consistent, TRT as a therapeutic treatment seems to be indicated and beneficial. Especially elderly men (aged over 60 years), treatment-resistant depressive men, men with major depression and HIV infection, hypogonadal men and dysthymic men are assumed to benefit from TRT.

Significant improvements in erectile function and SD were shown in an intervention study with 712 hypogonadal men (aged 18-99 years) following exogenous testosterone administration [188], replicating findings from previously conducted studies of beneficial effects with respect to several aspects of sexual functioning [e.g., 189]. As 20-30% of adult men have at least one manifest sexual dysfunction [106], the restoration of T levels, especially in hypogonadal men, promises substantial symptom relief. Therefore, men with ED and hypoactive SD should be screened for low T levels and receive treatment accordingly [190].

Intervention studies examining the effect of T supplementation on cognition have yielded mixed results, while existing RCT data in men with dementia indicate beneficial effects of T administration on cognition [140, 93].

Several intervention studies evaluating the effect of exogenously administered T on body composition and frailty have been conducted. A randomized double-blind placebo-controlled trial in 44 healthy men aged 60-75 reported that following TRT, significant dose-dependent increases in skeletal muscle mass and leg strength were observed [191]. In 122 men (mean age 70.8 years), TRT led to improvements in lean mass and muscle strength and reductions in body fat [192]. 209 men aged 65 or over with mobility limitations were randomly assigned to either a placebo or a TRT group [193]. The follow-up examination showed significant improvements in muscle strength and stair-climbing power. TRT in 131 men (mean age 77.1 years) with low endogenous T levels improved body composition and axial BMD [194]. Finkelstein and colleagues [113] reported that in men with chemically induced depletion of endogenous androgen levels and subsequent administration of different exogenous T doses in different groups, a dose-dependent improvement in muscle mass and strength was observed. Taken together, T shows a consistent association with body composition and frailty, and overall TRT in older men significantly improves body composition and reduces frailty.

Bassil [186] summarized the potential benefits of TRT as improvements in mood, energy, and quality of life, erectile function and SD, cognitive function, muscle mass and strength, and BMD.

Several risks of TRT have been reported. Although most studies show that atherosclerosis is mostly related to low T levels [195, 108], an increased risk of cardiovascular disorders with T administration has also been described [196]. Therefore, T supplementation in andropause patients with higher cardiovascular risk should only be undertaken with great caution, and TRT should not be initiated in patients with recent ischemic heart disease or heart failure [24]. Another concern of TRT in patients is polycythemia. As an increase in hematocrit in relation to TRT has been described, subsequent observation of hematocrit levels after a TRT is recommended [24, 197]. TRT has often been related to prostate cancer growth [198], but so far, there is not sufficient empirical evidence to support this assumption [199]. However, follow-up prostate-specific antigen levels should be continuously observed in TRT patients. Furthermore, TRT has been linked to benign and malignant hepatic tumors, intrahepatic cholestasis, hepatotoxicity, and liver failure, but only for oral forms of TRT [186]. Therefore, other delivery systems like transdermal gels or intramuscular injections are recommended. In addition, sleep apnea has been discussed to correlate with TRT in hypogonadal men, but empirical evidence in this regard is lacking [186]. Other TRT side effects such as edema, dyslipidemia, hypercalcemia, decreased testicular size, acne, azoospermia, gynecomastia and alopecia have been reported [186, 197]. Taken together, TRT is one possible strategy to counter symptoms of andropause, and the safety profile is considered to be rather good.

E2 is administered to men under special clinical conditions only. For example, the experience of hot flushes after prostate cancer treatment can be treated with E2 [200]. Falahti-Nini and colleagues [201] reported beneficial effects of E2 administration in hypogonadal elderly men in terms of the maintenance of bone formation. E2 administration in men is accompanied by several negative side effects. Reis and colleagues [202] reported findings of nipple/breast tenderness and gynecomastia, while other studies have reported ED, changes in blood coagulation and genotoxic changes in the liver [203]. E2 administration is not indicated as a therapeutic treatment against symptoms of age-related hormonal changes in men.

DHEA is seen as a youth-preserving hormone and its replacement is common in anti-aging medicine. When DHEA was administered in healthy adults aged between 40 and 70 years, restoring their DHEA levels to levels found in young adults, a significant improvement in perceived psychological well-being was found [204]. Schmidt and colleagues [205] reported a significant improvement in depressive symptoms of patients suffering from midlife-onset major and minor depression. Similar findings of beneficial DHEA supplementation were reported for midlife-onset dysthymia [206]. Therefore, DHEA could also be seen as a protective biological factor with the potential to improve mood and to buffer stress or promote resilience during stress.

In an intervention study, half of a group of 40 impotent patients aged 41-69 underwent a DHEA supplementation of 50mg/day over six months, while the other half received placebo. Significant improvements in erectile function, intercourse satisfaction, SD, overall satisfaction, and orgasmic function were reported [207]. In a six-week DHEA supplementation in depressed men and women aged 45-65 years, significant improvements in sexual functioning were documented [205]. However, in men with sexual dysfunction, Morales and colleagues [208] found that 50mg DHEA supplementation twice per day over 16 weeks had no influence on sexual function. A recent review concluded that controlled trials conducted to date have not definitively proven that DHEA administration is useful for improving male sexual function [209].

Several intervention studies in older men failed to find any significant improvements following DHEA(S) supplementation for different dimensions of cognition [210-213]. However, few studies reported beneficial effects of DHEA administration in cognitive dimensions such as episodic memory recollection in healthy young men [214]. Maggio and colleagues [34] concluded, after reviewing the literature on DHEA(S) and replacement therapy, that DHEA supplementation is not useful in improving or maintaining memory and other cognitive domains in older individuals. Therefore, DHEA seems to exert little effect as a therapeutic treatment against dementia or other forms of cognitive decline in elderly men, while higher endogenous DHEA(S) levels seem to be associated with better cognitive functioning in older men.

Most intervention studies on DHEA supplementation in older persons failed to report any significant improvements in muscle mass or strength [215, 216], but others found, with a daily dose of 100mg DHEA, a significant decrease in fat mass and an increase in knee muscle strength and lumbar back strength in 9 men aged 50-65 years [217]. Morley and Malmström [218] concluded that at present, DHEA does not play a main role in sarcopenia or frailty.

Reviewing the literature, Samaras and colleagues [24] concluded that potential benefits of DHEA replacement are the improvement of well-being, erectile function and SD, muscle mass and muscle strength, BMD and overall frailty, and a decrease in depression, risk of atherosclerosis and overall mortality.

Table 6. The effect of hormone replacement on different areas of life in men

Hormone Replacement	Testosterone (T)	Dehydroepiandrosterone (DHEA)	Estradiol (E2)
Psychological Consequences	Mood ↑ Depression ↓	Psychological well-being ↑ Depression ↓	
Sexual Consequences	Erectile function ↑ Sexual desire ↑	Erectile function ↑ Sexual desire ↑	Erectile function ↓
Cognitive Consequences	Cognitive function (↑) Dementia ↓	Cognitive function (↑)	
Physical Consequences	Muscle mass ↑ Fat mass ↓ Frailty ↓ BMD ↑	Muscle mass (↑) Fat mass (↓) Frailty (↓) BMD (↑)	Bone formation ↑
References	Amanatkar et al., 2014; Bassil, 2011; Borst et al., 2014; Buvat et al., 2010; Finkelstein et al., 2013; Kenny et al., 2010; Pexman-Fieth et al., 2013; Sattler et al., 2009; Storer et al., 2008; Wang et al., 2004; Yeap, 2014	Bloch et al., 1999; Kim & Morley, 2005; Kritz-Silverstein et al., 2008; Maggio et al., 2014; Morales et al., 1994; Morales et al., 1998; Percheron et al., 2003; Reiter et al., 1999; Samaras et al., 2014; Schmidt et al., 2005; Van Niekerk et al., 2001; Wolf et al., 1997; Wolf et al., 1998	Falahti-Nini et al., 2000; Kuebler et al., 2002; Reis et al., 2014

The reduction of a specific condition caused by hormone replacement is indicated by ↓; the increase of a specific condition caused by hormone replacement is indicated by ↑; no change is indicated by ≈; moderate effects are represented in brackets (e.g., (↓)).

Negative side effects of DHEA supplementation such as acne, seborrhea, facial hair growth, and ankle swelling have been reported, but these were rather mild and no relation to prostate cancer has been described [24]. However, so far, studies on long-term DHEA administration (more than 2 years) and its impact on health-related domains are lacking. Taken together, the general positive associations of higher DHEA levels in the aging male and the mild side effects of its supplementation indicate DHEA treatment as another possible strategy, besides TRT, to address symptoms of age-related hormonal changes. The effect of hormone replacement on different areas of life in men is summarized in table 6.

General Health and Successful Aging

For men, a continuous decline in circulating androgen levels (e.g., T and DHEA) with age has consistently been reported throughout the literature [20, 46]. Higher endogenous androgen levels were shown to be associated with various positive outcomes in older men, such as increased well-being [96], decreased symptoms of depression [140], increased erectile function and SD [109, 209], better overall cognition [140, 144], increased muscle mass [113, 98], and lower risk of frailty [108, 182].

Therefore, in older men, maintenance of androgen levels in a normal range (serum T levels 300-400 ng/dL) is a crucial aim in healthy aging. In the MMAS, apparent good health defined as absence of chronic illness, prescription medication, obesity, or excessive drinking, added 10-15% to the level of several androgens in the cross-sectional analysis. The authors conclude that incident poor health accelerates age-related decline in androgen levels and suggest a parallel track model, where not apparently healthy males are on a lower track, losing androgen levels at a greater rate compared to men with maintained good health [20]. Similar findings were replicated for obese versus non-obese men [219]. In 400 men aged 40-80, general health status modified the association of tT with age [220]. These results were extended by findings by Sartorius and colleagues [221], who examined 325 men with self-reported very good or excellent health. Repeated measures of serum T, Dihydrotestosterone and E2 levels showed no age-related decline [221]. Thus, age-related T decline and accompanying nonspecific symptoms in older men seem to be strongly associated with accumulating age-related comorbidities [221]. Although decreased health contributes to lower T levels, age remains a major independent factor for T decrease in older men [140].

The concept of successful aging has been proposed to reach beyond avoidance of disease and disability [222]. While Rowe and Khan [223] characterized successful aging as involving freedom from disability, and high cognitive, physical, and social functioning, Depp and colleagues [224] pointed out that successful aging is a multidimensional construct, with the biological age as an important dimension. Since the endocrine systems are strongly linked to the biological age, several lifestyle factors related to successful aging are considered to beneficially modify hormone concentrations in the aging male. Therefore, instead of implementing "routine" medical intervention programs offering long-term hormonal replacement therapy to restore or reduce hormone concentrations in older men, modification of specific lifestyle factors emerges as a crucial goal for healthy endocrine aging in men.

A recent review focusing on the maintenance of physical and cognitive function as well as the quality of life and independence in aging men identified crucial lifestyle factors [225]. Longevity with preserved good health and postponed disability were significantly associated with physical activity, sedentary behaviors, smoking, diet and alcohol consumption [225]. For example, a longitudinal study in 12,201 older men showed a positive effect of weekly physical activity on survival and successful aging [226]. Along with a direct relationship of different lifestyle factors on successful aging, studies demonstrate an association of several lifestyle factors and endocrine health (T, E2, DHEA and C), therefore also contributing on an indirect pathway to successful aging (Figure 1) [e.g., 227, 219, 110, 52].

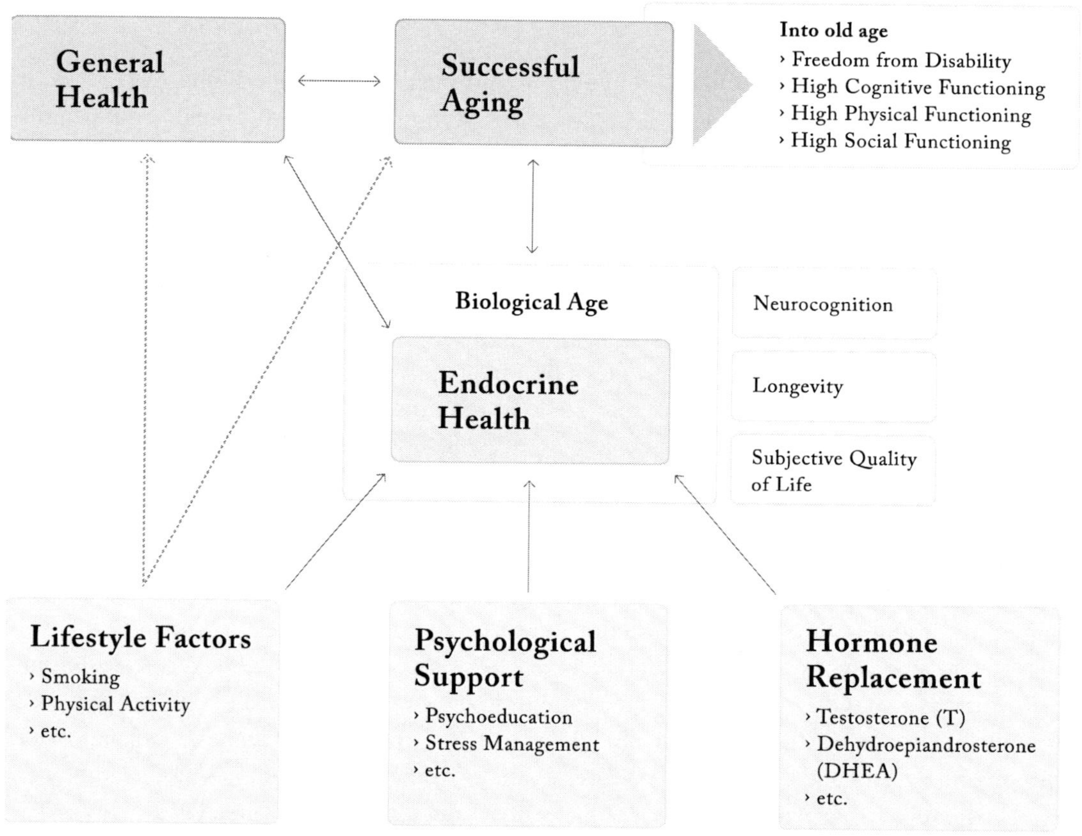

Figure 1. Lifestyle factors, psychological support and hormone replacement as possible targets to improve endocrine health and successful aging in men.

Prevention of disability, mental disorders, sexual dysfunction, and cognitive decline are of paramount interest for successful aging and can only be achieved within a multidimensional framework of intervention. Endocrine health plays a key role in preserving not only a functional state in various areas of life but also a high quality of life into old age, as shown in Figure 1.

Conclusion

Demographic changes are leading to a far larger proportion of elderly people in modern societies. Preserving good health and postponing disability into late life has been shown to be a vital goal for the future. Age-related hormonal changes in men are partially responsible for several clinical conditions in different areas of life. Psychological, sexual, cognitive and physical disorders may originate in declining T, DHEA, E2, or increasing C levels. Therefore, a greater understanding of the specific contribution of each hormone and its age-related change to the onset and progression of different clinical conditions such as depression, ED, dementia, or frailty is necessary.

Successful aging with less disease burden is strongly related to a healthy steroid profile. Since age-related hormonal changes occur as part of aging, strategies to restore or maintain healthy steroid levels are being established. TRT or the replacement of DHEA and E2 have been shown to significantly improve some clinical conditions related to the aging process, while they have not been found to improve others. A key role concerning age-related hormonal changes is played by lifestyle factors which contribute to a preserved healthy steroid profile, general health and successful aging (shown in Figure 1). Endogenous hormone concentrations in the optimal range seem to provide better protection against disease than supplemented hormone systems. Therefore, much more emphasis should be given to the prevention of age-related hormonal changes, by implementing psychoeducation about age-related hormonal changes in males at a population-based level. This would be one strategy to inform men about suitable healthy lifestyle habits, further options and to support them to counter these disorders. In individual cases, different psychological interventions, such as stress management training to maintain or restore optimal HPA axis functioning, could be applied to further improve endocrine health in aging men. However, as clinical conditions cannot be prevented for every male by maintaining good health and providing psychoeducation, hormone replacement represents a reasonable opportunity to overcome specific impairments caused by age-related steroid alterations. Men with severe symptoms caused by age-related hormonal changes should consider, besides consulting an endocrinologist, seeking psychological support for the successful management of the widespread effects.

References

[1] Kirkwood TBL: Understanding the odd science of aging. *Cell* 2005; 120:437-447.

[2] Hayflick L: Biological aging is no longer an unsolved problem. *Annals of the New York Academy of Science* 2007; 1100:1-13.

[3] Kirkwood TBL, Shanley DP: The connections between general and reproductive senescene and the revolutionary basis of menopause. *Ann. N.Y. Acad. Sci.* 2010; 1204:21-29.

[4] Baer, N., Schuler, D., Füglister-Dousse, S., & Moreau-Gruet, F. (2013). Depressionen in der Schweizer Bevölkerung. Daten zur Epidemiologie, Behandlung und sozial-beruflichen Integration (Obsan Bericht 56). Schweizerisches Gesundheits-observatorium, Neuchâtel.

[5] Weissman MM, Myers JK: Rates and risks of depressive symptoms in a United States urban community. *Acta psychiat. scand.* 1978; 57:219-231.

[6] Djukanovic I., Sorjonen K., Peterson U: Association between depressive symptoms and age, sex, loneliness and treatment among older people in Sweden. *Aging & Mental Health ahead-of-print* 2014:1-9.

[7] Johnson JM, Nachtigall LB, Stern TA: The effect of testosterone levels on mood in men: a review. *Psychosomatics* 2013; 54:509-514.

[8] Bertolote JM, Fleischmann A: A global perspective in the epidemiology of suicide. *Suicidologi* 2002; 7:6-8.

[9] Rüesch P, & Manzoni P: *Psychische Gesundheit in der Schweiz. Monitoring.* Neuchâtel: edition obsan 2003.

[10] Rosen R, Altwein J, Boyle P, Kirby RS, Kukacs B, Meuleman E, Leary MPO, Puppo P, Robertson C, Giuliano F: Lower urinary tract symptoms and male sexual dysfunction: a multinational survey of the aging male (MSAM-7). *European Urology* 2003; 44:637-649.

[11] Claassen J, Hu Z, Rohrbeck P: Fractures among active component, recruit trainees, and deployed service members, *U.S. Armed Forces,* 2003-2012. MSMR 2013; 21:2-19.

[12] Park DC, Bischof GN: The aging mind: neuroplasticity in response to cognitive training. *Dialogues in Clinical Neuroscience* 2013; 15:109-119.

[13] Kyle UG, Genton L, Slosman DO, Pichard C: Fat-free and fat mass percentiles in 5225 healthy subjects aged 15 to 98 years. *Nutrition* 2001; 17:534-541

[14] Bales CW, Ritchie CS: Sarcopenia, weight loss, and nutritional frailty in the elderly. *Annu. Rev. Nutr.* 2002; 22:309-323.

[15] Ensrud KE: Epidemiology of fracture risk with advancing age. *J Gerontol A Biol Sci Med Sci* 2013; 68:1236-1242.

[16] Shores MM, Smith NL, Forsberg CW, Anawalt BD, Matsumoto AM: Testosterone treatment and mortality in men with low testosterone levels. *J Clin Endocrinol Metab* 2012; 97:2050-2058.

[17] Lustig C, Shah P, Seidler R, Reuter-Lorenz PA: Aging, training, and the brain: a review and future directions. *Neuropsychol Rev* 2009; 19:504-522.

[18] Samaras N, Samaras D, Lang PO, Forster A, Pichard C, Frangos E, Meyer P: A review of geriatrics through hormones. What is the relation between andropause and well-known geriatric syndromes? *Maturitas* 2013; 74:213-219.

[19] Harman SM, Metter EJ, Tobin JD, Pearson J, Blackman MR: Longitudinal effects of aging on serum total and free testosterone levels in healthy men. *The Journal of Clinical Endocrinology and Metabolism* 2001; 86:724-731.

[20] Feldman HA, Longcope C, Derby CA, Johannes CB, Araujo AB, Coviello AD, Bremner WJ, McKinlay JB: Age trends in the level of serum testosterone and other hormones in middle-age men: longitudinal results from the Massachusetts Male Aging Study. *The Journal of Clinical Endocrinology and Metabolism* 2002; 87:589-598.

[21] Singh P: Andropause: Current concepts. Indian Journal of Endocrinology and Metabolism 2013; 17:621-629.

[22] Wu FCW, Tajar A, Beynon JM, Pye SR, Silman AJ, Finn JD, O`Neill TW, Bartfai G, Casanueva FF, Forti G, Giwercman A, Han TS, Kula K, Lean MEJ, Pendleton N, Punab M, Boonen S, Vanderschueren D, Labrie F, Huhtaniemi IT: Identification of

late-onset hypogonadism in middle-aged and elderly men. *N Engl J Med* 2010; 363:123-135.

[23] Nieschlag E: Andrologie: Grundlagen und Klinik der reproduktiven Gesundheit des Mannes. Edited by Nieschlag E, Behre HM. Heidelberg, *Springer Medizin Verlag* 2009, pp. 25-37.

[24] Samaras N, Papadopoulou MA, Samaras D, Ongaro F: Off-label use of hormones as an antiaging strategy: a review. *Clinical interventions in aging* 2014; 9:1175-1186.

[25] Gray A, Berlin JA, McKinlay JB, Longcope C: An examination of research design effects on the association of testosterone and male aging: results of a meta-analysis. *Journal of clinical epidemiology* 1991; 44:671-684.

[26] Vermeulen A, Kaufman JM, Giagulli VA: Influence of some biological indexes on sex hormone-binding globulin and androgen levels in aging or obese males. *The Journal of Clinical Endocrinology & Metabolism* 1996; 81:1821-1826.

[27] Chahal HS, Drake WM: The endocrine system and ageing. *J Pathol* 2007; 211:173-180.

[28] Migeon et al., 1957 Migeon CJ, Keller AR, Lawrence B, Shepard T: Dehydroepiandrosterone and androsterone levels in human plasma. Effect of age and sex; Day-to-day and diurnal variations*. *The Journal of Clinical Endocrinology & Metabolism 1957*; 17:1051-1062.

[29] Orentreich N, Brind JL, Rizer RL, Vogelman JH. 1984. Age changes and sex differences in serum dehydroepiandrosterone sulfate concentrations throughout adulthood. *The Journal of Clinical Endocrinology & Metabolism* 1984; 59:551-555.

[30] Orentreich N, Brind JL, Vogelman JH, Andres R, Baldwin H: Long-term longitudinal measurements of plasma dehydroepiandrosterone sulfate in normal men. *The Journal of Clinical Endocrinology & Metabolism* 1992, 75:1002-1004.

[31] Baulieu EE, Thomas G, Legrain S, Lahlou N, Roger M, Debuire B, Faucounau V, Girard L, Hervy MP, Latour F, Leaud MC, Mokrane A, Pitti-Ferrandi H, Trivalle C, de Lacharrière O, Nouveau S, Rakoto-Arison B, Souberielle JC, Raison J, Le Bouc Y, Raynaud A, Girerd X, Forette F: Dehydroepiandrosterone (DHEA), DHEA sulfate, and aging: contribution of the DHEAge Study to a sociobiomedical issue. *Proceedings of the National Academy of Sciences* 2000; 97:4279-4284.

[32] Parker CR: Dehydroepiandrosterone and dehydroepiandrosterone sulfate production in the human adrenal during development and aging. *Elsevier Science Inc.* 1999; 64:640-647.

[33] Genazzani AD, Lanzoni C, Genazzani AR: Might DHEA be Considered a Beneficial Replacement Therapy in the Elderly? *Drugs Aging* 2007; 24:173-185.

[34] Maggio M, De Vita F, Fisichella A, Colizzi E, Provenzano S, Lauretani F, ... Ceda GP: DHEA and cognitive function in the elderly. *The Journal of steroid biochemistry and molecular biology* 2014.

[35] Vermeulen A, Kaufman JM, van Pottelberg G, van Pottelberg I: Estradiol in elderly men. *The aging male* 2002; 5:98-102.

[36] Hemsell DL, Grodin JM, Brenner PF, Siiteri PK, MacDonald PC: Plasma Precursors of Estrogen. II. Correlation of the Extent of Conversion of Plasma Androstenedione to Estrone with Age 1. *The Journal of Clinical Endocrinology & Metabolism* 1974; 38:476-479.

[37] Kyle UG, Genton L, Slosman DO, Pichard C: Fat-free and fat mass percentiles in 5225 healthy subjects aged 15 to 98 years. *Nutrition* 2001; 17:534-541.

[38] Longcope C, Pratt JH, Stephen HS, Fineberg SE: Aromatization of Androgens by Muscle and Adipose Tissue in Vivo*. *The Journal of Clinical Endocrinology & Metabolism* 1978; 46:146-152.

[39] Ferrini RL, Barrett-Connor E: Sex hormones and age: a cross-sectional study of testosterone and estradiol and their bioavailable fractions in community-dwelling men. *American Journal of Epidemiology* 1998; 147:750-754.

[40] Khosla S, Melton III LJ, Atkinson EJ, O'fallon WM: Relationship of serum sex steroid levels to longitudinal changes in bone density in young versus elderly men. *The Journal of Clinical Endocrinology & Metabolism* 2001; 86:3555-3561.

[41] Van den Beld AW, de Jong FH, Frobbee DE, Pols HAP, Lamberts SWJ: Measures of bioavailable serum testosterone and estradiol and their relationships with muscle strength, bone density, and body composition in elderly men. The Journal of Clinical Endocrinology and Metabolism 2000; 85:3276-3282.

[42] Orwoll E, Lambert LC, Marshall LM, Phipps K, Blank J, Barrett-Connor E, ... Cummings S: Testosterone and estradiol among older men. *The Journal of Clinical Endocrinology & Metabolism* 2006; 91:1336-1344.

[43] Yeap BB, Alfonso H, Chubb SP, Handelsman DJ, Hankey GJ, Norman PE, Flicker L: Reference ranges and determinants of testosterone, dihydrotestosterone, and estradiol levels measured using liquid chromatography-tandem mass spectrometry in a population-based cohort of older men. *The Journal of Clinical Endocrinology & Metabolism* 2012; 97:4030-4039.

[44] Lewerin C, Nilsson-Ehle H, Jacobsson S, Johansson H, Sundh V, Karlsson MK, ... Mellström D: Serum estradiol associates with blood hemoglobin in elderly men; The MrOS Sweden Study. *The Journal of Clinical Endocrinology & Metabolism* 2014; (2014).

[45] Jasuja GK, Travison TG, Davda M, Murabito JM, Basaria S, Zhang A, ... Bhasin S: Age trends in estradiol and estrone levels measured using liquid chromatography tandem mass spectrometry in community-dwelling men of the Framingham Heart Study. *The Journals of Gerontology Series A: Biological Sciences and Medical Sciences* 2013; 68:733-740.

[46] Kozloski MJ, Schumm LP, McClintock MK: The Utility and Dynamics of Salivary Sex Hormone Measurements in the National Social Life, Health, and Aging Project, Wave 2. *The Journals of Gerontology Series B: Psychological Sciences and Social Sciences* 2014; 69:S215-S228.

[47] Ehlert U: Enduring psychobiological effects of childhood adversity. *Psychoneuroendocrinology* 2013; 38:1850-1857.

[48] Van Cauter E, Leproult R, Kupfer DJ: Effects of gender and age on the levels and circadian rhythmicity of plasma cortisol. *The Journal of Clinical Endocrinology & Metabolism* 1996; 81:2468-2473.

[49] Van Cauter E, Leproult R, Plat L: Age-related changes in slow wave sleep and REM sleep and relationship with growth hormone and cortisol levels in healthy men. *Jama* 2000; 284:861-868.

[50] Kern W, Dodt C, Born J, Fehm HL: Changes in cortisol and growth hormone secretion during nocturnal sleep in the course of aging. *Journal of Gerontology* 1996; 51A:M3-M9.

[51] Karlamangla AS, Friedman EM, Seeman TE, Stawksi RS, Almeida DM: Daytime trajectories of cortisol: demographic and socioeconomic differences – Findings from the National Study of Daily Experiences. *Psychoneuroendocrinology* 2013; 38:2585-2597.

[52] Feller S, Vigl M, Bergmann MM, Boeing H, Kirschbaum C, Stalder T: Predictors of hair cortisol concentrations in older adults. *Psychoneuroendocrinology* 2014; 39:132-140.

[53] Oettel M, Mukhopadhyay AK: Progesterone: the forgotten hormone in men? The aging male 2004; 7:6-257.

[54] Basaria S: Testosterone thearpy in older men with late-onset hypogonadism: a counter-rationale. *Endocrine Practice* 2013; 19:853-863.

[55] Kaler LW, Neaves WB: Attrition of the human leydig cell population with advancing age. *Anat. Rec.* 1978; 192:513-518.

[56] Neaves WB, Johnson L, Petty CS: Age-related change in numbers of other interstitial cells in testes of adult men: evidence bearing on the fate of leydig cells lost with increasing age. *Biology of Reproduction* 1985; 33:259-269.

[57] Rubens R, Dhont M, Vermeulen A: Further studies on Leydig cell function in old age. *The Journal of Clinical Endocrinology & Metabolism* 1974; 39:40-45.

[58] Longcope C: The effect of human chorionic gonadotropin on plasma steroid levels in young and old men. *Steroids* 1973; 21:583-592.

[59] Harman SM, Tsitouras PD: Reproductive hormones in aging men. I. Measurement of sex steroids, basal luteinizing hormone, and Leydig cell response to human chorionic gonadotropin. *The Journal of Clinical Endocrinology & Metabolism* 1980; 51:35-40.

[60] Nankin HR, Lin T, Murono EP, Osterman J: The aging leydig cell: III. Gonadotropin Stimulation in Men. *J Androl* 1981; 2:181-189.

[61] Vermeulen, A. (1991). Androgens in the aging male. *The Journal of Clinical Endocrinology & Metabolism,* 73(2), 221-224.

[62] Vermeulen A, Kaufman JM: Ageing of the hypothalamo-pituitary-testicular axis in men. *Horm Res.* 1995; 43:25-28.

[63] Iranmanesh A, Mulligan T, Veldhuis JD: Age in men does not determine gonadotropin-releasing hormone's dose-dependent stimulation of luteinizing hormone secretion under an exogenous testosterone clamp. *The Journal of Clinical Endocrinology & Metabolism* 2010; 95: 2877-2884.

[64] Kaufman JM, Deslypere JP, Giri M, Vermeulen A: Neuroendocrine regulation of pulsatile luteinizing hormone secretion in elderly men. *The Journal of steroid biochemistry and molecular biology* 1990; 37:421-430.

[65] Tenover JS: Effects of testosterone supplementation in the aging male. *The Journal of Clinical Endocrinology & Metabolism* 1992; 75: 1092-1098.

[66] Vermeulen A, Deslypere JP, Kaufman JM: Influence of Antiopioids on Luteinizing Hormone Pulsatility in Aging Men*. *The Journal of Clinical Endocrinology & Metabolism* 1989; 68:68-72.

[67] Urban RJ, Veldhuis JD, Blizzard RM, Dufau ML: Attenuated release of biologically active luteinizing hormone in healthy aging men. *Journal of Clinical Investigation* 1988; 81:1020.

[68] Mulligan T, Iranmanesh A, Kerzner R, Demers LW, Veldhuis JD: Two-week pulsatile gonadotropin releasing hormone infusion unmasks dual (hypothalamic and Leydig cell)

defects in the healthy aging male gonadotropic axis. *European Journal of Endocrinology* 1999; 141:257-266.

[69] Kaufman JM, Taelman P, Vermeulen A, Vandeweghe M: Bone mineral status in growth hormone-deficient males with isolated and multiple pituitary deficiencies of childhood onset. *The Journal of Clinical Endocrinology & Metabolism* 1992; 74:118-123.

[70] Schneider HJ, Sieverst C, Klotsche J, Böhler S, Pittrow D, Lehnert H, Wittchen HU, Stalla GK: Prevalence of low male testosterone levels in primary care in Germany: cross-sectional results from the DETECT study. *Clinical Endocrinology* 2009; 70:446-454.

[71] Parker CR, Mixon RL, Brissie RM, Grizzle WE: Aging alters zonation in the adrenal cortex of men. Journal of Clinical Endocrinology and Metabolism 1997; 82:3898-3901.

[72] Ferrari M, Mantero F: Male aging and hormones: the adrenal cortex. *Journal of endocrinological investigation* 2004; 28(suppl 11):92-95.

[73] Parker JR CR, Carr BR, Winkel CA, Casey ML, Simpson ER, Macdonald PC: Hypercholesterolemia due to Elevated Low Density Lipoprotein-Cholesterol in Newborns with Anencephaly and Adrenal Atrophy*. *The Journal of Clinical Endocrinology & Metabolism* 1983; 57:37-43.

[74] Van Cauter E, Leproult R, Plat L: Age-related changes in slow wave sleep and REM sleep and relationship with growth hormone and cortisol levels in healthy men. *Jama* 2000; 284:861-868.

[75] Veldhuis JD, Sharma A, Roelfsema F: Age-dependent and gender-dependent regulation of hypothalamic-adrenocorticotropic-adrenal axis. *Endocrinology and Metabolism Clinics of North America* 2013; 42:201-225.

[76] Yau JL, Hibberd C, Noble J, Seckl JR: The effect of chronic fluoxetine treatment on brain corticosteroid receptor mRNA expression and spatial memory in young and aged rats. *Molecular Brain Research* 2002; 106:117-123.

[77] Zhou J-N, Swaab DF: Activation and degeneration during aging: a morphometric study of the human hypothalamus. *Microscopy Research and Technique* 1999; 44:36-48.

[78] Deuschle M, Gotthardt U, Schweiger U, Weber B, Körner A, Schmider J, Standhardt H, Lammers C-H, Heuser I: With aging in humans the activity of the hypothalamus-pituitary-adrenal system increases and its diurnal amplitude flattens. *Life Sciences* 1997; 61:2239-2246.

[79] Ferrari E, Magri F, Dori D, Migliorati G, Nescis T, Molla G, Fioravanti M, Solerte SB: Neuroendocrine correlates of the aging brain in humans. *Neuroendocrinology* 1995; 61:464-470.

[80] Kudielka BM, Schommer NC, Hellhammer DH, Kirschbaum C: Acute HPA axis responses, heart rate, and mood changes to psychosocial stress (TSST) in humans at different times of day. *Psychoneuroendocrinology* 2004; 29:983-992.

[81] Patel RS, Shaw SR, Maclntyre H, McGarry GW, Wallace AM: Production of gender-specific morning salivary cortisol reference intervals using internationally accepted procedures. *Clinical Chemistry and Laboratory Medicine* 2004; 42:1424-1429.

[82] Keenan DM, Roelfsema F, Carroll BJ, Iranmanesh A, Veldhuis JD: Sex defines the age dependence of endogenous ACTH-cortisol dose responsiveness. *Am J Physiol Regul Integr Comp* Physiol 2009; 297:R515-R523.

[83] Barrett-Connor E, von Mühlen DG, Kritz-Silverstein D: Bioavailable testosterone and depressed mood in older men: the Rancho Bernardo Study. *The Journal of Clinical Endocrinology & Metabolism* 1999; 84:573-577.

[84] Seidman SN, Araujo AB, Roose SP, Devanand DP, Xie S, Cooper TB, McKinlay JB: Low testosterone levels in elderly men with dysthymic disorder. *American Journal of Psychiatry* 2002; 159:456-459.

[85] Almeida OP, Yeap BB, Hankey GJ, Jamrozik K, Flicker L: Low free testosterone concentration as a potentially treatable cause of depressive symptoms in older men. *Archives of general psychiatry* 2008; 65:283-289.

[86] Shores MM, Sloan KL, Matsumoto AM, Moceri VM, Felker B, Kivlahan DR: Increased incidence of diagnosed depressive illness in hypogonadal older men. *Arch Gen Psychiatry* 2004; 61:162-167.

[87] Joshi D, van Schoor NM, de Ronde W, Schaap LA, Comijs HC, Beekman ATF, Lips P: Low free testosterone levels are associated with prevalence and incidence of depressive symptoms in older men. *Clinical Endocrinology* 2010; 72:232-240.

[88] Colangelo LA, Shap L, Kopp P, Scholtens D, Chiu BC-H, Liu K, Gapstur SM: Total testosterone, androgen receptor polymorphism, and depressive symptoms in young black and white men: the CARDIA male hormone study. *Psychoneuroendocrinology* 2007; 32:951-958.

[89] Amiaz R, Seidman SN: Testosterone and depression in men. *Curr Opin Endocrinol Diabetes Obes* 2008; 15:278-283.

[90] Schmidt PJ, Nieman L, Danaceau MA, Tobin MB, Roca CA, Murphy JH, Rubinow DR: Estrogen replacement in perimenopause-related depression: a preliminary report. *Am J Obstet Gynecol 2000*; 183:414-420.

[91] De Novaes Soares C, Almeida OP, Joffe H, Cohen LS: Efficacy of estradiol for the treatment of depressive disorders in perimenopausal women. *Arch Gen Psychiatry* 2001; 58:529-534.

[92] Almeida OP, Waterreus A, Spry N, Flicker L, Martins RN: One year follow-up study of the association between chemical castration, sex hormones, beta-amyloid, memory and depression in men. *Psychoneuroendocrinology* 2004; 29:1071-1081.

[93] Borst SE, Yarrow JF, Fernandez C, Conover CF, Ye F, Meuleman JR, Morrow M, Zou B, Shuster JJ: Cognitive effects of testosterone and finasteride administration in older hypogonadal men. *Clinical Interventions in Aging* 2014; 9:1327-1333.

[94] Castanho TC, Moreira PS, Portugal-Nunes C, Novais A, Costa PS, Palha JA, Sousa N, Santos NC: The role of sex and sex-related hormone in cognition, mood and well-being in older men and women. *Biological Psychology* 2014; 103:158-166.

[95] Basar MM, Aydin G, Mert HC, Keles I, Caglayan O, Orkun S, Batislam E: Relationship between serum sex steroids and aging male symptoms score and international index of erectile function. *Urology* 2005; 66:597-601.

[96] Davis SR, Panjari M, Stanczyk FZ: DHEA replacement for postmenopausal women. *J Clin Endocrinol Metab* 2011; 96:1642-1653.

[97] Michael A, Jenaway A, Paykel ES, Herbert J: Altered salivary dehydroepiandrosterone levels in major depression in adults. *Biol Psychiatry* 2000; 48:989-995.

[98] Abraham PA, Kazman JB, Zeno SA, Poth M, Deuster PA: Age-related decline in salivary dehydroepiandrosterone sulfate and associated health risks among African Americans. *Ethnicity and Disease* 2013; 23: 149-154.

[99] Morgan CA, Rasmusson A, Pietrzak RH, Coric V, Southwick SM: Relationships among plasma dehydroepiandrosterone and dehydroepiandrosterone sulfate, cortisol, symptoms of dissociation, and objective performance in humans exposed to underwater navigation stress. *Biol Psychiatry* 2009; 66:334-340.

[100] Izawa S, Sugaya N, Shirotsuki K, Yamada KC, Ogawa N, Ouchi Y, Nagano Y, Suzuki K, Nomura S: Salivary dehydroepiandrosterone secretion in response to acute psychosocial stress and its correlations with biological and psychological changes. *Biological Psychology* 2008; 79:294-298.

[101] Stetler C, Miller GE: Depression and hypothalamic-pituitary-adrenal activation: a quantitative summary of four decades of research. *Psychosomatic Medicine* 2011; 73:114-126.

[102] Pariante CM: Risk factors for development of depression and psychosis. Glucocorticoids and Mood: *Ann. N.Y. Acad. Sci* 2009; 1179:144-152.

[103] McKay MS, Zakzanis KK: The impact of treatment on HPA axis activity in unipolar major depression. *Journal of Psychiatric Research* 2010; 44:183-192.

[104] Hsiao FH, Jow GM, Lai YM, Chen YT, Wang KC, Ng SM, Ho RTH, Chan CLW, Yang TT: The long-term effects of psychotherapy added to pharmacotherapy on morning to evening diurnal cortisols patterns in outpatients with major depression. *Psychother Psychosom* 201; 80:166-172.

[105] Lok A, Mocking RJT, Ruhé HG, Visser I, Koeter MWJ, Assies J, Bockting CLH, Olff M, Schene AH: Longitudinal hypothalamic-pituitary-adrenal axis trait and state effects in recurrent depression. *Psychoneuroendocrinology* 2012; 37:892-902.

[106] Lewis RW, Fugl-Meyer KS, Bosch R, Fugl-Meyer AR, Laumann EO, Lizza E, Martin-Morales A: Epidemiology/risk factors of sexual dysfunction. *The journal of sexual medicine* 2004; 1:35-39.

[107] Martin S, Atlantis E, Wilson D, Lange K, Haren MT, Taylor A, Wittert G: Clinical and biopsychosocial determinants of sexual dysfunction in middle-aged and older Australian men. *The Journal of Sexual Medicine* 2012; 9:2093-2103.

[108] Hyde Z, Flicker L, Almeida OP, Hankey GJ, McCaul KA, Chubb SP, Yeap BB: Low free testosterone predicts frailty in older men: the health in men study. *The Journal of Clinical Endocrinology and Metabolism* 2012; 95:3165-3172.

[109] Shi MD, Chao JK, Ma MC, Hao LJ, Chao I: Factors associated with sex hormones and erectile dysfunction in male Taiwanese participants with obesity. *The Journal of Sexual Medicine* 2014; 11:230-239.

[110] Jastrzębska S, Walczak-Jędrzejowska R, Kramek E, Marchlewska K, Oszukowska E, Filipiak E, Kula K, Słowikowska-Hilczer J: Relationship between sexual function, body mass index and levels of sex steroid hormones in young men. *Endokrynologia Polska* 2014; 65:203-209.

[111] El-Sakka AI: Impact of the association between elevated oestradiol and low testosterone levels on erectile dysfunction severity. *Asian journal of andrology* 2013; 15:492-496.

[112] Vignozzi L, Filippi S, Comeglio P, Cellai I, Morelli A, Marchetta M, Maggi M: Estrogen mediates metabolic syndrome-induced erectile dysfunction: a study in the rabbit. *J Sex Med* 2014; 11:2890-2902.

[113] Finkelstein JS, Lee H, Burnett-Bowie SAM, Pallais JC, Yu EW, Borges LF, Jones BF, Barry CV, Wulczyn KE, Thomas BJ, Leder BZ: Gonadal steroids and body

composition, strength, and sexual function in men. *New England Journal of Medicine* 2013; 369:1011-1022.

[114] Ramasamy R, Scovell JM, Kovac JR, Lipshultz LI: Elevated serum estradiol is associated with higher libido in men on testosterone supplementation therapy. *European urology* 2014; 65:1224-1225.

[115] Costa IC, Carvalho HN, Pacheco-Figueiredo L, Tomada I, Tomada N: Hormonal modulation in aging patients with erectile dysfunction and metabolic syndrome. *International Journal of Endocrinology* 2013; 2013.

[116] Levy J: Impotence and its medical and psychosocial correlates: results of the Massachusetts Male Aging Study. *Journal of Urology* 1994; 151:54-61.

[117] Bodenmann G, Ledermann T, Blattner D, Galluzzo C: Associations among everyday stress, critical life events, and sexual problems. *The Journal of nervous and mental disease 2006*; 194:494-501.

[118] Hamilton LD, Julian AM: The relationship between daily hassles and sexual function in men and women. *Journal of Sex and Marital Therapy* 2014; 40:379-395.

[119] Starkman MN, Schteingart DE: Neuropsychiatric manifestations of patients with Cushing`s Syndrome: relationship to cortisol and adrenocorticotropic hormone levels. *Arch Intern Med* 1981; 141:215-219.

[120] Ückert S, Fuhlenriede MH, Becker AJ, Stief CG, Scheller F, Knapp WH, Jonas U: Is there an inhibitory role of cortisol in the mechanism of male sexual arousal and penile erection? *Urol Res* 2003; 31:402-406.

[121] Derouet H, Lehmann J, Stamm B, Lühl C, Römer D, Georg T, Isenberg E, Gebhardt T, Stoeckle: Age dependent secretion of LH and ACTH in healthy men and patients with erectile dysfunction. *European Urology* 2002; 41:114-154.

[122] Gray A, Feldman HA, McKinlay JB, Longcope C: Age, disease, and changing sex hormone levels in middle-aged men: results of the Massachusetts Male Aging Study. *The Journal of Clinical Endocrinology and Metabolism* 1991; 73:1016-1025.

[123] Kalaitzidou I, Venetikou MS, Konstadinidis K, Artemiadis AK, Chrousos G, Darviri C: Stress management and erectile dysfunction: a pilot comparative study. *Andrologia* 2013; 46:698-702.

[124] Kobori Y, Koh E, Sugimoto K, Izumi K, Narimoto K, Maeda Y, Konaka H, Mizokami A, Matsushita T, Iwamoto T, Namiki M: The relationship of serum and salivary cortisol levels to male sexual dysfunction as measured by the International Index of Erectile Function. *International Journal of Impotence Research* 2009; 21:207-212.

[125] Chrousos GP: Stress and disorders of the stress system. *Nature Reviews Endocrinology* 2009; 5:374-381.

[126] Yaffe K, Lui LY, Zmuda J, Cauley J: Sex hormones and cognitive function in older men. *Journal of the American Geriatrics Society* 2002; 50:707-712.

[127] Muller M, Aleman A, Grobbee DE, De Haan EHF, van der Schouw YT: Endogenous sex hormone levels and cognitive function in aging men. Is there an optimal level? *Neurology* 2005; 64:866-871.

[128] Thilers PP, MacDonald SW, Herlitz A: The association between endogenous free testosterone and cognitive performance: a population-based study in 35 to 90 year-oldmen and women. *Psychoneuroendocrinology* 2006; 31:565-576.

[129] Yonker JE, Eriksson E, Nilsson LG, Herlitz A: Negative association of testosterone on spatial visualization in 35 to 80 year old men. *Cortex* 2006; 42:376-386.

[130] Yeap BB, Almeida OP, Hyde Z, Chubb SA, Hankey GJ, Jamrozik K, Flicker L: Higher serum free testosterone is associated with better cognitive function in older men, while total testosterone is not. The health in men study. *Clinical endocrinology* 2008; 68:404-412.

[131] Fonda SJ, Bertrand R, O'Donnell A, Longcope C, McKinlay JB: Age, hormones, and cognitive functioning among middle-aged and elderly men: cross-sectional evidence from the Massachusetts Male Aging Study. *The Journals of Gerontology Series A: Biological Sciences and Medical Sciences* 2005; 60:385-390.

[132] World Health Organization (2012) *Dementia: a public health priority.* WHO, Genf.

[133] Hogervorst E, Bandelow S, Combrinck M, Smith AD: Low free testosterone is an independent risk factor for Alzheimer's disease. *Experimental Gerontology* 2004; 39:1633-1639.

[134] Moffat SD, Zonderman AB, Metter EJ, Kawas C, Blackman MR, Harman SM, Resnick SM: Free testosterone and risk for Alzheimer disease in older men. *Neurology* 2004; 62:188-193.

[135] Rosario ER, Chang L, Head EH, Stanczyk FZ, Pike CJ: Brain levels of sex steroid hormones in men and women during normal aging and in Alzheimer's disease. *Neurobiology of Aging* 2011; 32:604-613.

[136] LeBlanc ES, Wang PY, Janowsky JS, Neiss MB, Fink HA, Yaffe K, Marshall LM, Lapidust JA, Stefanick ML, Orwoll ES: Association between sex steroids and cognition in elderly men. *Clinical endocrinology* 2010; 72:393-403.

[137] Muller M, van den Beld AW, Grobbee DE, de Jong FH, Lamberts SW: Sex hormones and cognitive decline in elderly men. *Psychoneuroendocrinology* 2009; 34:27-31.

[138] Geerlings MI, Strozyk D, Masaki K, Remaley AT, Petrovitch H, Ross GW, White LR, Launer LJ: Endogenous sex hormones, cognitive decline, and future dementia in old men. *Annals of Neurology* 2006; 60:346-355.

[139] Zimmerman ME, Lipton RB, Santoro N, McConnell DS, Derby CA, Katz MJ, Baigi K, Saunders-Pullman R: Endogenous estradiol is associated with verbal memory in nondemented older men. *Brain and Cognition* 2011; 76:158-165.

[140] Yeap BB: Hormonal changes and their impact on cognition and mental health of ageing men. *Maturitas* 2014; 79:227-235.

[141] Sorwell KG, Urbanski HF: Dehydroepiandrosterone and age-related cognitive decline. *Age* 2010; 32:61-67.

[142] Valenti G, Ferrucci L, Lauretani F, Ceresini G, Bandinelli S, Luci M, Ceda G, Maggio M, Schwartz RS: Dehydroepiandrosterone sulfate and cognitive function in the elderly: The InCHIANTI Study. *Journal of Endocrinological Investigation* 2009; 32:766-772.

[143] Haren MT, Banks WA, Perry III HM, Patrick P, Malmstrom TK, Miller DK, Morley JE: Predictors of serum testosterone and DHEAS in African-American men. International *Journal of Andrology* 2008; 31:50-59.

[144] Hildreth KL, Gozansky WS, Jankowski CM, Grigsby J, Wolfe P, Kohrt WM: Association of serum dehydroepiandrosterone sulfate and cognition in older adults: Sex steroid, inflammatory, and metabolic mechanisms. *Neuropsychology* 2013; 27:356-363.

[145] Goldman N, Glei DA: Sex differences in the relationship between DHEAS and health. *Experimental Gerontology* 2007; 42:979-987.

[146] Moffat SD, Zonderman AB, Harman SM, Blackman MR, Kawas C, Resnick SM: The relationship between longitudinal declines in dehydroepiandrosterone sulfate

concentrations and cognitive performance in older men. *Archives of internal medicine* 2000; 160:2193-2198.

[147] Sanders JL, Cappola AR, Arnold AM. Boudreau RM, Chaves PH, Robbins J, Cushman M, Newman AB: Concurrent change in dehydroepiandrosterone sulfate and functional performance in the oldest old: results from the Cardiovascular Health Study All Stars study. *The Journals of Gerontology Series A: Biological Sciences and Medical Sciences* 2010; 65A:976-981.

[148] Sapolsky RM, Krey LC, McEwen BS: The neuroendocrinology of stress and aging: the glucocorticoid cascade hypothesis. *Endocrine Reviews* 1986; 7:284-301.

[149] Sapolsky RM, Romero LM, Munck AU: How do glucocorticoids influence stress responses? Integrating permissive, suppressive, stimulatory, and preparative actions. *Endocrine reviews* 2000; 21:55-89.

[150] Lupien S, Lecours AR, Lussier I, Schwartz G, Nair NP, Meaney MJ: Basal cortisol levels and cognitive deficits in human aging. *The Journal of Neuroscience* 1994; 14:2893-2903.

[151] Lupien SJ, De Leon M, De Santi S, Convit A, Tarshish C, Nair NPV, Thakur M, McEwen BS, Hauger RL, Meaney MJ: Cortisol levels during human aging predict hippocampal atrophy and memory deficits. *Nature Neuroscience* 1998; 1:69-73.

[152] Elgh E, Lindqvist Åstot A, Fagerlund M, Eriksson S, Olsson T, Näsman B: Cognitive dysfunction, hippocampal atrophy and glucocorticoid feedback in Alzheimer's disease. *Biological Psychiatry* 2006; 59:155-161.

[153] Hartmann A, Veldhuis JD, Deuschle M, Standhardt H, Heuser I: Twenty-four hour cortisol release profiles in patients with Alzheimer's and Parkinson's disease compared to normal controls: ultradian secretory pulsatility and diurnal variation. *Neurobiology of Aging* 1997; 18:285-289.

[154] Rasmuson S, Andrew R, Näsman B, Seckl JR, Walker BR, Olsson T: Increased glucocorticoid production and altered cortisol metabolism in women with mild to moderate Alzheimer's disease. *Biological Psychiatry* 2001; 49:547-552.

[155] Csernansky J, Dong H, Fagan A, Wang L, Xiong C, Holtzman D, Morris J: Plasma cortisol and progression of dementia in DAT subjects. *American Journal of Psychiatry* 2006; 163:2164-2169.

[156] Karlamangla AS, Singer BH, Chodosh J, McEwen BS, Seeman TE: Urinary cortisol excretion as a predictor of incident cognitive impairment. *Neurobiology of aging* 2005; 26:80-84.

[157] Schrijvers EM, Direk N, Koudstaal PJ, Kirschbaum C, Hofman A, Tiemeier H, Breteler MM: Associations of serum cortisol with cognitive function and dementia: the Rotterdam Study. *Journal of Alzheimer's Disease* 2011; 25:671-677.

[158] Beluche I, Carrière I, Ritchie K, Ancelin ML: A prospective study of diurnal cortisol and cognitive function in community-dwelling elderly people. *Psychological medicine* 2010; 40:1039-1049.

[159] Fiocco AJ, Wan N, Weekes N, Pim H, Lupien SJ: Diurnal cycle of salivary cortisol in older adult men and women with subjective complaints of memory deficits and/or depressive symptoms: relation to cognitive functioning. Stress. *The International Journal on the Biology of Stress* 2006; 9:143-152.

[160] Stawski RS, Almeida DM, Lachman ME, Tun PA, Rosnick CB, Seeman T: Associations between cognitive function and naturally occurring daily cortisol during

middle adulthood: timing is everything. *The Journals of Gerontology Series B: Psychological Sciences and Social Sciences* 2011, 66:i71-i81.

[161] Franz CE, O'Brien RC, Hauger RL, Mendoza SP, Panizzon MS, Prom-Wormley E, Eaves LJ, Jacobson K, Lyons MJ, Lupien S, Hellhammer D, Xian H, Kremen WS: Cross-sectional and 35-year longitudinal assessment of salivary cortisol and cognitive functioning: the Vietnam Era twin study of aging. *Psychoneuroendocrinology* 2011; 36:1040-1052.

[162] Lee B, Glass T, Wand G, McAtee M, Bandeen-Roche K, Bolla K, Schwartz B: Apolipoprotein e genotype, cortisol, and cognitive function in community-dwelling older adults. *American Journal of Psychiatry* 2008; 165:1456-1464.

[163] Gerritsen L, Comijs HC, Deeg DJ, Penninx BW, Geerlings MI: Salivary cortisol, APOE-ε4 allele and cognitive decline in a prospective study of older persons. *Neurobiology of Aging* 2011; 32:1615-1625.

[164] Singh-Manoux A, Dugravot A, Elbaz A, Shipley M, Kivimaki M, Kumari M: No evidence of a longitudinal association between diurnal cortisol patterns and cognition. *Neurobiology of Aging* 2014; 35:2239-2245.

[165] Moriarty AS, Bradley AJ, Anderson KN, Watson S, Gallagher P, McAllister-Williams RH: Cortisol awakening response and spatial working memory in man: a U-shaped relationship. *Human Psychopharmacology: Clinical and Experimental* 2014; 29:295-298.

[166] Pulopulos MM, Hidalgo V, Almela M, Puig-Perez S, Villada C, Salvador A: Hair cortisol and cognitive performance in healthy older people. *Psychoneuroendocrinology* 2014; 44:100-111.

[167] Lippuner K, Johansson H, Kanis JA, Rizzoli R: Remaining lifetime and absolute 10-year probabilities of osteoporotic fracture in Swiss men and women. *Osteoporosis International* 2009; 20:1131-1140.

[168] Kanis JA, Johnell O, Oden A, Sernbo I, Redlund-Johnell I, Dawson A, De Laet C, Jonsson B: Long-term risk of osteoporotic fracture in Malmö. *Osteoporosis International* 2000; 11:669-674.

[169] Melton 3rd LJ, Khosla S, Crowson CS, O'Connor MK, O'Fallon WM, Riggs BL: Epidemiology of sarcopenia. *Journal of the American Geriatrics Society* 2000; 48:625-630.

[170] Baumgartner RN, Waters DL, Gallagher D, Morley JE, Garry PJ: Predictors of skeletal muscle mass in elderly men and women. *Mechanisms of Ageing and Development* 1999; 107:123-136.

[171] Roy TA, Blackman MR, Harman SM, Tobin JD, Schrager M, Metter EJ: Interrelationships of serum testosterone and free testosterone index with FFM and strength in aging men. *American Journal of Physiology-Endocrinology and Metabolism* 2002; 283:E284-E294.

[172] Yeap BB, Alfonso H, Chubb SP, Handelsman DJ, Hankey GJ, Almeida OP, Golledge J, Norman PE, Flicker L: In older men an optimal plasma testosterone is associated with reduced all-cause mortality and higher dihydrotestosterone with reduced ischemic heart disease mortality, while estradiol levels do not predict mortality. *The Journal of Clinical Endocrinology & Metabolism* 2014; 99:E9-E18.

[173] Mellström D, Vandenput L, Mallmin H, Holmberg AH, Lorentzon M, Odén A, Johansson H, Orwoll ES, Labrie F, Karlsson MK, Ljunggren Ö, Ohlsson C: Older men

with low serum estradiol and high serum SHBG have an increased risk of fractures. *Journal of Bone and Mineral Research* 2008; 23:1552-1560.

[174] Khosla S, Melton III LJ, Atkinson EJ, O'fallon WM: Relationship of serum sex steroid levels to longitudinal changes in bone density in young versus elderly men. *The Journal of Clinical Endocrinology & Metabolism* 2001; 86:3555-3561.

[175] Hoppé E, Morel G, Biver E, Borg S, Chopin F, Legrand E: Male osteoporosis: do sex steroids really benefit bone health in men? *Joint Bone Spine* 2011; 78:S191-S196.

[176] Amin S, Zhang Y, Sawin CT, Evans SR, Hannan MT, Kiel DP, Wilson PWF, Felson DT: Association of hypogonadism and estradiol levels with bone mineral density in elderly men from the framing study. *Ann Intern Med* 2000; 133:951-963.

[177] Szulc P, Munoz F, Claustrat B, Garnero P, Marchand F, Duboeuf F, Delmas PD: Bioavailable estradiol may be an important determinant of osteoporosis in men: the MINOS Study 1. *The Journal of Clinical Endocrinology & Metabolism* 2001; 86:192-199.

[178] Van Pottelbergh I, Goemaere S, Kaufman JM: Bioavailable estradiol and an aromatase gene polymorphism are determinants of bone mineral density changes in men over 70 years of age. *The Journal of Clinical Endocrinology & Metabolism* 2003; 88:3075-3081.

[179] Mellström D, Johnell O, Ljunggren Ö, Eriksson AL, Lorentzon M, Mallmin H, Homberg A, Redlund-Johnell I, Orwoll E, Ohlsson C: Free testosterone is an independent predictor of BMD and prevalent fractures in elderly men: MrOS Sweden. *Journal of Bone and Mineral Research* 2006; 21:529-535.

[180] Nguyen HT, von Schoultz B, Nguyen TV, Thang TX, Chau TT, Duc PT, Hirschberg AL: Sex hormone levels as determinants of bone mineral density and osteoporosis in Vietnamese women and men. *Journal of Bone and Mineral Metabolism* 2014:1-8.

[181] Voznesensky M, Walsh S, Dauser D, Brindisi J, Kenny AM: The association between dehydroepiandosterone and frailty in older men and women. Age and Ageing 2009; 38:401-406.

[182] Baylis D, Bartlett DB, Syddall HE, Ntani G, Gale CR, Cooper C, Lord JM, Sayer AA: Immune-endocrine biomarkers as predictors of frailty and mortality: a 10-year longitudinal study in community-dwelling older people. *Age* 2013; 35:963-971.

[183] Warriner AH, Saag KG: Glucocorticoid-related bone changes from endogenous or exogenous glucocorticoids. Current Opinion in Endocrinology, *Diabetes and Obesity* 2013; 20:510-516.

[184] Holanda CMDA, Guerra RO, Nóbrega PVDN, Costa HF, Piuvezam MR, Maciel ÁCC: Salivary cortisol and frailty syndrome in elderly residents of long-stay institutions: A cross-sectional study. *Archives of Gerontology and Geriatrics* 2012; 54:e146-e151.

[185] Zouboulis CC, Makrantonaki E: Hormonal therapy of intrinsic aging. *Rejuvenation Research* 2012; 15:302-312.

[186] Bassil N: Late-onset hypogonadism. *Medical Clinics of North America* 2011; 95:507-523.

[187] Amanatkar HR, Chibnall JT, Seo BW, Manepalli JN, Grossberg GT: Impact of exogenous testosterone on mood: a systematic review and meta-analysis of randomized placebo-controlled trials. Annals of Clinical Psychiatry: *Official Journal of the American Academy of Clinical Psychiatrists* 2014; 26:19-32.

[188] Pexman-Fieth C, Behre HM, Morales A, Kan-Dobrosky N, Miller MG: A 6-month observational study of energy, sexual desire, and body proportions in hypogonadal men treated with a testosterone 1% gel. *The Aging Male* 2014; 17:1-11.

[189] Wang C, Cunningham G, Dobs A, Iranmanesh A, Matsumoto AM, Snyder PJ., ... Swerdloff RS: Long-term testosterone gel (AndroGel) treatment maintains beneficial effects on sexual function and mood, lean and fat mass, and bone mineral density in hypogonadal men. *The Journal of Clinical Endocrinology & Metabolism* 2004; 89:2085-2098.

[190] Buvat J, Maggi M, Gooren L, Guay AT, Kaufman J, Morgentaler A, Schulman C, Tan HM, Torres LO, Yassin A, Zitzmann M: Endocrine aspects of male sexual dysfunctions. *J Sex Med* 2010; 7:1627-1656.

[191] Storer TW, Woodhouse L, Magliano L, Singh AB, Dzekov C, Dzekov J, Bhasin S: Changes in muscle mass, muscle strength, and power but not physical function are related to testosterone dose in healthy older men. *Journal of the American Geriatrics Society* 2008; 56:1991-1999.

[192] Sattler FR, Castaneda-Sceppa C, Binder EF, Schroeder ET, Wang Y, Bhasin S, Kawakubo M, Stewart Y, Yarasheski KE, Ullor J, Colletti P, Roubenoff R, Azen SP: Testosterone and growth hormone improve body composition and muscle performance in older men. *The Journal of Clinical Endocrinology & Metabolism* 2009; 94:1991-2001.

[193] Travison TG, Basaria S, Storer TW, Jette AM, Miciek R, Farwell WR, Choong K, Lakshman K, Mazer NA, Coviello AD, Knapp PE, Ulloor J, Zhang A, Brooks B, Nguyen A-H, Eder R, LeBrasseur N, Elmi A, Appleman E, Hede-Brierley L, Bhasin G, Bhatia A, Lazzari A, Davis S, Ni P, Collins L, Bhasin S: Clinical meaningfulness of the changes in muscle performance and physical function associated with testosterone administration in older men with mobility limitation. *J Gerontol A BioL Sci Med Sci* 2011; 66A:1090-1099.

[194] Kenny AM, Kleppinger A, Annis K, Rathier M, Browner B, Judge JO, McGee D: Effects of transdermal testosterone on bone and muscle in older men with low bioavailable testosterone levels, low bone mass, and physical frailty. *Journal of the American Geriatrics Society* 2010; 58:1134-1143.

[195] Ohlsson C, Barrett-Connor E, Bhasin S, Orwoll E, Labrie F, Karlsson MK, Ljunggren Ö, Vandenput L, Mellström D, Tivesten Å: High serum testosterone is associated with reduced risk of cardiovascular events in elderly men: The MrOS (Osteoporotic Fractures in Men) study in Sweden. *Journal of the American College of Cardiology* 2011; 58:1674-1681.

[196] Vigen R, O'Donnell CI, Barón AE, Grunwald GK, Maddox TM, Bradley SM, Barqawi A, Woning G, Wierman ME, Plomondon ME, Rumsfeld JS, Ho PM: Association of testosterone therapy with mortality, myocardial infarction, and stroke in men with low testosterone levels. *Jama* 2013; 310:1829-1836.

[197] Ullah MI, Riche DM, Koch CA: Transdermal testosterone replacement therapy in men. *Drug design, development and therapy* 2014; 8:101-112.

[198] Pierorazio PM, Ferrucci L, Kettermann A, Longo DL, Metter EJ, Carter HB: Serum testosterone is associated with aggressive prostate cancer in older men: results from the Baltimore Longitudinal Study of Aging. *BJU International* 2010; 105:824-829.

[199] Shabsigh R, Crawford ED, Nehra A, Slawin KM: Testosterone therapy in hypogonadal men and potential prostate cancer risk: a systematic review. *International Journal of Impotence Research* 2008; 21:9-23.

[200] Frisk J: Managing hot flushes in men after prostate cancer - a systematic review. Maturitas 2010; 65:15-22.

[201] Falahati-Nini A, Riggs BL, Atkinson EJ, O'Fallon WM, Eastell R, Khosla S: Relative contributions of testosterone and estrogen in regulating bone resorption and formation in normal elderly men. *The Journal of Clinical Investigation* 2000; 106:1553-1560.

[202] Reis LO, De Mendonça GB, Carneiro BD, Schneider E, Gewehr EV, Meirelles A, Denardi F, Gugliotta A: Diethylstilbestrol 1 mg in the treatment of acute urinary retention due to prostatic obstruction in the ederly: a preliminary study. *Advances in Urology* 2014; (2014).

[203] Kuebler JF, Jarrar D, Toth B, Bland KI, Rue L, Wang P, Chaudry ICH: Estradiol administration improves splanchnic perfusion following trauma-hemorrhage and sepsis. *Archives of Surgery* 2002; 137:74-79.

[204] Morales AJ, Nolan JJ, Nelson JC, Yen SS: Effects of replacement dose of dehydroepiandrosterone in men and women of advancing age. *The Journal of Clinical Endocrinology and Metabolism* 1994; 78:1360-1367.

[205] Schmidt PJ, Daly RC, Bloch M, Smith MJ, Danaceau MA, Simpson St. Clair L, Murphy JH, Haq N, Rubinow DR: Dehydroepiandrosterone monotherapy in midlife-onset major and minor depression. *Arch Gen Psychiatry* 2005; 62:154-162.

[206] Bloch M, Schmidt PJ, Danaceau MA, Adams LF, Rubinow DR: Dehydroepiandrosterone treatment of midlife dysthymia. *Biol Psychiatry* 1999; 45:1533-1541.

[207] Reiter WJ, Pycha A, Schatzl G, Pokorny A, Gruber DM, Huber JC, Marberger M: Dehydroepiandrosterone in the treatment of erectile dysfunction: a prospective, double-blind, randomized, placebo-controlled study. *Urology* 1999: 53:590-594.

[208] Morales A, Black A, Emerson L, Barkin J, Kuzmarov I, Day A: Androgens and sexual function: a placebo-controlled, randomized, double-blind study of testosterone vs. dehydroepiandrosterone in men with sexual dysfunction and androgen deficiency. *The Aging Male* 2009; 12:104-112.

[209] Maggi M, Buvat J, Corona G, Guay A, Torres LO: Hormonal causes of male sexual dysfunctions and their management (hyperprolactinemia, thyroid disorders, GH disorders, and DHEA). *The Journal of Sexual Medicine* 2013; 10:661-677.

[210] Wolf OT, Naumann E, Hellhammer DH, Kirschbaum C: Effects of dehydroepiandrosterone replacement in elderly men on event-related potentials, memory, and well-being. *The Journals of Gerontology Series A: Biological Sciences and Medical Sciences* 1998; 53:M385-M390.

[211] van Niekerk JK, Huppert FA, Herbert J: Salivary cortisol and DHEA: association with measures of cognition and well-being in normal older men, and effects of three months of DHEA supplementation. *Psychoneuroendocrinology* 2001; 26:591-612.

[212] Wolf OT, Neumann O, Hellhammer DH, Geiben AC, Strasburger CJ, Dressendörfer RA, ... Kirschbaum C: Effects of a Two-Week Physiological Dehydroepiandrosterone Substitution on Cognitive Performance and Well-Being in Healthy Elderly Women and Men 1. *The Journal of Clinical Endocrinology & Metabolism* 1997; 82:2363-2367.

[213] Kritz-Silverstein D, Von Mühlen D, Laughlin GA, Bettencourt R: Effects of Dehydroepiandrosterone Supplementation on Cognitive Function and Quality of Life: The DHEA and Well-Ness (DAWN) Trial. *Journal of the American Geriatrics Society* 2008; 56:1292-1298.

[214] Alhaj HA, Massey AE, McAllister-Williams RH: Effects of DHEA administration on episodic memory, cortisol and mood in healthy young men: a double-blind, placebo-controlled study. *Psychopharmacology* 2006; 188:541-551.

[215] Kim MJ, Morley JE: The hormonal fountains of youth: myth or reality? *J Endocrinol Invest* 2005; 28(Suppl 11):5-14.

[216] Percheron G, Hogrel JY, Denot-Ledunois S, Fayet G, Forette F, Baulieu EE, ... Marini JF: Effect of 1-year oral administration of dehydroepiandrosterone to 60-to 80-year-old individuals on muscle function and cross-sectional area: a double-blind placebo-controlled trial. *Archives of internal medicine* 2003; 163:720-727.

[217] Morales AJ, Haubrich RH, Hwang JY, Asakura H, Yen SS: The effect of six months treatment with a 100 mg daily dose of dehydroepiandrosterone (DHEA) on circulating sex steroids, body composition and muscle strength in age-advanced men and women. *Clinical endocrinology* 1998; 49:421-432.

[218] Morley JE, Malmstrom TK: Frailty, sarcopenia, and hormones. *Endocrinology and metabolism clinics of North America* 2013; 42:391-405.

[219] Travison TG, Araujo AB, O'Donnell AB, Kupelian V, McKinlay JB: A population-level decline in serum testosterone levels in American men. *The Journal of Clinical Endocrinology & Metabolism* 2007; 92:196-202.

[220] Muller M, den Tonkelaar I, Thijssen JH, Grobbee DE, van der Schouw YT: Endogenous sex hormones in men aged 40-80 years. European *Journal of Endocrinology* 2003; 149:583-589.

[221] Sartorius G, Spasevska S, Idan A, Turner L, Forbes E, Zamojska A, ... Handelsman DJ: Serum testosterone, dihydrotestosterone and estradiol concentrations in older men self-reporting very good health: the healthy man study. *Clinical endocrinology* 2012; 77:755-763.

[222] *Active Ageing a Policy Framework.* Geneva, World Health Organization, 2003.

[223] Rowe JW, Kahn RL: Human aging: usual and successful. *Science* 1987; 237:143-149.

[224] Depp CA, Glatt SJ, Jeste DV: Recent advances in research on successful or healthy aging. *Current psychiatry reports* 2007; 9:7-13.

[225] Södergren M: Lifestyle predictors of healthy ageing in men. *Maturitas* 2013; 75:113-117.

[226] Almeida OP, Khan KM, Hankey GJ, Yeap BB, Golledge J, Flicker L: 150 minutes of vigorous physical activity per week predicts survival and successful ageing: a population-based 11-year longitudinal study of 12 201 older Australian men. *British journal of sports medicine* 2014; 48:220-225.

[227] Steptoe A, Ussher M: Smoking, cortisol and nicotine. *International Journal of Psychophysiology* 2006; 59:228-235.

In: Neurobiology of Men's Mental Health
Editors: Leo Sher and Timothy R. Rice
ISBN: 978-1-63483-191-8
© 2015 Nova Science Publishers, Inc.

Chapter 25

FUTURE RESEARCH AND CLINICAL DIRECTIONS IN THE FIELD OF MEN'S MENTAL HEALTH: THE MADRID DECLARATION

Leo Sher*[*,1,2], MD, Zoltan Rihmer[3], MD, PhD, Javier Didia-Attas[4], MD, Jose de Leon[5], MD, Shih-Ku Lin[6,7], MD, Carlos Roncero[8], MD, Nestor Szerman[9], MD and Timothy Rice[1], MD

[1]Department of Psychiatry,
Icahn School of Medicine at Mount Sinai, New York, NY, US
[2]James J. Peters Veterans' Administration Medical Center,
Bronx, New York, US
[3]Semmelweis University, Budapest, Hungary
[4]Hospital Italiano, Buenos Aires, Argentina
[5]University of Kentucky, Lexington, KY, US
[6]Taipei Medical University, Taipei, Taiwan
[7]Psychiatric Center, Taipei City Hospital, Taipei, Taiwan
[8]University Hospital Vall Hebron,
Universidad Autónoma de Barcelona, Barcelona, Spain
[9]Complutense University of Madrid, Madrid, Spain

ABSTRACT

The research and clinical directions in the field of Men's Mental Health is rapidly developing. This chapter is the product of an international collaboration under the auspices of the World Federation of Biological Psychiatry's Task Force on Men's Mental Health to identify these clinical directions and areas for further research. The chapter

---

[*] Corresponding author: Leo Sher, M.D. James J. Peters Veterans' Administration, Medical Center, 130 West Kingsbridge Road, Bronx, New York 10468, USA, Tel: 1-718-584-9000 x 6821, Fax: 1-718-741-4703, E-mail: Leo.Sher@mssm.edu

identifies help-seeking behavior, impulsivity, alcohol and drug abuse in men, work related stress, bereavement, hormones, pharmacokinetics, and social awareness as areas of promise in the furhter development of the field of Men's Mental Health.

Keywords: men's mental health, gender, biological psychiatry, impulsivity, help-seeking

INTRODUCTION

The members of the World Federation of Biological Psychiatry's Task Force on Men's Mental Health met in Madrid in September 2014 to discuss the research and clinical directions in the field of Men's Mental Health. Leo Sher, M.D. (USA), Zoltan Rihmer, M.D., Ph.D. (Hungary), Javier Didia-Attas, M.D. (Argentina), Jose de Leon, M.D. (USA), Shih-Ku Lin, M.D. (Taiwan), Carlos Roncero, M.D. (Spain), and Nestor Szerman, M.D. (Spain) participated in the meeting. The following consensus recommendations were made.

HELP-SEEKING BEHAVIOR

Men seek help less frequently than women [1, 2]. Young men in particular receive less health care and are more often uninsured [3]. Without routine preventative care, risk factors, and warning signs for mental health concerns can go unobserved and unaddressed. Among men, and young men, in particular, masculinity is directly associated with receiving less healthcare and particularly mental health services [1, 4, 5]. Research, clinical work, and public health interventions targeting misconceptions among males that a lifestyle involving poor self-care and lack of help seeking is masculine are needed.

IMPULSIVITY

Men are more impulsive and aggressive than women [6, 7]. This leads to suicidality [6–8], substance abuse [9], and aggressive violence and homicidality [7, 10]. There are defined neural correlates for this association [11]. The focus of research and clinical work must be how to continue to further define these neural correlates, how to diagnose and treat impulsivity, and how to reduce access to lethal means.

ALCOHOL AND DRUG ABUSE IN MEN

There is a consistent and greater male use of alcohol, benzodiazepines, and illegal drugs of abuse [2, 12]. This is particularly true with alcohol, especially in select cultures, including those of Eastern Europe and the Americas. Patterns of use and behavior additionally diverge. Research and clinical work should be focused on targeted interventions to reduce male substance abuse and its associated disruptive behavior, suicidality, and aggression.

WORK RELATED STRESS

Employment and income expectations and concerns are traditionally of more importance to men [2, 13, 14]. In Western cultures, men are expected to provide for their families. For men, it is more difficult to lose a social and economic status than for women. The widespread youth unemployment in many countries worldwide may have particularly negative impact upon young men [15]. This needs to be studied in the context of stress reduction to prevent depression and suicide and to preserve function to help unemployed men to find new jobs.

BEREAVEMENT

The process of bereavement after widowhood may be more difficult for husbands [16, 17], though recent data suggests many complex factors may underlie earlier evidence that supported this claim [18]. As researchers and clinicians, we must explore the particular factors, which may make bereavement more difficult, and we must ask what kind of psychotherapy can be performed that may specifically target male vulnerability to complications of bereavement.

HORMONES

Testosterone levels in adults have been directly linked to impulsivity, aggression, desire to dominate, depression, and criminality [19, 20]. Testosterone may produce this effect via organizational and activational effects. Organizational effects refer to the effects which testosterone produces on the developing central nervous system during pregnancy and adolescence, whereas activational effects refer to those which serum testosterone levels directly produce. Clinical and research work must define and develop the understanding of both these pathways.

PHARMACOKINETICS

Male and female sex hormones exert divergent effects upon the cytochrome P450 system and other metabolic enzymes [21, 22]. Female hormones are inhibitors of CYP1A2, and smoking, a CYP1A2 inducer, is usually more frequent in men. Males, thus, on average, have greater CYP1A2 activity than females and may need higher doses of some psychiatric medications, including clozapine. There is more need for research in these gendered pharmacokinetic differences that, once well-established, need to be incorporated by clinicians in their work.

Social Awareness

There should be an increased public health awareness of differences in behavior and mental health between men and women. Public health interventions including education and primary preventative efforts focused on men are needed.

Men's mental health needs more attention from clinicians, researchers, and health policy makers. This declaration serves as a preliminary communication of the World Federation of Biological Psychiatry's Task Force on Men's Mental Health concerning future directions for research, clinical care, and public health policy.

Conflict of Interest Statement

The authors declare that the research was conducted in the absence of any commercial or financial relationships that could be construed as a potential conflict of interest.

References

[1] Rutz W, Rihmer Z. Suicidality in men – practical issues, challenges, solutions. *J. Mens. Health Gend.* (2000) 4:393–401. doi: 10.1016/j.jmhg.2007.07.046

[2] Möller-Leimkühler AM. The gender gap in suicide and premature death or: why are men so vulnerable? *Eur. Arch. Psychiatry Clin Neurosci* (2003) 253:1–8. doi:10.1007/s00406-003-0397-6

[3] Adams SH, Newacheck PW, Park MJ, Brindis CD, Irwin CE. Health insurance across vulnerable ages: patterns and disparities from adolescence to the early 30s. *Pediatrics* (2007) 119:e1033–9. doi:10.1542/peds.2006-1730

[4] Marcell AV, Ford CA, Pleck JH, Sonenstein FL. Masculine beliefs, parental communication, and male adolescents' health care use. *Pediatrics* (2007) 119:e966–75. doi:10.1542/peds.2006-1683

[5] Vogel DL, Heimerdinger-Edwards SR, Hammer JH, Hubbard A. Boys don't cry: examination of the links between endorsement of masculine norms, self-stigma, and help-seeking attitudes for men from diverse backgrounds. *J. Couns Psychol.* (2011) 58:368–82. doi:10.1037/a0023688

[6] Archer J. Sex differences in aggression in real-world settings: a meta-analytic review. *Rev. Gen. Psychol.* (2004) 8:291–322. doi:10.1037/1089-2680.8.4.291

[7] Strüber D, Lück M, Roth G. Sex, aggression and impulse control: an integrative account. *Neurocase* (2008) 14:93–121. doi:10.1080/13554790801992743

[8] Gvion Y, Apter A. Aggression, impulsivity, and suicide behavior: a review of the literature. *Arch. Suicide Res.* (2011) 15:93–112. doi:10.1080/13811118.2011.565265

[9] Perry JL, Carroll ME. The role of impulsive behavior in drug abuse. *Psychopharmacology* (Berl) (2008) 200:1–26. doi:10.1007/s00213-008-1173-0

[10] Sánchez-Martín JR, Azurmendi A, Pascual-Sagastizabal E, Cardas J, Braza F, Braza P, et al., Androgen levels and anger and impulsivity measures as predictors of physical,

verbal and indirect aggression in boys and girls. *Psychoneuroendocrinology* (2011) 36:750–60. doi:10.1016/j.psyneuen.2010.10.011

[11] Meyer-Lindenberg A, Buckholtz JW, Kolachana B, R Hariri A, Pezawas L, Blasi G, et al. Neural mechanisms of genetic risk for impulsivity and violence in humans. *Proc. Natl. Acad. Sci. USA* (2006) 103:6269–74. doi:10.1073/pnas.0511311103

[12] Substance Abuse and Mental Health Services Administration. Addressing the specific behavioral health needs of men. Treatment Improvement Protocol (TIP) Series 56. HHS Publication No. (SMA) 13-4736. Rockville, MD: Substance Abuse and Mental Health Services Administration. (2013). p.1–245.

[13] Stansfeld S, Candy B. Psychosocial work environment and mental health – a meta-analytic review. *Scand. J. Work Environ. Health* (2006) 32:443–62. doi:10.5271/sjweh.1050

[14] Sher L. Per capita income is related to suicide rates in men but not in women. *J. Mens. Health Gend.* (2006) 3:39–42. doi:10.1371/journal.pone.0098405

[15] Hammarström A. Health consequences of youth unemployment – review from a gender perspective. *Soc. Sci. Med.* (1994) 38:699–709. doi:10.1016/0277-9536(94)90460-X

[16] Bowling A, Windsor J. Death after widow(er)hood: an analysis of mortality rates up to 13 years after bereavement. *Omega* (1995) 31:35–49. doi:10.1093/ageing/afp108

[17] Stroebe MS. The broken heart phenomenon: an examination of the mortality of bereavement. *J. Community Appl. Soc. Psychol.* (1994) 4:47–61. doi:10.1002/casp.2450040110

[18] Sasson I, Umberson DJ. Widowhood and depression: new light on gender differences, selection, and psychological adjustment. *J. Gerontol. B. Psychol. Sci. Soc. Sci.* (2014) 69:135–45. doi:10.1093/geronb/gbt058

[19] Mehta PH, Beer J. Neural mechanisms of the testosterone-aggression relation: the role of orbitofrontal cortex. *J. Cogn. Neurosci.* (2010) 22:2357–68. doi:10.1162/jocn.2009.21389

[20] Mazur A. The role of testosterone in male dominance contests that turn violent. *Soc. Biol.* (2006) 53:24–9. doi:10.1080/19485565.2006.9989114

[21] Parkinson A, Mudra DR, Johnson C, Dwyer A, Carroll KM. The effects of gender, age, ethnicity, and liver cirrhosis on cytochrome P450 enzyme activity in human liver microsomes and inducibility in cultured human hepatocytes. *Toxicol. Appl. Pharmacol.* (2004) 199:193–209. doi:10.1016/j.taap.2004.01.010

[22] Rodenburg EM, Eijgelsheim M, Geleijnse JM, Amin N, van Duijn CM, Hofman A, et al. CYP1A2 and coffee intake and the modifying effect of sex, age, and smoking. *Am. J. Clin. Nutr.* (2012) 96:182–7. doi:10.3945/ajcn.111.027102

The article, "Future research and clinical directions in the field of Men's Mental Health: The Madrid Declaration" has been published previously under the terms of the Creative Commons Attribution License (Sher L, Rihmer Z, Didia-Attas J, de Leon J, Lin S, Roncero C, Szerman N and Rice T (2014). Future research and clinical directions in the field of Men's Mental Health: The Madrid Declaration. Front. Public Health 2:242. doi: 10.3389/fpubh.2014.00242). Therefore, we can include this article as a chapter in the book.

Index

#

A

B

C

D

E

Q

R

S

T

U

V

W

Y

Γ